IET ELECTROMAGNETIC WAVES SERIES 51

Series Editors: Professor P.J.B. Clarricoats
Professor E.V. Jull

Asymptotic and Hybrid Methods in Electromagnetics

Other volumes in this series:

Volume 1 **Geometrical theory of diffraction for electromagnetic waves, 3rd edition** G.L. James
Volume 10 **Aperture antennas and diffraction theory** E.V. Jull
Volume 11 **Adaptive array principles** J.E. Hudson
Volume 12 **Microstrip antenna theory and design** J.R. James, P.S. Hall and C. Wood
Volume 15 **The handbook of antenna design, volume 1** A.W. Rudge, K. Milne, A.D. Oliver and P. Knight (Editors)
Volume 16 **The handbook of antenna design, volume 2** A.W. Rudge, K. Milne, A.D. Oliver and P. Knight (Editors)
Volume 18 **Corrugated horns for microwave antennas** P.J.B. Clarricoats and A.D. Oliver
Volume 19 **Microwave antenna theory and design** S. Silver (Editor)
Volume 21 **Waveguide handbook** N. Marcuvitz
Volume 23 **Ferrites at microwave frequencies** A.J. Baden Fuller
Volume 24 **Propagation of short radio waves** D.E. Kerr (Editor)
Volume 25 **Principles of microwave circuits** C.G. Montgomery, R.H. Dicke and E.M. Purcell (Editors)
Volume 26 **Spherical near-field antenna measurements** J.E. Hansen (Editor)
Volume 28 **Handbook of microstrip antennas, 2 volumes** J.R. James and P.S. Hall (Editors)
Volume 31 **Ionospheric radio** K. Davies
Volume 32 **Electromagnetic waveguides: theory and application** S.F. Mahmoud
Volume 33 **Radio direction finding and superresolution, 2nd edition** P.J.D. Gething
Volume 34 **Electrodynamic theory of superconductors** S.A. Zhou
Volume 35 **VHF and UHF antennas** R.A. Burberry
Volume 36 **Propagation, scattering and diffraction of electromagnetic waves** A.S. Ilyinski, G. Ya.Slepyan and A. Ya.Slepyan
Volume 37 **Geometrical theory of diffraction** V.A. Borovikov and B.Ye. Kinber
Volume 38 **Analysis of metallic antenna and scatterers** B.D. Popovic and B.M. Kolundzija
Volume 39 **Microwave horns and feeds** A.D. Olver, P.J.B. Clarricoats, A.A. Kishk and L. Shafai
Volume 41 **Approximate boundary conditions in electromagnetics** T.B.A. Senior and J.L. Volakis
Volume 42 **Spectral theory and excitation of open structures** V.P. Shestopalov and Y. Shestopalov
Volume 43 **Open electromagnetic waveguides** T. Rozzi and M. Mongiardo
Volume 44 **Theory of nonuniform waveguides: the cross-section method** B.Z. Katsenelenbaum, L. Mercader Del Rio, M. Pereyaslavets, M. Sorella Ayza and M.K.A. Thumm
Volume 45 **Parabolic equation methods for electromagnetic wave propagation** M. Levy
Volume 46 **Advanced electromagnetic analysis of passive and active planar structures** T. Rozzi and M. Farinai
Volume 47 **Electromagnetic mixing formulas and applications** A. Sihvola
Volume 48 **Theory and design of microwave filters** I.C. Hunter
Volume 49 **Handbook of ridge waveguides and passive components** J. Helszajn
Volume 50 **Channels, propagation and antennas for mobile communications** R. Vaughan and J. Bach-Anderson
Volume 51 **Asymptotic and hybrid methods in electromagnetics** F. Molinet, I. Andronov and D. Bouche
Volume 52 **Thermal microwave radiation: applications for remote sensing** C. Matzler (Editor)
Volume 53 **Principles of planar near-field antenna measurements** S. Gregson, J. McCormick and C. Parini
Volume 502 **Propagation of radiowaves, 2nd edition** L.W. Barclay (Editor)

Asymptotic and Hybrid Methods in Electromagnetics

F. Molinet, I. Andronov
and D. Bouche

The Institution of Engineering and Technology

Published by The Institution of Engineering and Technology, London, United Kingdom

First published 2005

The Institution of Engineering and Technology
Michael Faraday House
Six Hills Way, Stevenage
Herts, SG1 2AY, United Kingdom

www.theiet.org

British Library Cataloguing in Publication Data
Molinet, Frederic
Asymptotic and hybrid methods in electromagnetics
1. Electromagentism 2. Diffraction
I. Title II. Andronov, I.V. (Iran. V.) III. Bouche, Daniel
1958– IV. Institution of Electrical Engineers

ISBN (10 digit) 0 86341 447 8
ISBN (13 digit) 978-0-86341-447-3

Typeset in India by Newgen Imaging Systems (P) Ltd, Chennai

Contents

Preface

In the last few years, progress in the field of numerical methods for diffraction problems has gained immense importance. Among a wealth of new methods and improvements of classical techniques, one can outline two major breakthroughs. First, the emergence and improvement of multipole methods for solving integral equations has provided a reliable order $N \log(N)$ method for diffraction problems. Second, progress in parallel computers has been very fast and machines with one or several teraflops peak power are now available to many engineers and scientists. As a result, it is now possible to perform computations with more than ten million unknowns, whereas in 1997, a problem with a hundred thousand was considered as big.

However, there is still room for asymptotic methods. First, they provide considerable physical insight and understanding of diffraction mechanisms and are therefore very useful in the design of electromagnetic devices such as radar targets and antennas. Second, some objects are still too large in terms of wavelength to fall in the realm of numerical methods. Third, very low radar cross-section objects, such as the NASA almond with antireflection coating, are sometimes difficult to compute using multipole methods. Moreover, objects that are very large in terms of wavelength, but with complicated details, are still a challenge both for asymptotic and numerical methods. The best, but not widely explored, solution for these problems is to combine in some way the above cited methods in the so-called hybrid methods.

Therefore, the demand for understanding and learning how to apply asymptotic and hybrid methods to solve diffraction problems is still important in the engineering and mathematical community. To this end, this book presents the state of the art in the field of asymptotic and hybrid methods. It does not intend to provide a complete coverage of the field, as a number of dedicated survey papers or books are available for self study, some of which appear in the references. The purpose is rather to give a detailed presentation of subjects where recent progress has been achieved. Most of these results are scattered in papers, or in unpublished internal reports. To the best of our knowledge, some of the results presented are new.

Chapter 1 gives a unified presentation of the geometrical theory of diffraction. By using a generalised Fermat principle, geometrical optics, edge or tip diffracted and creeping rays are written in a common framework. Ray methods give precise solutions but fail in transition zones, in the vicinity of diffracting bodies in the shadow zones and more generally in boundary layers appearing in diffraction problems. To compute the field in these zones, the boundary-layer method is a versatile and efficient tool. The foundation of the boundary-layer method with a number of illustrative applications to diffraction by smooth convex bodies, whispering gallery waves, transparent bodies, to cite a few, is presented in the second part of the chapter. The goal of this section is to provide the reader with a working knowledge of the method. Thus, we choose to deal with scalar equations and two-dimensional examples to keep the algebra as simple as possible. The precision of the method is demonstrated by comparison to numerical computations in the last part of the chapter.

Chapter 2 is devoted to electromagnetic creeping waves on curved objects. The first part of the chapter deals with diffraction of electromagnetic waves by a three-dimensional (3D) body. The algebra is somewhat more involved than in the previous chapter, because the equations are vectorial, and the geometry is 3D, but remains accessible to the reader acquainted with the boundary-layer method in Chapter 1. A complete solution for creeping waves, including the anisotropic impedance case and caustic of creeping rays is provided. Generalised creeping waves on strongly elongated objects are examined in the second part. It is shown that some of these waves exhibit very low attenuation. The third part is devoted to excitation and propagation of waves running along curvilinear interfaces between materials with different index. Different kinds of waves, depending on the order of magnitude of the index difference, can exist. For the small contrast case, waves that are a hybrid of creeping and whispering gallery waves appear.

Chapter 3 is a synthesis of the present knowledge on hybrid diffraction coefficients. These coefficients are useful to solve a number of diffraction problems, such as diffraction of creeping rays by edge or curvature discontinuities, creeping wave launching at discontinuities in curved surface, as well as diffraction and observation near grazing incidence.

Chapter 4 explains how to compute the currents on an obstacle. The purpose of this chapter is twofold. First, these so-called asymptotic currents extend the well-known physical optics and PTD approximations to smooth surfaces presenting edge discontinuities, with special emphasis on concave surfaces, for which results equivalent to those of previous chapters for convex surfaces are provided. Explicit formulae are given both in illuminated and shadow zone. Second, asymptotic currents are an essential ingredient for hybrid methods, which are the subject of Chapter 5.

A number of hybrid methods have been proposed in the literature. After a brief survey of these methods, an in-depth presentation of the so-called regularisation method is provided. This method is well suited to the computation of diffraction by objects that are large in terms of wavelengths, but with small details of complicated shape whose contribution is outside the scope of asymptotic methods. The diffraction problem is divided in two parts, one large regular part namely the object without the detail, which is computed with asymptotic methods, the other small but of complicated shape, whose contribution is computed with numerical methods, using as incident field the field diffracted by the regular part. The method is especially efficient if the field first diffracted by the detail and then by the main object is small as compared to the total diffracted field. However, if this condition is not satisfied, it is possible to enhance the precision of the result by using an iterative procedure, providing a systematic way of improving the solution.

1 JAMES, G. L.: 'Geometrical theory of diffraction for electromagnetic waves' (IEE Electromagnetic Waves Series, vol. 1, 1986)
2 MCNAMARA, D. A., PISTORIUS, C. W. I., MALHERBE, J. A. G. 'Introduction to the uniform geometrical theory of diffraction' (Artech House, 1990)
3 BABICH, V. M., and BULDYREV, V. S.: 'Short wave length diffraction theory: asymptotic methods' (Springer-Verlag, Berlin, 1991)
4 BOUCHE, D., MOLINET, F., and MITTRA, R.: 'Asymptotic and hybrid techniques for electromagnetic scattering', *Proceedings of the IEEE*, 1993, **81** (12)
5 KINBER, B. Y., and BOROVIKOV, V. A.: 'Geometrical theory of diffraction' (IEE Electromagnetic Waves Series, vol. 37, 1994)
6 BOUCHE, D., MOLINET, F., and MITTRA, R.: 'Asymptotic methods in electromagnetics' (Springer-Verlag, Berlin, 1997)
7 PATHAK, P. H., and BURKHOLDER, R.: 'High frequency methods', in PIKE, R., and SABATIER, P. (Eds): 'Scattering' (Academic Press, New York, 2002)

Chapter 1
Asymptotic theory of diffraction

1.1 Introduction to the geometrical theory of diffraction

1.1.1 General overview of the theory and basic concepts

The geometrical theory of diffraction (GTD) was conceived by Keller in the mid-1950s (1953–57), and published only some time later [1–4]. He showed that diffraction phenomena can be incorporated into a geometrical strategy and phrased in geometrical terms by introducing diffracted rays. These rays have paths determined by a generalisation of Fermat's principle.

The concept of diffracted rays was developed by Keller from the asymptotic evaluation (as the wave number k tends to infinity) of the known exact solution to scattering from simple shapes, referred to as the canonical problems of GTD. There exists a direct relationship between ray representations and the asymptotics of the solution of the Helmholtz equation $\Delta u + k^2 u = 0$ (or a system of Maxwell's equations) first outlined by Sommerfeld and Runge [5] in 1911 for the geometrical optics (GO) rays. It has given the basis to a formal technique called the ray method for constructing asymptotic expansions with respect to the inverse power of k of GO solutions of diffraction problems by smooth objects [6]. Construction of a high-frequency asymptotic solution by the ray method is possible only if the field of rays is regular. This condition, which will be mathematically specified in Section 1.1.4, is normally satisfied away from caustics and shadow boundaries.

In the Russian literature (see Reference 7), Keller's technique is known as the model or etalon problem method. The etalon problem method is based on the principle that similar ray geometry leads to similar asymptotic formulae (as $k \to \infty$) for wave fields. It is the simplest problem in which the field of rays has the same singularities as in the original problem (caustics, shadow boundaries). The simplest etalon problem means that it can be solved exactly, usually by the method of separation of the variables. The exact solution of the etalon (or canonical) problem is investigated as $k \to \infty$ and from this an asymptotic expansion is obtained which describes the field in the region of interest. The analytical expression for the field found in the investigation of the etalon problem is taken over to the original problem. This procedure permits us to start the resolution of the original problem by stating a general analytic form of the solution usually called an 'Ansatz'. The coefficients of the asymptotic series are determined recursively by substituting this analytical expression into the Helmholtz equation (or Maxwell's equations) and boundary conditions that correspond to the original problem.

The canonical (or model or etalon) problem method is the basis of GTD. A GTD solution is normally thought of in two stages. The first stage covers the construction of a system of rays corresponding to both GO and diffraction fields and the evaluation of transition regions, where the field of rays is singular. At the second stage the ray skeleton is covered by the flesh of computed field amplitudes (expression first used by Borovikov and Kinber [8]). The GO reflected and refracted waves are determined by the Fresnel and Snell's (Descartes) laws. The amplitudes of diffracted waves are calculated by the specific GTD laws based on model problem solutions.

For the second stage it is convenient to introduce the concept of asymptotic expansion. As we mentioned before, a direct relationship exists between ray representations and asymptotics of the solutions of the Helmholtz equation, hence GTD may be viewed also as an asymptotic ($k \to \infty$) theory of the solutions of the Helmholtz equation (or system of Maxwell's equations), that is, as a mathematical discipline. A GTD solution is thought of in the form of an asymptotic expansion of the solution with $k \to \infty$. This strategy, which allows easy derivation of – with rigorous mathematical proofs – the specific GTD laws from the model problem solutions, will be adopted in our presentation.

Although the model problem method is an extension of the method of comparison equations, extensively used for deriving asymptotic expansions for solutions of ordinary differential equations, it is not a rigorous mathematical procedure since it uses assumptions that are supplementary to the mathematical formulation of the problem. A rigorous mathematical proof at the level of theorems that the formulae of the ray method underlying the model problem method are asymptotic expansions of solutions of the given boundary-value problems does not exist in the general case, but such proofs have been demonstrated in a number of special cases especially for two-dimensional problems. Moreover, the asymptotic formulae derived by the model problem method agree with the physics of short-wave propagation and have been satisfactorily confirmed by experiment.

In summary, the GTD comprises three main steps:

- the generalisation of Fermat's principle
- the representation of the solution in the form of an asymptotic series, the analytical form of which is given by the solution of a canonical or model problem
- the calculation of the field amplitudes.

We will now develop each of these steps and derive the general form of the GTD solutions.

A number of general reviews of asymptotic methods in diffraction theory are available. We mention in particular the article by Kouyoumjian [9] written in 1965, which lays emphasis on the physical concepts underlying GTD and gives additional insight into the foundation of this theory.

1.1.2 Fermat principle

In 1654, the French mathematician Fermat postulated that regardless of to what kind of reflection or refraction a ray is subjected, it travels from a point to another in such a way that the time taken is minimum. This so-called Fermat principle was later put on a firmer mathematical basis by Hamilton.

In a medium with refractive index $n(\mathbf{r})$, the optical path length from a point A to a point B is the integral:

$$L(T) = \int_T n(\mathbf{r})\, ds,$$

where T is a path connecting the points A and B. A ray is defined as a trajectory satisfying the Fermat principle which selects it from all the curves from A to B as the one rendering the integral from A to B stationary.

The generalised Fermat principle consists in a generalisation of the concept of rays to include any extremal path satisfying constraints inferred from the environment.

It has been first applied to the reflection of a ray on a smooth surface or to the refraction of a ray into a different media, through a smooth interface, where the ray is treated as an extremal path constraint having a point on the smooth surface or interface.

In studying the diffraction by a wedge, Keller introduced diffracted rays that are defined by an extremum path subject to the constraints that it must include a point somewhere on the edge (Figure 1.1).

He also extended the principle to encompass diffraction by smooth convex surfaces in the shadow region.

In this case, the ray is defined by an extremum path subject to the constraint that it must include an arc on the surface (Figure 1.2).

1.1.2.1 Conditions for a path to be a ray

A generalisation of the Fermat principle including all types of interactions can be carried out in the following way. Let us consider a path T connecting the points

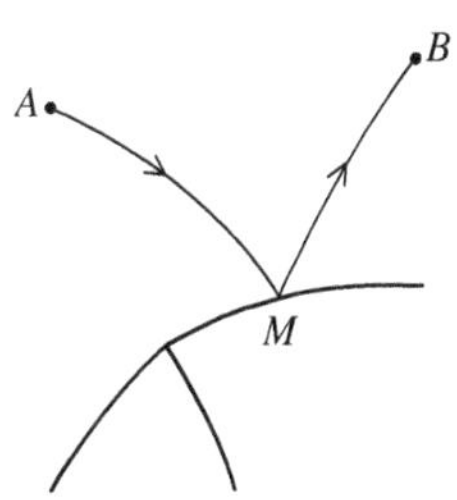

Figure 1.1 Constraints for edge diffraction

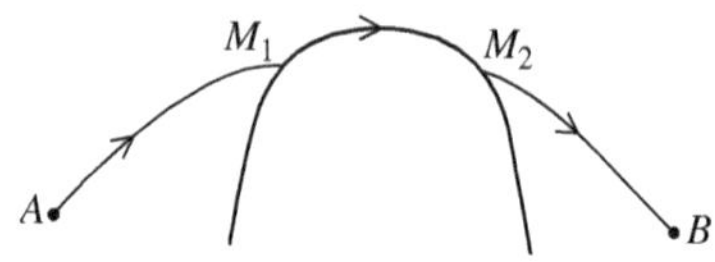

Figure 1.2 *Constraints for the diffraction by a smooth convex surface at an observation point in the shadow region*

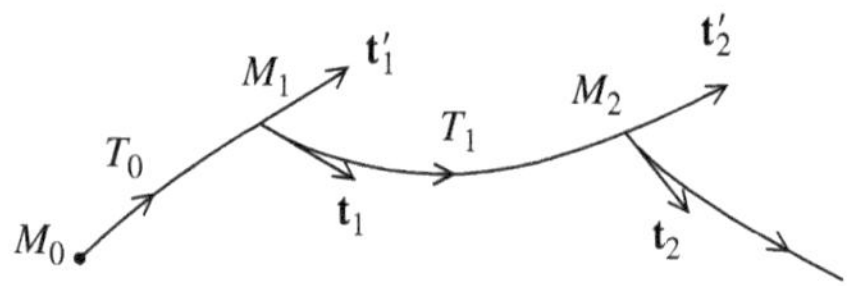

Figure 1.3 *General path between the two extremities of a ray*

M_0 and M_{N+1}. Let T be composed of N regular segments T_i and let the connection points be M_i, $i = 1, \ldots, N$, where either the direction or the curvature of T changes. The segments T_i may reside either in the space outside the object or be located on the surface. The points M_i, in turn, are located either on the surface, on edges, or on the tips (see Figure 1.3). N represents the total number of interactions of the path T with the object.

Let us designate $\mathbf{t}_i$, as the tangent at M_i, to the segment T_i and let $\mathbf{t}'_i$ be the tangent at M_i to the segment T_{i-1}, where $\mathbf{t}_i$ is not equal to $\mathbf{t}'_i$, in general. When the object is located in free space, the optical path is simply the length of the path T

$$L(T) = \int_T ds.$$

This is a functional expression defined on the entire set of paths, compatible with the connections on the surface, that links the points M_0 to M_{N+1} in a way such that the intermediate points M_i located on the surface, the edge or the tip, remain there and the segments T_i on the object remain on the surface of the object.

Fermat's generalised principle is stated as follows: 'T is a ray if and only if the length of T is stationary for all paths satisfying the connections on the surface'.

Applying the technique of the calculus of variations to express the infinitesimal variation $\delta(L(T))$ when each point of the curve undergoes an incremental displacement $\delta\mathbf{M}$ compatible with the connections on the surface, we get for an object located in free space:

$$\delta(L(T)) = \int_T \delta(ds) = \int_T \mathbf{t} \cdot d(\delta\mathbf{M}). \tag{1.1}$$

Since

$$\delta(ds) = \delta\left(\sqrt{|d\mathbf{M}|^2}\right) = \frac{d\mathbf{M}}{ds} \cdot \delta(d\mathbf{M}) = \mathbf{t} \cdot d(\delta\mathbf{M}),$$

where $\mathbf{t} = d\mathbf{M}/ds$ is the unit vector of the tangent to T at point M, integrating by parts, (1.1) becomes

$$\delta(L(T)) = \sum_{i=0}^{N} (\mathbf{t}'_i - \mathbf{t}_i) \cdot \delta\mathbf{M}_i - \sum_{i=1}^{N} \int_{T_i} \delta\mathbf{M} \cdot d\mathbf{t}.$$

We can now define a ray as follows: 'T is a ray if $\delta(L(T)) = 0$, for all $\delta\mathbf{M}$ compatible with the constraints imposed on the ray segments'. This definition leads to the following two types of conditions:

1 $N + 1$ conditions characterising the $N + 1$ elementary segments T_i

$$\int_{T_i} \delta\mathbf{M} \cdot d\mathbf{t} = 0. \tag{1.2}$$

2 N conditions of transition associated with the diffraction points

$$(\mathbf{t}'_i - \mathbf{t}_i) \cdot \delta\mathbf{M}_i = 0. \tag{1.3}$$

When the ray crosses at some connection point $\mathbf{M}_q$ the interface between two media with constant, but different, refractive indexes n_1 and n_2, the condition (1.2) still applies, however, the condition (1.3) at $\mathbf{M}_q$ is modified and replaced by

$$(n_1\mathbf{t}'_q - n_2\mathbf{t}_q) \cdot \delta\mathbf{M}_q = 0. \tag{1.4}$$

At all other points $\mathbf{M}_i$ connecting two paths T_i and T_{i+1} lying in the same medium, condition (1.3) holds.

As we will see below, conditions (1.2), (1.3) and (1.4) embody all the laws of ray propagation and enable us to construct the optical as well as the diffracted rays.

1.1.2.2 Applications of conditions (1.2) and (1.3) or (1.4) to specific problems

Segments in free space. In free space, $\delta\mathbf{M}$ is arbitrary and has three degrees of freedom. According to (1.2), $d\mathbf{t} = 0$, hence $\mathbf{t} = \mathbf{t}_i = \mathbf{t}'_i =$ constant, and as a result (1.3) is also satisfied. This leads us to the first law of GO, which is simply stated as: 'In free space, the rays are straight lines'.

Reflection from a smooth surface. For a smooth surface, both the incoming and outgoing rays corresponding to a reflection point M are straight lines. Let us denote the unit vector along the incident ray as $\mathbf{t}'_i = \mathbf{i}$ and the corresponding unit vector along the reflected ray as $\mathbf{t}_i = \mathbf{r}$. Since the reflection point is on the surface S, $\delta\mathbf{M}$ must be a vector in the tangent plane P at the reflection point M. From (1.3), it appears that $\mathbf{i} - \mathbf{r}$ must be normal to S:

$$\mathbf{i} - \mathbf{r} = \lambda\mathbf{n}, \tag{1.5}$$

where λ is a scalar and $\mathbf{n}$ is the unit vector normal to S at M, pointing outward. Let us denote by θ_i, the incident angle. Then,

$$\mathbf{i} \cdot \mathbf{n} = -\cos\theta_i. \tag{1.6}$$

In view of (1.5) and utilising the fact that $\mathbf{r}$ is a unit vector, we obtain

$$\mathbf{r} = \mathbf{i} + 2\cos\theta_i\,\mathbf{n}. \tag{1.7}$$

Equation (1.7) represents the law of reflection. It is often stated as follows: 'The reflected ray is in the plane of incidence, which is defined by the surface normal and the propagation vector of the incident ray, and the angle of reflection is equal to the angle of incidence'.

Transmission through a smooth interface between different media with constant refractive indexes. According to (1.2), both the incident and refracted rays are straight lines. Let us denote by n_1 and n_2, the refractive indexes of media 1 and 2 and by $\mathbf{t}'_q = \mathbf{i}$ and $\mathbf{t}_q = \mathbf{t}$ the unit vectors along the incident ray in medium 1 and along the refracted ray in medium 2, respectively. Since the refraction point $\mathbf{M}_q$ is constrained to be on the interface S, it appears from (1.4) that

$$n_1\mathbf{i} - n_2\mathbf{t} = \lambda\mathbf{n}, \tag{1.8}$$

where λ is a scalar and $\mathbf{n}$ is the unit vector normal to S at $\mathbf{M}_q$, pointing towards medium 1.

In view of (1.8), vector $\mathbf{t}$ is in the plane T defined by $\mathbf{i}$ and $\mathbf{n}$. Moreover, if we denote the incident angle as θ_i and the refracted angle as θ_t, then by multiplying both terms of (1.8) by the unit vector $\mathbf{n}$ tangent to the interface S and lying in the plane T, we get

$$n_1 \sin\theta_i = n_2 \sin\theta_t. \tag{1.9}$$

Equations (1.8) and (1.9) define the law of refraction which may be stated as follows: 'The refracted ray is in the plane of incidence, which is defined by the normal to the interface and the propagation vector of the incident ray, and the angle of refraction is related to the angle of incidence by the law (1.9)'.

Diffraction by an edge. Again, both incoming and outgoing rays corresponding to a diffraction point M on the edge, are straight lines. Since the diffraction point is on the edge, $\delta\mathbf{M}$ is along the tangent $\mathbf{s}$ to the diffracting edge. Denoting the unit vector on the diffracted ray as $\mathbf{d}$, we obtain from (1.3)

$$(\mathbf{i} - \mathbf{d}) \cdot \mathbf{s} = 0. \tag{1.10}$$

Equation (1.10) defines a cone of diffracted rays, the axis of the cone being tangential to the edge, and its semi-angle β at the vertex of the cone being the angle between $\mathbf{i}$ and $\mathbf{s}$. This cone of diffraction is called the Keller's cone (see Figure 1.4).

The law of diffraction from an edge can be simply stated as 'All of the diffracted rays originating from a diffraction point must reside on the Keller's cone'. The law stated above is valid for any line discontinuity of the tangent plane, curvature or higher order derivatives of the surface.

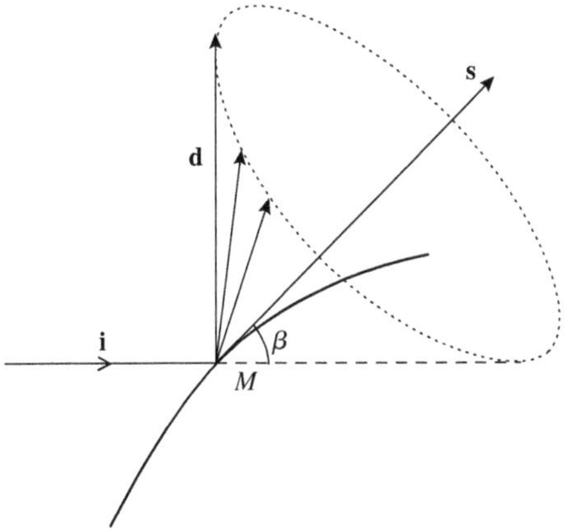

Figure 1.4 Keller's cone

Surface rays. If T_i is a surface ray, $\delta\mathbf{M}$ lies in the plane tangential to the surface. Then, according to (1.2), $d\mathbf{t} \cdot \delta\mathbf{M} = 0$ and consequently:

$$\frac{d\mathbf{t}}{ds} = \lambda\mathbf{n},$$

where **n** is the normal to the surface.

Since according to Frenet's formula:

$$\frac{d\mathbf{t}}{ds} = \frac{\mathbf{h}}{R(s)},$$

where **h** is the principal normal to the surface ray, we see that **h** and **n** coincide. This property is characteristic of the geodesics on a surface. This leads to the following rule governing the propagation of rays on a surface: 'The surface rays follow the geodesics of the surface'. In the case of a convex surface, the surface rays are also referred to as 'creeping rays'.

Creeping rays launched by space rays on a smooth convex surface. At the intersection of the two rays, the point M can move in the plane tangential to the surface. Let $\mathbf{t}_i' = \mathbf{i}$ be the unit vector along the space ray and $\mathbf{t}_i = \mathbf{r}$ along the creeping ray. Then from (1.3), we have:

$$\mathbf{i} - \mathbf{r} = \lambda\mathbf{n}.$$

Since $\mathbf{r} \cdot \mathbf{n} = 0$, it follows that $\mathbf{i} = \mathbf{r}$. As a consequence, $\mathbf{i} \cdot \mathbf{n} = 0$ and from (1.6) it follows that the angle of incidence $\theta_i = \pi/2$. This implies that **i** is along the direction of grazing incidence and consequently, the point M is located on the light–shadow boundary (Figure 1.5).

This leads us to the following rule: 'The space rays launch creeping rays at the light–shadow boundaries and the tangent to the creeping rays is along the incident space rays'.

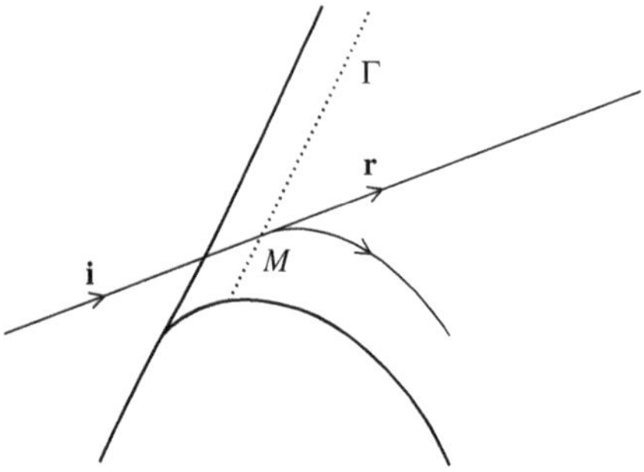

Figure 1.5 Creeping ray launched at the light–shadow boundary Γ

Once we have the rays, the next step consists in calculating the field. Before doing this, we need to introduce some elementary definitions and properties of asymptotic expansions.

1.1.3 Fundamentals of asymptotic expansions

The terminology 'asymptotic expansion' was first introduced by Poincaré in 1892 [10]. However, the underlying concept was known much earlier by all those attempting to solve a physical problem, using a perturbation procedure. An application of the perturbation method to differential equations leads, in general, to a solution expressed in a series expansion in terms of integer powers of a small parameter ε. The series in question satisfies the equation

$$f(\varepsilon) - \sum_{n=0}^{N} a_n \varepsilon^n = o(\varepsilon^N), \tag{1.11}$$

where $f(0)$ is the solution of the unperturbed problem and $o(\,)$ is the Landau symbol, small 'o' having the following definition.

Definition. We say $f = o(g)$ as ε tends to 0 if given any number $\eta > 0$, as small as we want, there exists a neighbourhood R_0 of the origin such that $|f| \leq \eta |g|$ for all $\varepsilon \in R_0$.

Equation (1.11) means that if ε is sufficiently small, the N-terms expansion represents an approximation to the solution. As ε tends towards 0, the sum (1.11) converges to the exact solution of the problem.

A representation of the type (1.11), which is limited to a finite number of terms, is called an asymptotic expansion.

1.1.3.1 Asymptotic sequence

Consider a sequence $\{\nu_n(\varepsilon)\}$, $n = 1, 2, \ldots$ of real and positive functions of ε defined and continuous in a neighbourhood R_0 of the origin. Such a sequence is called an asymptotic sequence if:

$$\nu_{n+1}(\varepsilon) = o(\nu_n(\varepsilon)), \qquad \text{as } \varepsilon \to 0. \tag{1.12}$$

If the sequence is infinite and if (1.12) is valid uniformly in n which means that the choice of η and R_0 in the definition of small 'o' does not depend on n, then the sequence is said to be uniform in n.

It is possible to extend the concept of asymptotic expansions by using in (1.11) the sequence $\nu_n(\varepsilon)$ instead of the sequence ε^n. Indeed, owing to (1.12), we have:

$$f(\varepsilon) - \sum_{n=0}^{N} a_n \nu_n(\varepsilon) = o(\nu_N(\varepsilon)), \qquad \text{as } \varepsilon \to 0. \tag{1.13}$$

Generally $f(\varepsilon)$ depends on other variables, say x, belonging to a given domain D. The expansion:

$$F(x,\varepsilon) = \sum_{n=0}^{N} a_n \nu_n(\varepsilon),$$

is said to be uniform, if:

$$f(\varepsilon) - F(x,\varepsilon) = o(\nu_N(\varepsilon)), \qquad \text{as } \varepsilon \to 0,$$

is uniformly valid in all of the domain D. This definition implies that the error in f remains of the order of $\nu_N(\varepsilon)$ not only when ε tends to zero for a fixed x, but also for all possible variations of $x = x(\varepsilon)$ provided that x continues to remain in the domain D.

1.1.3.2 Compatible asymptotic sequence

It is, in principle, possible to determine the coefficients of the asymptotic sequence by taking the limits:

$$\begin{aligned}
a_0 &= \lim_{\varepsilon\to 0} \frac{f(\varepsilon)}{\nu_0(\varepsilon)},\\
a_1 &= \lim_{\varepsilon\to 0} \frac{f(\varepsilon) - a_0\nu_0(\varepsilon)}{\nu_1(\varepsilon)},\\
&\vdots\\
a_K &= \lim_{\varepsilon\to 0} \frac{f(\varepsilon) - \sum_{n=0}^{K-1} a_n \nu_n(\varepsilon)}{\nu_K(\varepsilon)}.
\end{aligned} \tag{1.14}$$

However, it may happen that these limits are all equal to zero or infinity. In such a case, we say that the sequence $\nu_n(\varepsilon)$ is not compatible with the function $f(\varepsilon)$.

Example

$$f(\varepsilon) = \cos\varepsilon,$$
$$\nu_n(\varepsilon) = \varepsilon^{n+1/2}, \qquad n = 0, 1, 2, \ldots .$$

This implies that it is necessary to impose certain restrictions to the choice of the asymptotic sequence $\nu_n(\varepsilon)$.

An asymptotic sequence $\nu_n(\varepsilon)$ is compatible with the function $f(\varepsilon)$ if it contains a sequence $\mu_n(\varepsilon)$ defined as follows, where Ord is an abbreviation for order:

$$\mathrm{Ord}[\mu_0(\varepsilon)] = \mathrm{Ord}[f(\varepsilon)],$$

$$\mathrm{Ord}[\mu_1(\varepsilon)] = \mathrm{Ord}[f(\varepsilon) - b_0\mu_0(\varepsilon)],$$

$$\vdots$$

$$\mathrm{Ord}[\mu_K(\varepsilon)] = \mathrm{Ord}\left[f(\varepsilon) - \sum_{n=0}^{K-1} b_n\mu_n(\varepsilon)\right],$$

where $b_0, b_1, \ldots, b_n$ denote the non-vanishing coefficients of (1.14).

$$b_0 = \lim_{\varepsilon\to 0} \frac{f(\varepsilon)}{\mu_0(\varepsilon)},$$

$$\vdots$$

$$b_n = \lim_{\varepsilon\to 0} \frac{f(\varepsilon) - \sum_{q=0}^{n-1} b_q\mu_q(\varepsilon)}{\mu_n(\varepsilon)}.$$

We give below the mathematical definition of the Landau symbols large 'O' and Ord.

Definitions

We write $f = O(g)$ in R_1 if there exists a constant A independent of ε such that $|f| \leq A|g|$ for all $\varepsilon \in R_1$.
We say that f is of the order of g, which we write $f = \mathrm{Ord}(g)$ if we have simultaneously $f = O(g)$ and $g = O(f)$ where 'O' is the Landau symbol large 'O'.

1.1.3.3 Properties of an asymptotic expansion

1. Given a compatible asymptotic sequence, the asymptotic expansion is unique.
2. An asymptotic expansion represents an approximation to the solution when ε is small.

According to (1.13), for a fixed value of N, say $N = N_1$, and for $\eta > 0$ arbitrarily small, there exists $\varepsilon = \varepsilon_1$ such that:

$$|f(\varepsilon_1) - F_{N_1}(\varepsilon_1)| < \eta,$$

where

$$F_{N_1}(\varepsilon) = \sum_{n=0}^{N_1} a_n\nu_n(\varepsilon).$$

Conversely, if $\varepsilon = \varepsilon_1$ is fixed and if we augment N from zero to infinity, the difference passes through a value smaller than η, when N passes through N_1, and then augments indefinitely if the series $F_N(\varepsilon_1)$ diverges, or tends to zero if the series converges.

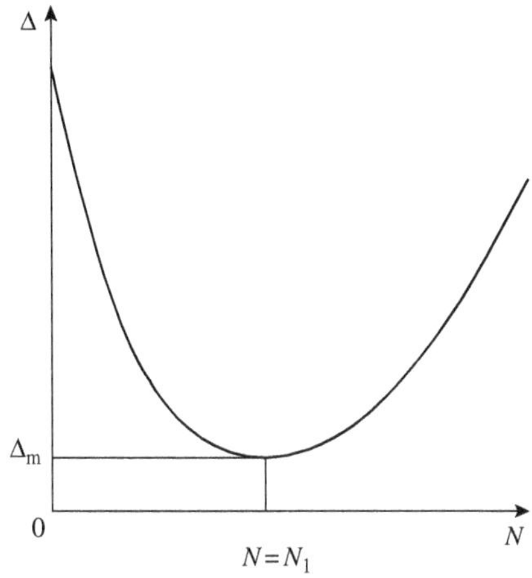

Figure 1.6 Typical shape giving the variation of the error Δ as a function of N

In general, the series $F_N(\varepsilon_1)$ diverges and a typical shape of the curve giving the variation of error Δ with N is shown in Figure 1.6.

The minimum value Δ_{m} is a function of ε and tends to zero, when $\varepsilon \to 0$. Hence, for a given value of ε, there exists an optimal value of N for which the difference between the exact solution and its asymptotic expansion is the smallest. Unfortunately, there does not exist a general rule by which we can predict the value of this optimum and a certain amount of experience is needed to determine the desired number of terms to be retained in the asymptotic expansion. More information concerning asymptotic expansions can be found in References 10–14.

1.1.4 Asymptotic solution of Maxwell's equations in a source-free region

1.1.4.1 Derivation of the asymptotic expansion of the solution

We first consider the scalar Helmholtz equation in free space:

$$(\Delta + k^2)U = 0. \tag{1.15}$$

To derive a solution using the perturbational approach, the small parameter being $1/k$, we divide the Helmholtz equation by k^2 to get:

$$\left(\frac{1}{k^2}\Delta + 1\right) U = 0.$$

The term $k^{-2}\Delta U$ could be viewed as a perturbation, but the solution of the unperturbed equation would then simply be a trivial solution $U = 0$. To avoid this pitfall, we perform the transformation:

$$U(\mathbf{r}) = e^{ikS(\mathbf{r})}u(\mathbf{r}), \tag{1.16}$$

known as the quasi-optics Ansatz, which was introduced by Sommerfeld and Runge in 1911 [5] who derived for the first time the laws of GO from the wave equation. This Ansatz can also be stated in accordance to the asymptotic solution of the model problem of a point source in free space. If we introduce (1.16) in (1.15) and order the different terms according to the parameter k^{-1}, which is assumed to be small, the wave equation is converted into a new form as follows:

$$(1 - (\nabla S)^2)u + \frac{i}{k}(\Delta S + 2\nabla S \cdot \nabla u) + \frac{1}{k^2}\Delta u = 0.$$

If we neglect the last term, we are led to the eikonal equation:

$$|\nabla S|^2 = 1$$

and the transport equation:

$$u\Delta S + 2\nabla S \cdot \nabla u = 0,$$

which contains, as we will see later, all the laws of the GO for the scalar waves.

Consider now the case of Maxwell's equations in free space:

$$\begin{cases} \text{curl}\,\mathbf{E} = ik\eta\mathbf{H}, \\ \\ \text{curl}\,\mathbf{H} = -\dfrac{ik}{\eta}\mathbf{E}, \end{cases} \qquad \eta = \sqrt{\frac{\mu_0}{\varepsilon_0}}.$$

An equivalent form of these equations is given by:

$$\begin{aligned} &\Delta\mathbf{E} + k^2\mathbf{E} = 0 && \text{vector Helmholtz equation,} \\ &\text{div}\,\mathbf{E} = 0 && \text{Gauss law,} \\ &\mathbf{H} = \frac{1}{ik\eta}\,\text{curl}\,\mathbf{E}. \end{aligned} \tag{1.17}$$

According to the procedure followed by Sommerfeld and Runge starting with the quasi-optics Ansatz (1.16), we seek a solution for the Maxwell equations of the form:

$$\begin{aligned} \mathbf{E}(\mathbf{r}) &= e^{ikS(\mathbf{r})}\sum_{n=0}^{N}(ik)^{-n}\mathbf{e}_n(\mathbf{r}) + o(k^{-N}), \\ \mathbf{H}(\mathbf{r}) &= e^{ikS(\mathbf{r})}\sum_{n=0}^{N}(ik)^{-n}\mathbf{h}_n(\mathbf{r}) + o(k^{-N}), \end{aligned} \tag{1.18}$$

which is an asymptotic expansion in terms of the asymptotic sequence k^{-n}, where n is an integer.

Inserting (1.18) in (1.17), we obtain the following set of equations:

$$|\nabla S|^2 = 1, \tag{1.19}$$

$$(\Delta S + 2\nabla S \cdot \nabla)\mathbf{e}_n = -\Delta \mathbf{e}_{n-1}, \tag{1.20}$$

$$\nabla S \cdot \mathbf{e}_n = -\nabla \cdot \mathbf{e}_{n-1}, \tag{1.21}$$

$$\mathbf{h}_n = \frac{1}{\eta}(\nabla S \times \mathbf{e}_n + \nabla \times \mathbf{e}_{n-1}). \tag{1.22}$$

Since $\mathbf{e}_{-1} = 0$, we see that $\mathbf{e}_0$ verifies the transport equation:

$$(\Delta S + 2\nabla S \cdot \nabla)\mathbf{e}_0 = 0.$$

In addition, we have from (1.21) and (1.22):

$$\nabla S \cdot \mathbf{e}_0 = 0,$$

$$\mathbf{h}_0 = \frac{1}{\eta}\nabla S \times \mathbf{e}_0,$$

which shows that $\mathbf{e}_0$ is orthogonal to $\mathbf{h}_0$ and to ∇S.

1.1.4.2 Resolution of the eikonal equation

In a Cartesian system of coordinates (x_1, x_2, x_3) a point in space is represented by $\mathbf{r} = \mathbf{r}(x_1, x_2, x_3)$ and the eikonal equation (1.19) may be written as

$$\left(\frac{\partial S}{\partial x_1}\right)^2 + \left(\frac{\partial S}{\partial x_2}\right)^2 + \left(\frac{\partial S}{\partial x_3}\right)^2 - 1 = 0.$$

We see that the eikonal equation is a first-order differential equation, the so-called Hamilton–Jacobi equation which has the general form $F(x_i, \xi_i) = 0$, where $\xi_i = \partial S/\partial x_i$. An equation of this type is usually solved by using the method of characteristics which consists in writing the total derivative of F:

$$dF = \frac{\partial F}{\partial x_i}dx_i + \frac{\partial F}{\partial \xi_i}d\xi_i = 0,$$

leading to:

$$dx_i\left(\frac{\partial F}{\partial \xi_i}\right)^{-1} = -d\xi_i\left(\frac{\partial F}{\partial x_i}\right)^{-1} = ds,$$

or equivalently:

$$\frac{d\xi_i}{ds} = -\frac{\partial F}{\partial x_i}, \qquad \frac{dx_i}{ds} = \frac{\partial F}{\partial \xi_i}. \tag{1.23}$$

The system of equations (1.23) are the parametric equations of the characteristic curves. In a homogeneous medium:

$$\frac{dx_i}{ds} = 2\xi_i, \qquad \frac{d\xi_i}{ds} = 0.$$

The first equation shows that the tangent to a characteristic curve (or a ray) is directed along the gradient of the phase:

$$\frac{d\mathbf{r}}{ds} = 2\nabla S, \tag{1.24}$$

whereas the second equation tells us that this tangent is constant along a characteristic curve (or a ray): $d(\nabla S)/ds = 0$. The characteristic curves (or rays) are therefore straight lines directed along the gradient of the phase.

By change of the metric of the space (element of length) we want $|d\mathbf{r}/d\sigma| = 1$, where σ is the curvilinear coordinate along the characteristic curve. According to (1.24), we see that $d\sigma = 2\,ds$.

1.1.4.3 Properties of the characteristic curves

1 The characteristic curves are orthogonal to the surfaces $S(\mathbf{r}) = \text{const}$.
Indeed we have:

$$\frac{d\mathbf{r}}{d\sigma} = \boldsymbol{\sigma} = \frac{\nabla S}{|\nabla S|},$$

where $d\mathbf{r}/d\sigma$ is a tangent to the characteristic curve and $\nabla S/|\nabla S|$ is a unit vector orthogonal to the surface $S(\mathbf{r}) = \text{const}$.

2 Through each point $\mathbf{r}_0$ on $S(\mathbf{r}) = \text{const}$, passes one characteristic curve. Since a point of S is defined by two parameters, the characteristic curves form a family of curves depending on two parameters: this is called a congruence of curves (here straight lines). A mathematical property is: a congruence of curves has always an envelope which in general is a surface. In some cases, the surface can be degenerated into a line or a point. In GO, the following vocabulary is used: The surface $S(\mathbf{r}) = \text{const}$ is called a wave front, the characteristic curves are called the rays and the envelope of the characteristic curves is called a caustic (see Figure 1.7).

3 The phase variation from $\mathbf{r}_0$ to $\mathbf{r}_1$ is given by:

$$k[S(\mathbf{r}_1) - S(\mathbf{r}_0)] = k\int_0^{\sigma} \frac{\partial S}{\partial \sigma}\, d\sigma = k\int_0^{\sigma} \nabla S \cdot \frac{d\mathbf{r}}{d\sigma}\, d\sigma$$

$$= k\int_0^{\sigma} d\sigma = k\sigma = \mathbf{k} \cdot (\mathbf{r}_1 - \mathbf{r}_0).$$

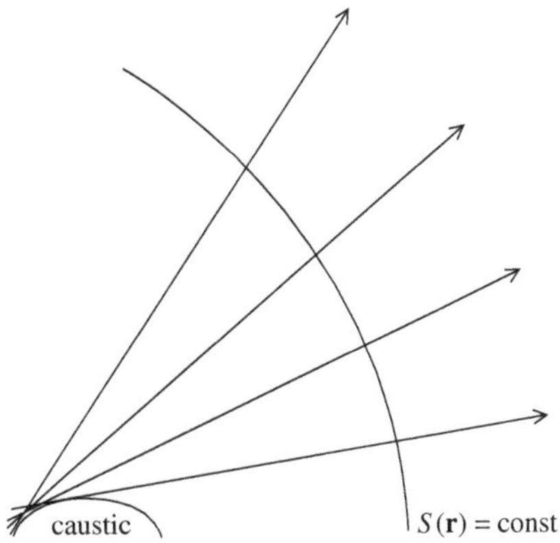

Figure 1.7 *Wave front, rays and caustic*

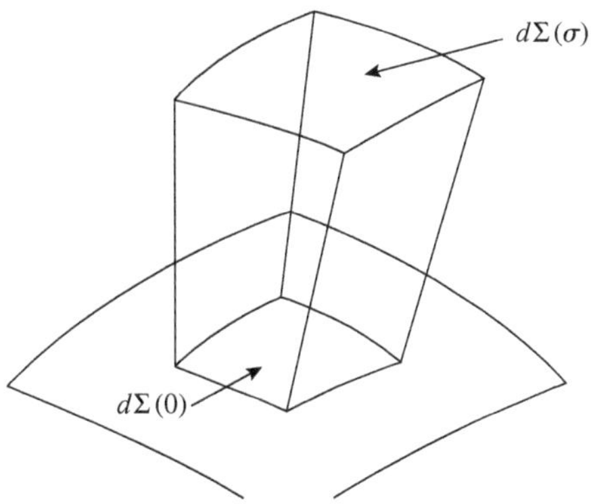

Figure 1.8 *Ray tube*

1.1.4.4 Resolution of the transport equation

In the scientific literature we can find different methods of resolution of this equation. The most familiar one concerns the equation for $\mathbf{e}_0$. If u is a Cartesian component of this vector, we have to solve the equation:

$$u\Delta S + 2\nabla S \cdot \nabla u = 0. \tag{1.25}$$

By multiplying it by u, we obtain:

$$\nabla \cdot (u^2 \nabla S) = 0, \tag{1.26}$$

which shows that $u^2 \nabla S$ has a vanishing divergence and consequently its flux in a tube of rays is conserved. The flux on the walls formed by the rays vanishes since $u^2 \nabla S$ is parallel to the rays (Figure 1.8).

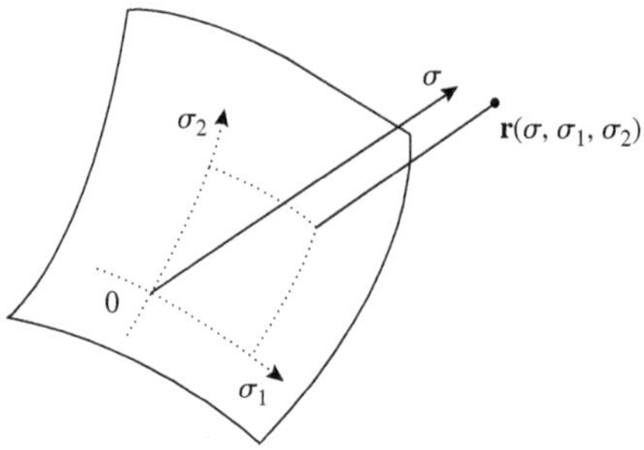

Figure 1.9 Curvilinear (ray) coordinate system

The flux on the walls comprising the wave fronts is given by

$$u^2(\sigma)d\Sigma(\sigma) = u^2(0)d\Sigma(0), \tag{1.27}$$

and since (1.26) and (1.27) are verified by all the three components of $\mathbf{e}_0$, we see that the flux of the square of the field amplitude is conserved in a tube of rays:

$$|\mathbf{e}_0(\sigma)|^2 d\Sigma(\sigma) = |\mathbf{e}_0(0)|^2 d\Sigma(0). \tag{1.28}$$

This result corresponds to the second postulate of GO. However, Equation (1.28) does not give any information on the direction of the field.

The transport Equation (1.25) can also be solved in a curvilinear coordinate system $(\sigma, \sigma_1, \sigma_2)$ where σ is along a ray and σ_1, σ_2 are coordinates along the principal directions of the wave front (see Figure 1.9).

This coordinate system is orthogonal. The general expression for the Laplacian is:

$$\Delta u = \frac{1}{h_1 h_2 h}\left\{\frac{\partial}{\partial\sigma}\left(\frac{h_1 h_2}{h}\frac{\partial u}{\partial\sigma}\right) + \frac{\partial}{\partial\sigma_1}\left(\frac{h_2 h}{h_1}\frac{\partial u}{\partial\sigma_1}\right) + \frac{\partial}{\partial\sigma_2}\left(\frac{h h_1}{h_2}\frac{\partial u}{\partial\sigma_2}\right)\right\}.$$

Here

$$h_1 = 1 + \frac{\sigma}{R_1}, \qquad h_2 = 1 + \frac{\sigma}{R_2}, \qquad h = 1,$$

where R_1 and R_2 are, respectively, the principal radii of curvature along the coordinate curves σ_1 and σ_2. Since S is constant on a wave front, we have:

$$\frac{\partial S}{\partial\sigma_1} = \frac{\partial S}{\partial\sigma_2} = 0,$$

and hence:

$$\Delta S = \frac{1}{R_1 + \sigma} + \frac{1}{R_2 + \sigma}. \tag{1.29}$$

Along a ray, we have:

$$\nabla S \cdot \nabla u = \sigma \cdot \nabla u = \frac{\partial u}{\partial\sigma}, \tag{1.30}$$

and since σ_1 and σ_2 are constant along a ray, we can replace $\partial u/\partial\sigma$ by $du/d\sigma$.

According to (1.29) and (1.30), the transport Equation (1.25) can be written as

$$2\frac{du}{d\sigma} + \left(\frac{1}{R_1 + \sigma} + \frac{1}{R_2 + \sigma}\right) u = 0. \tag{1.31}$$

Hence

$$\frac{u(\sigma)}{u(0)} = \sqrt{\frac{R_1 R_2}{(R_1 + \sigma)(R_2 + \sigma)}}$$

and

$$\mathbf{e}_0(\sigma) = \sqrt{\frac{R_1 R_2}{(R_1 + \sigma)(R_2 + \sigma)}}\,\mathbf{e}_0(0). \tag{1.32}$$

Equation (1.32) shows that the direction of the field $\mathbf{e}_0$ is invariant along a ray. As $d\Sigma(\sigma) \sim (R_1 + \sigma)(R_2 + \sigma)$, one can easily check that (1.32) implies (1.28). This result corresponds to the third postulate of GO.

Combining all the properties we have found leads us to the following important result: 'The GO field is the first term of the asymptotic expansion (1.18) also called the Luneberg–Kline expansion'.

However, as we will see, the laws of GO do not apply to the higher-order terms as there is no power conservation in a tube of rays because of the right-hand side of the transport Equation (1.20), which is different from zero for $n \neq 0$.

In order to solve the transport Equation for $n \neq 0$, it is convenient to rewrite Equation (1.31) in another form by introducing the function

$$\frac{J(\sigma)}{J(0)} = \frac{(R_1 + \sigma)(R_2 + \sigma)}{R_1 R_2},$$

which verifies

$$\frac{1}{J(\sigma)}\frac{dJ(\sigma)}{d\sigma} = \frac{1}{R_1 + \sigma} + \frac{1}{R_2 + \sigma}. \tag{1.33}$$

Inserting (1.33) in the transport Equation for the component u_n of $\mathbf{e}_n$, this equation reduces to the differential equation along the ray:

$$2\frac{du_n}{d\sigma} + \frac{u_n}{J}\frac{dJ}{d\sigma} = \frac{2}{\sqrt{|J|}}\frac{d}{d\sigma}(\sqrt{|J|}u_n) = -\Delta u_{n-1},$$

which can be integrated immediately to give

$$u_n(\sigma) = \sqrt{\left|\frac{J(0)}{J(\sigma)}\right|}u_n(0) - \frac{1}{2}\int_0^\sigma \sqrt{\left|\frac{J(\sigma')}{J(\sigma)}\right|}\Delta u_{n-1}(\sigma')\,d\sigma'. \tag{1.34}$$

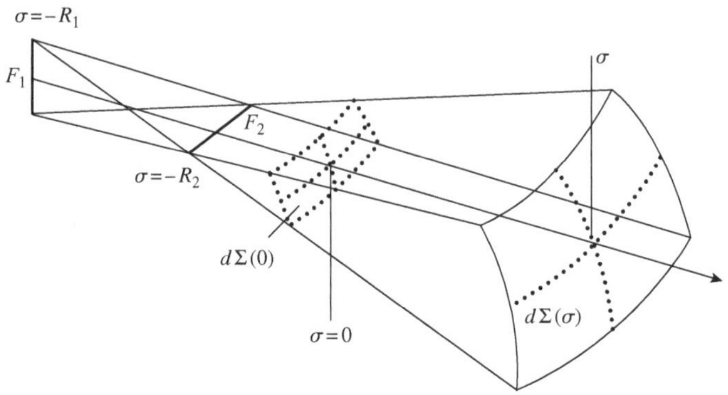

Figure 1.10 Astigmatic pencil of rays

This equation is verified by each component of the electric field and consequently:

$$\mathbf{e}_n(\sigma) = \sqrt{\left|\frac{J(0)}{J(\sigma)}\right|}\mathbf{e}_n(0) - \frac{1}{2}\int_0^{\sigma}\sqrt{\left|\frac{J(\sigma')}{J(\sigma)}\right|}\Delta\mathbf{e}_{n-1}(\sigma')\,d\sigma'. \tag{1.35}$$

The function $|J(\sigma)|$ is the Jacobian along a given ray. It is related to the cross section of a narrow pencil of rays $d\Sigma(\sigma)$ by the well-known formula:

$$d\Sigma(\sigma) = |J(\sigma)|d\sigma_1 d\sigma_2.$$

In Figure 1.10, four rays are shown defining an astigmatic pencil of rays with curvilinear cross section. From elementary geometry, we find:

$$\frac{d\Sigma(0)}{d\Sigma(\sigma)} = \left|\frac{R_1 R_2}{(R_1+\sigma)(R_2+\sigma)}\right| = \left|\frac{J(0)}{J(\sigma)}\right|. \tag{1.36}$$

The radii of curvature are taken positive if the rays emanating from the corresponding focus are divergent and negative if the rays are convergent. When $R_2 < 0$, the focus line F_2 lies ahead of the wave front. In other words, when we progress along the ray pencil from 0 to σ in the direction of wave propagation, we cross a focal line. Then a change of $-\pi/2$ has to be introduced in the phase to take into account the crossing of a caustic surface (crossing means here passing through a point of tangency of a ray with a caustic surface). This condition is verified if the absolute value in (1.36) is removed and if the following convention for the square roots in (1.36), and also in (1.32), is adopted: $\sqrt{R_{1,2}+\sigma}$ takes positive real, negative imaginary or zero values. The selection of the correct square root can be justified by the behaviour of the field near a caustic which must be analysed with a different Ansatz (see Section 1.2.9).

Let us now come back to (1.35). The formula enables one to continue $\mathbf{e}_n$ along a given ray.

For the zeroth-order amplitude $\mathbf{e}_0$, it is only necessary to know one initial value $\mathbf{e}_0$ at a reference point in order to carry out this continuation. We see also that the direction $\mathbf{e}_0$ which defines the polarisation of the zeroth-order field, remains constant along a ray. In addition, it is found from Equations (1.21) and (1.22) that:

$$\nabla S \cdot \mathbf{e}_0 = 0, \qquad \mathbf{h}_0 = \frac{1}{\eta}\nabla S \times \mathbf{e}_0,$$

which means that $(\mathbf{e}_0, \mathbf{h}_0, \nabla S)$ form a right-handed system of vectors. Thus the leading term of the asymptotic expansion (1.20) describes a local plane wave field. At a caustic where $\sigma = -R_1$ or $\sigma = -R_2$ vector $\mathbf{e}_0$ becomes infinite, hence (1.32) fails.

For a higher-order amplitude $\mathbf{e}_n$ $(n > 0)$, more information is needed in order to continue $\mathbf{e}_n$ along a given ray. In addition to the initial value $\mathbf{e}_n(0)$ function $\Delta\mathbf{e}_{n-1}(\sigma')$ must be known for all σ' in the range $0 < \sigma' < \sigma$. For $n = 1$, for instance, one needs $\Delta\mathbf{e}_0(\sigma')$ which implies the knowledge of the first and the second derivatives of R_1 and R_2 with respect to the transverse coordinates (σ_1, σ_2) on the initial wave front.

For $n > 0$, the direction of $\mathbf{e}_n$ remains constant along the ray, but it is no longer orthogonal to the ray. The corresponding Poynting vector is therefore not directed along the ray, but makes an angle with it which accounts for energy flow transverse to the ray. According to the hypothesis underlying the perturbation method, this energy flow and hence the corresponding angle, are assumed to be small.

The formulae which have been obtained in this chapter are not valid near the singularities of the ray field which are given by the zeros of the Jacobian of the transformation $(x_1, x_2, x_3) \to (\sigma_1, \sigma_2, \sigma)$:

$$J = \frac{D(x_1, x_2, x_3)}{D(\sigma_1, \sigma_2, \sigma)} = \det\left(\frac{\partial x_i}{\partial \sigma_j}\right), \qquad \text{with } \sigma_3 = \sigma,$$

where (x_1, x_2, x_3) are the Cartesian coordinates of a point in space and $(\sigma_1, \sigma_2, \sigma)$ are its 'ray coordinates' defined in Figure 1.9. The ray method can be applied only if the Cartesian coordinates are smooth functions of the ray coordinates and the Jacobian differs from zero. When this condition is satisfied, the field of rays is called regular. In the presence of an obstacle, discontinuities of the ray coordinates appear at shadow boundaries and caustics.

1.1.5 Field reflected by a smooth object

The model problem for the reflection by a smooth object is the reflection by an infinite plane. We know from the solution to this problem that the reflected field has the same asymptotic structure as the incident field. It can therefore be represented by an asymptotic expansion having the same asymptotic sequence.

We assume that the incident field is represented by the asymptotic expansion:

$$\begin{aligned}\mathbf{E}^i(\mathbf{r}) &= e^{ikS^i(\mathbf{r})}\sum_{n=0}^{N}(ik)^{-n}\mathbf{e}^i_n(\mathbf{r}) + o(k^{-N}),\\ \mathbf{H}^i(\mathbf{r}) &= e^{ikS^i(\mathbf{r})}\sum_{n=0}^{N}(ik)^{-n}\mathbf{h}^i_n(\mathbf{r}) + o(k^{-N}),\end{aligned} \tag{1.37}$$

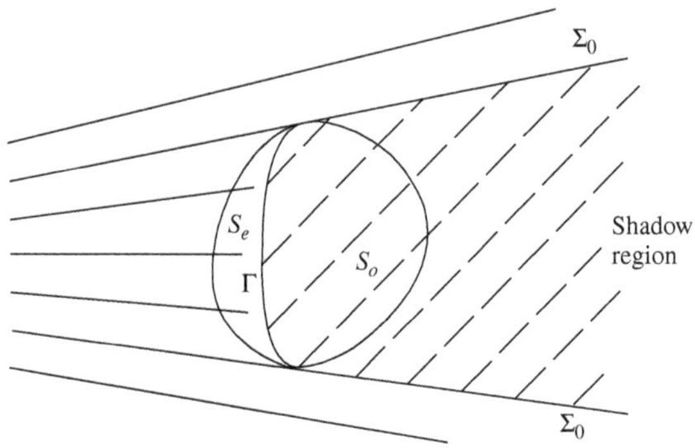

Figure 1.11 Incident ray field intercepted by a smooth object

where the phase $S^i(\mathbf{r})$ verifies the eikonal equation and where the amplitudes $\mathbf{e}_n^i(\mathbf{r})$, $\mathbf{h}_n^i(\mathbf{r})$ verify Equations (1.20)–(1.22).

A field of the form (1.37) is called a ray field. If it is limited to its zero-order term, it is a GO field.

We have seen that in a homogeneous medium, the rays are straight lines in the direction of ΔS. Some of them intercept the surface of the target and divide the space into an illuminated region and a shadow region separated by a surface Σ_0, which is called the shadow boundary of the incident field (see Figure 1.11).

The shadow region is tangential to the surface S of the object along the curve Γ separating the lit region S_e of S from the shadow region S_0. In the lit region, the incident field gives rise to an extended reflected field $(\mathbf{E}^R, \mathbf{H}^R)$, which we represent away from the shadow boundary and possible caustics by an asymptotic expansion similar to (1.37):

$$\mathbf{E}^R(\mathbf{r}) = e^{ikS^R(\mathbf{r})} \sum_{n=0}^{N} (ik)^{-n} \mathbf{e}_n^R(\mathbf{r}) + o(k^{-N}),$$
$$\mathbf{H}^R(\mathbf{r}) = e^{ikS^R(\mathbf{r})} \sum_{n=0}^{N} (ik)^{-n} \mathbf{h}_n^R(\mathbf{r}) + o(k^{-N}), \tag{1.38}$$

where the phase $S^R(\mathbf{r})$ and the amplitudes $(\mathbf{e}_n^R, \mathbf{h}_n^R)$ verify, respectively, the eikonal equation and the system (1.20)–(1.22) which are a consequence of Maxwell's equations.

As for the incident field, the characteristics or rays of the eikonal equation are orthogonal to the wave front $S^R(\mathbf{r}) = \text{const}$, and form a congruence in $\mathbb{R}^3$. They have therefore an envelope or caustic which can be located outside (real caustic) or inside (virtual caustic) of the target.

The asymptotic expansions (1.37) and (1.38) are valid at every point in the illuminated region of $\mathbb{R}^3$ with the exception of those points located on, or in the vicinity of, the shadow boundaries and caustics. It is therefore possible to use these expansions at observation points located on the lit side S_e of S not too close to Γ and apply the boundary conditions at those points.

If $\mathbf{n}$ is the unit normal to S oriented to the outside of the volume delimited by S, we have:

For a perfectly conducting body:

$$\mathbf{n} \times \mathbf{E}_t = 0, \qquad \mathbf{r} \in S. \tag{1.39}$$

For an imperfectly conducting body characterised by an impedance Z:

$$\mathbf{E}_t - (\mathbf{n} \cdot \mathbf{E}_t)\mathbf{n} = Z\mathbf{n} \times \mathbf{H}_t, \qquad \mathbf{r} \in S, \tag{1.40}$$

where $(\mathbf{E}_t, \mathbf{H}_t)$ is the total field on S.

The asymptotic expansions (1.37) and (1.38) do not give the total field on S, but only an approximation of it. We will see that other diffraction phenomena occur such as creeping waves for a regular object or waves diffracted by edges if the surface S is not regular. But since the asymptotic expansions that are associated with these diffraction phenomena are defined with respect to different asymptotic sequences (other than k^{-n}) the boundary conditions (1.39) and (1.40) are separately verified by each species of wave. This property holds on for multiple reflected rays, since in such a case, the phase function of the asymptotic expansion is different which implies that the boundary conditions must be applied separately for each order of interaction.

Accordingly, we impose the conditions:

$$\mathbf{n} \times \mathbf{E}_a = 0, \qquad \mathbf{r} \in S \tag{1.41}$$

or

$$\mathbf{E}_a - (\mathbf{n} \cdot \mathbf{E}_a)\mathbf{n} = Z\mathbf{n} \times \mathbf{H}_a, \qquad \mathbf{r} \in S, \tag{1.42}$$

where

$$\mathbf{E}_a = \mathbf{E}^i + \mathbf{E}^R, \qquad \mathbf{H}_a = \mathbf{H}^i + \mathbf{H}^R,$$

with $(\mathbf{E}^i, \mathbf{H}^i)$ and $(\mathbf{E}^R, \mathbf{H}^R)$ given by (1.37) and (1.38), respectively.

The conditions (1.41) and (1.42) involve the continuity of the phase at every point on S_e:

$$S^i(\mathbf{r}) = S^R(\mathbf{r}), \qquad \mathbf{r} \in S_e. \tag{1.43}$$

The surface gradient of the eikonal is simply the projection of $\nabla S^i = \mathbf{s}^i$ of the incident ray on the tangent plane to the surface. According to (1.43), the projection of ∇S^i must be equal to the projection of $\nabla S^R = \mathbf{s}^R$. The direction $\mathbf{s}^R$ of the reflected field is pointed outward from the surface. Hence:

$$\mathbf{s}^R = \mathbf{s}^i - 2\mathbf{n}(\mathbf{n} \cdot \mathbf{s}^i),$$

which is the law of reflection in the lit region. In the shadow region:

$$\mathbf{s}^R = \mathbf{s}^i.$$

Thus, the reflected field in the shadow region is the negative of the incident field and the total field there vanishes. On Γ, vector $\mathbf{s}^i$ is tangent to S and consequently we are on the caustic of the reflected field.

On the surface S_e, the amplitudes $\mathbf{e}_n$ verify:

$$\mathbf{n} \times (\mathbf{e}_n^i + \mathbf{e}_n^R) = 0 \quad \text{or} \quad \mathbf{e}_n^i + \mathbf{e}_n^R - [\mathbf{n} \cdot (\mathbf{e}_n^i + \mathbf{e}_n^R)]\mathbf{n} = Z\mathbf{n} \times (\mathbf{h}_n^i + \mathbf{h}_n^R), \tag{1.44}$$

hence on a perfectly conducting surface, the projection of the incident electric field on the tangent plane of the surface at a point Q is equal to the opposite of the projection of the reflected electric field.

For $n = 0$, Equation (1.32) holds for $\mathbf{e}_0^R(\sigma)$

$$\mathbf{e}_0^R(\sigma) = \sqrt{\frac{\rho_1^R \rho_2^R}{(\rho_1^R + \sigma)(\rho_2^R + \sigma)}}\, \mathbf{e}_0^R(0), \tag{1.45}$$

where ρ_1^R, ρ_2^R are the principal radii of curvature of the reflected wave front at the point of reflection.

If we decompose $\mathbf{e}_0^i$ into its components respectively parallel and perpendicular to the plane of incidence, we get:

$$\mathbf{e}_0^i = (\mathbf{e}_0^i \cdot \mathbf{e}_\parallel^i)\mathbf{e}_\parallel^i + (\mathbf{e}_0^i \cdot \mathbf{e}_\perp^i)\mathbf{e}_\perp^i, \tag{1.46}$$

where $\mathbf{e}_\perp^i$ is a unit vector perpendicular to the plane of incidence and $\mathbf{e}_\parallel^i$ is a unit vector parallel to the plane of incidence (Figure 1.12) so that:

$$\mathbf{e}_\parallel^i = \mathbf{e}_\perp^i \times \mathbf{s}^i. \tag{1.47}$$

The same decomposition is used for the reflected field $\mathbf{e}_0^R$ with

$$\mathbf{e}_0^R = (\mathbf{e}_0^R \cdot \mathbf{e}_\parallel^R)\mathbf{e}_\parallel^R + (\mathbf{e}_0^R \cdot \mathbf{e}_\perp^R)\mathbf{e}_\perp^R, \tag{1.48}$$

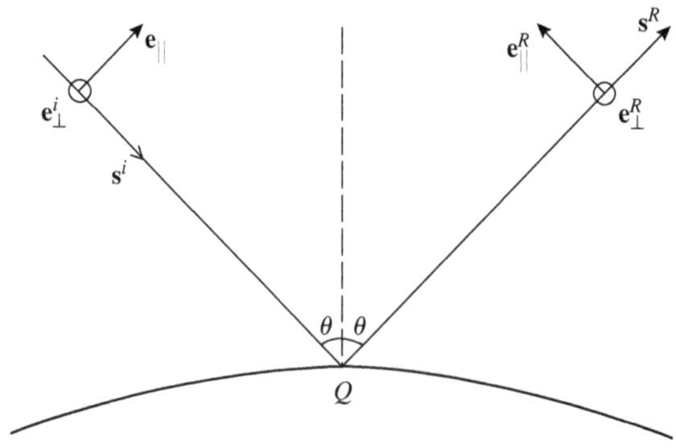

Figure 1.12 Ray fixed coordinate system for reflection

and

$$\mathbf{e}_{\parallel}^{R} = \mathbf{e}_{\perp}^{R} \times \mathbf{s}^{R}.$$

If (1.46) and (1.47) are inserted in the first equation of (1.44), we obtain:

$$\mathbf{e}_0^R(0) = \mathbf{e}_0^i(0) \cdot \mathcal{R}, \tag{1.49}$$

where $\mathcal{R}$ is the reflection dyadic given by:

$$\mathcal{R} = R_{\mathrm{TE}}\mathbf{e}_{\parallel}^{i}\mathbf{e}_{\parallel}^{R} + R_{\mathrm{TM}}\mathbf{e}_{\perp}^{i}\mathbf{e}_{\perp}^{R}, \tag{1.50}$$

with $R_{\mathrm{TE}} = -1$ and $R_{\mathrm{TM}} = 1$. Note that $\mathbf{e}_{\perp}^{i} = \mathbf{e}_{\perp}^{R} = \mathbf{e}_{\perp}$, unit vector orthogonal to the plane of incidence.

For an imperfectly conducting surface, we obtain from (1.46), (1.47) and the second equation of (1.44), the same relation (1.49) and (1.50) with:

$$R_{\mathrm{TM}} = \frac{Z_0 \cos\theta - Z}{Z_0 \cos\theta + Z}, \qquad R_{\mathrm{TE}} = \frac{Z \cos\theta - Z_0}{Z \cos\theta + Z_0},$$

where Z_0 is the impedance of vacuum.

Inserting (1.49) into (1.45) and in the first term of the asymptotic expansion of $\mathbf{E}^R$, we obtain the general formula for the GO reflected field at an observation point P:

$$\mathbf{E}_{\mathrm{GO}}^{R}(P) = \mathbf{E}^{i}(Q) \cdot \mathcal{R}\sqrt{\frac{\rho_1^R \rho_2^R}{(\rho_1^R + \sigma)(\rho_2^R + \sigma)}}\, e^{ik\sigma}, \tag{1.51}$$

where Q is the point of reflection, $\sigma = |QP|$ and:

$$\mathbf{E}^{i}(Q) = \mathbf{e}_0^{i}(Q) e^{ikS^{i}(Q)}.$$

General formulae for ρ_1^R and ρ_2^R are given in the literature [15–17].

For the calculation of the higher-order terms ($n > 0$), (1.45) is replaced by:

$$\mathbf{e}_n^R(\sigma) = \sqrt{\left|\frac{J(0)}{J(\sigma)}\right|}\mathbf{e}_n^R(0) - \frac{1}{2}\int_0^{\sigma} \sqrt{\left|\frac{J(\sigma')}{J(\sigma)}\right|}\Delta \mathbf{e}_{n-1}^R(\sigma')\, d\sigma'.$$

To this equation, we must also add the Gauss law:

$$\nabla S^R \cdot \mathbf{e}_n^R(\sigma) = -\nabla \cdot \mathbf{e}_{n-1}^R(\sigma).$$

Explicit expressions for an arbitrary convex object have only been established for $n = 1$ [18].

1.1.6 Field transmitted through a smooth interface between two different media with constant refractive indexes

The model problem for the transmission of an incident wave through a smooth interface separating two different media with constant refractive indexes is the transmission of an incident plane wave through a planar interface between two different media. We know from the solution to this problem that the transmitted field has the same asymptotic structure as the incident field. It can therefore be represented by an asymptotic expansion similar to (1.37), but with k replaced by the wave number of the medium in which the field is propagating.

Let us denote by k_1 and k_2 the wave numbers of media 1 and 2, respectively, and by n_1 and n_2 the corresponding refractive indexes.

We assume that the incident field is represented by the asymptotic expansion (1.37) with k replaced by k_1. Away from the shadow boundaries, the transmitted field is represented by:

$$\begin{aligned}\mathbf{E}^t(\mathbf{r}) &= e^{ik_2 S^t(\mathbf{r})} \sum_{n=0}^{N} (ik_2)^{-n} \mathbf{e}_n^t(\vec{r}) + o(k_2^{-N}),\\ \mathbf{H}^t(\mathbf{r}) &= e^{ik_2 S^t(\mathbf{r})} \sum_{n=0}^{N} (ik_2)^{-n} \mathbf{h}_n^t(\vec{r}) + o(k_2^{-N}),\end{aligned} \tag{1.52}$$

where the phase $S^t(\mathbf{r})$ and the amplitudes $(\mathbf{e}_n^t, \mathbf{h}_n^t)$ verify, respectively, the eikonal equation and the system (1.20)–(1.22) applied to medium 2. If $\mathbf{n}$ is the unit vector normal to the interface S between media 1 and 2 oriented towards medium 1, the boundary conditions impose the continuity of the tangential components of the total electric and magnetic field through S:

$$\mathbf{n} \times \mathbf{E}_1 = \mathbf{n} \times \mathbf{E}_2, \quad \mathbf{n} \times \mathbf{H}_1 = \mathbf{n} \times \mathbf{H}_2, \qquad \mathbf{r} \in S, \tag{1.53}$$

where $(\mathbf{E}_1, \mathbf{H}_1)$ and $(\mathbf{E}_2, \mathbf{H}_2)$ are the total field in media 1 and 2, respectively. As in the case of reflection by an impenetrable object, we can replace these fields by their asymptotic approximations:

$$\begin{aligned}\mathbf{E}_1 &= \mathbf{E}^i + \mathbf{E}^R, \qquad \mathbf{E}_2 = \mathbf{E}^t,\\ \mathbf{H}_1 &= \mathbf{H}^i + \mathbf{H}^R, \qquad \mathbf{H}_2 = \mathbf{H}^t,\end{aligned} \tag{1.54}$$

where $(\mathbf{E}^i, \mathbf{H}^i)$, $(\mathbf{E}^R, \mathbf{H}^R)$ and $(\mathbf{E}^t, \mathbf{H}^t)$ are, respectively, given by the asymptotic expansions (1.37), (1.38) and (1.52). These asymptotic expansions do not give the total field on both sides of the interface, but only an approximation of it. Waves propagating along the interface, for instance, are not included. However, as we mentioned in Section 1.1.5, the boundary conditions (1.53) are separately verified by each species of wave. Accordingly, we impose the conditions:

$$\mathbf{n} \times (\mathbf{E}^i + \mathbf{E}^R) = \mathbf{n} \times \mathbf{E}^t, \quad \mathbf{n} \times (\mathbf{H}^i + \mathbf{H}^R) = \mathbf{n} \times \mathbf{H}^t, \qquad \mathbf{r} \in S. \tag{1.55}$$

Since k_1 and k_2 are supposed to be large but arbitrary, Equations (1.55) imply that:

$$k_1 S^i(\mathbf{r}) = k_1 S^R(\mathbf{r}) = k_2 S^t(\mathbf{r}), \qquad \mathbf{r} \in S. \tag{1.56}$$

The surface gradient of the eikonal is simply the projection of $\nabla S = \mathbf{s}$ on the tangent plane to the interface. Hence, according to (1.56), the vectors $\mathbf{s}^r$ and $\mathbf{s}^t$ are in the incident plane defined by $\mathbf{s}^i$ and $\mathbf{n}$ and

$$k_1 \sin\theta^i = k_1 \sin\theta^R = k_2 \sin\theta^t. \tag{1.57}$$

Thus, the transmitted wave satisfies the law of refraction:

$$n_1 \sin\theta^i = n_2 \sin\theta^t. \tag{1.58}$$

In order to derive the formulae for the amplitudes $\mathbf{e}_n$ from (1.55), we express the asymptotic expansion in (1.52) with respect to the asymptotic sequence $(ik_1)^{-n}$ by using the identity:

$$(ik_2)^{-n} = (ik_1)^{-n}\left(\frac{n_2}{n_1}\right)^{-n}. \tag{1.59}$$

Inserting (1.59) into (1.52) and applying (1.55), we obtain:

$$\begin{aligned}\mathbf{n}\times(\mathbf{e}_n^i+\mathbf{e}_n^R) &= \left(\frac{n_2}{n_1}\right)^{-n}(\mathbf{n}\times\mathbf{e}_n^t),\\ \mathbf{n}\times(\mathbf{h}_n^i+\mathbf{h}_n^R) &= \left(\frac{n_2}{n_1}\right)^{-n}(\mathbf{n}\times\mathbf{h}_n^t).\end{aligned} \tag{1.60}$$

For $n = 0$, we know from the results of Section 1.1.4 that the field $(\mathbf{e}_0^t, \mathbf{h}_0^t)$ is a GO field. It can therefore be expanded in the sum of a transverse electric (TE) and a transverse magnetic (TM) field with respect to the plane of refraction. By using the ray fixed coordinate system (see Figure 1.13) defined by the unit vectors $\mathbf{e}_\perp^t$ and $\mathbf{e}_\parallel^t$ respectively perpendicular and parallel to the plane of refraction which is identical to the plane of incidence, and the unit vector $\mathbf{s}^t$ along the refracted ray verifying

$$\mathbf{e}_\parallel^t = \mathbf{e}_\perp^t \times \mathbf{s}^t. \tag{1.61}$$

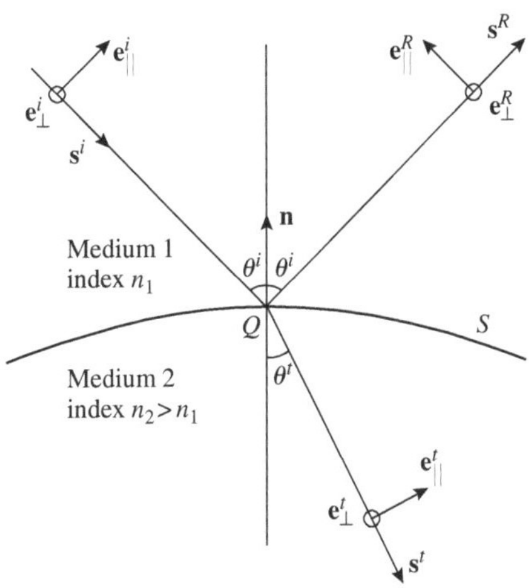

Figure 1.13 Ray fixed coordinate system for transmission. Note that $\mathbf{e}_\perp^i = \mathbf{e}_\perp^R = \mathbf{e}_\perp^t = \mathbf{e}_\perp$

Field, in medium 2 can be written

$$\mathbf{e}_0^t = (\mathbf{e}_0^t \cdot \mathbf{e}_\parallel^t)\mathbf{e}_\parallel^t + (\mathbf{e}_0^t \cdot \mathbf{e}_\perp^t)\mathbf{e}_\perp^t. \tag{1.62}$$

The first term on the right-hand side of (1.62) is a TM field whereas the second term is a TE field. Hence, by using a similar decomposition for $\mathbf{e}_0^i$ on the interface (see (1.46)), we have:

$$\mathbf{e}_0^t = \mathbf{e}_0^i \cdot T, \qquad \mathbf{r} \in S, \tag{1.63}$$

where T is the transmission dyadic given by ($\mathbf{e}_\perp^i = \mathbf{e}_\perp^t = \mathbf{e}_\perp$):

$$T = T_{\mathrm{TM}}\mathbf{e}_\parallel^i\mathbf{e}_\parallel^t + T_{\mathrm{TE}}\mathbf{e}_\perp\mathbf{e}_\perp. \tag{1.64}$$

In (1.64), T_{TM} and T_{TE} are the transmission coefficients from medium 1 to medium 2 for a TM and a TE field, respectively.

A similar decomposition to (1.62) holds for $\mathbf{e}_0^R$ (see (1.48)) and leads to (1.49) and (1.50) in which R_{TM} and R_{TE} are the reflection coefficients on the interface of a field incident from medium 1.

In order to derive explicit expressions for the transmission and reflection coefficients, we insert (1.62) and its homologues for $\mathbf{e}_0^R, \mathbf{h}_0^i, \mathbf{h}_0^R, \mathbf{h}_0^t$ into the boundary conditions (1.60) for $n = 0$. Then we replace the terms

$$(\mathbf{e}_0^\alpha \cdot \mathbf{e}_\parallel^\alpha)\mathbf{e}_\parallel^\alpha, \qquad (\mathbf{h}_0^\alpha \cdot \mathbf{e}_\parallel^\alpha)\mathbf{e}_\parallel^\alpha, \tag{1.65}$$

respectively by their TM and TE counterparts:

$$\frac{1}{\eta_\alpha}(\mathbf{e}_0^\alpha \cdot \mathbf{e}_\parallel^\alpha)\mathbf{e}_\perp^\alpha, \qquad \eta_\alpha(\mathbf{h}_0^\alpha \cdot \mathbf{e}_\parallel^\alpha)\mathbf{e}_\perp^\alpha, \tag{1.66}$$

where α stands for i, R or t and η_α is the intrinsic impedance of the corresponding medium. Owing to (1.66) the following relations resulting from the definitions of the transmission and reflection coefficients, hold:

$$\begin{aligned} (\mathbf{e}_0^t \cdot \mathbf{e}_\parallel^t) &= T_{\mathrm{TM}}(\mathbf{e}_0^i \cdot \mathbf{e}_\parallel^i), & \mathbf{e}_0^R \cdot \mathbf{e}_\parallel^R &= R_{\mathrm{TM}}(\mathbf{e}_0^i \cdot \mathbf{e}_\parallel^i), \\ \eta_2(\mathbf{h}_0^t \cdot \mathbf{e}_\parallel^t) &= T_{\mathrm{TE}}\eta_1(\mathbf{h}_0^i \cdot \mathbf{e}_\parallel^i), & \mathbf{h}_0^R \cdot \mathbf{e}_\parallel^R &= R_{\mathrm{TE}}(\mathbf{h}_0^i \cdot \mathbf{e}_\parallel^i). \end{aligned} \tag{1.67}$$

Inserting relations (1.67) into (1.65) and taking into account that

$$\mathbf{n} \times \mathbf{e}_\parallel^i = -\cos\theta^i\mathbf{e}_\perp, \qquad \mathbf{n} \times \mathbf{e}_\parallel^R = \cos\theta^i\mathbf{e}_\perp, \qquad \mathbf{n} \times \mathbf{e}_\parallel^t = -\cos\theta^t\mathbf{e}_\perp$$

leads to:

$$\begin{aligned} R_{\mathrm{TM}} &= \frac{\eta_1\cos\theta^i - \eta_2\cos\theta^t}{\eta_1\cos\theta^i + \eta_2\cos\theta^t}, & R_{\mathrm{TE}} &= \frac{(1/\eta_1)\cos\theta^i - (1/\eta_2)\cos\theta^t}{(1/\eta_1)\cos\theta^i + (1/\eta_2)\cos\theta^t}, \\ T_{\mathrm{TM}} &= \frac{2\eta_2\cos\theta^i}{\eta_1\cos\theta^i + \eta_2\cos\theta^t}, & T_{\mathrm{TE}} &= \frac{(2/\eta_1)\cos\theta^i}{(1/\eta_1)\cos\theta^i + (1/\eta_2)\cos\theta^t}. \end{aligned} \tag{1.68}$$

Away from the interface, (1.32) holds for $\mathbf{e}_0^R(\sigma)$ and $\mathbf{e}_0^t(\sigma)$.

Hence, the general formula for the GO refracted field at an observation point P is given by:

$$\mathbf{E}^t_{\mathrm{GO}}(P) = \mathbf{E}^i(Q) \cdot T \sqrt{\frac{\rho_1^t \rho_2^t}{(\rho_1^t + \sigma)(\rho_2^t + \sigma)}} e^{ik\sigma}, \tag{1.69}$$

where Q is the point of refraction, $\sigma = |QP|$ and ρ_1^t, ρ_2^t are the principal radii of curvature of the refracted wave front at Q.

General formulae for ρ_1^t and ρ_2^t are also given in References 15–17.

The formula for the reflected field is identical to (1.51). However, the reflection coefficients in (1.50) are those given by (1.68).

Higher-order terms ($n > 0$) can also be calculated by using (1.35) together with the Gauss law (1.21) and the boundary conditions (1.60).

1.1.7 Field diffracted by the edge of a curved wedge

We consider now an object, the surface of which has a discontinuity line in the tangent plane forming a sharp edge which may be straight or curved (see Figure 1.14). Away from this discontinuity line, the surface is assumed to be smooth with the radii of curvature large in terms of the wavelength.

An incident ray field defined by the asymptotic expansion (1.37) hits the boundary surface of the object giving rise to a scattered field $\mathbf{E}^S$ which can be written as the sum of a general reflected field $\mathbf{E}^R$ given by (1.38) and an additional term $\mathbf{E}^d$, due to the presence of the edge, called the diffracted field:

$$\mathbf{E}^S(\mathbf{r}) = \mathbf{E}^R(\mathbf{r}) + \mathbf{E}^d(\mathbf{r}). \tag{1.70}$$

We suppose that $\mathbf{E}^d$ has the general form:

$$\mathbf{E}^d(\mathbf{r}) = \frac{1}{\sqrt{k}} e^{ikS^d(\mathbf{r})} \sum_{n=0}^{N} (ik)^{-n} \mathbf{e}_n^d(\mathbf{r}) + o(k^{-N}). \tag{1.71}$$

The Ansatz (1.71) together with the decomposition (1.70) are suggested by the solution of the diffraction problem of an incident plane wave by a straight wedge which is

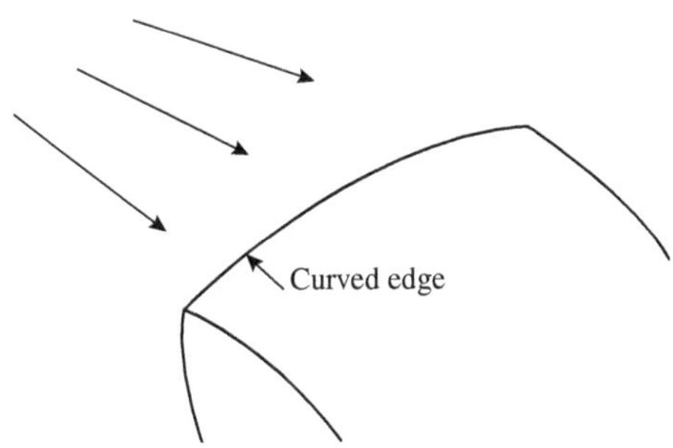

Figure 1.14 General shape of a curved wedge

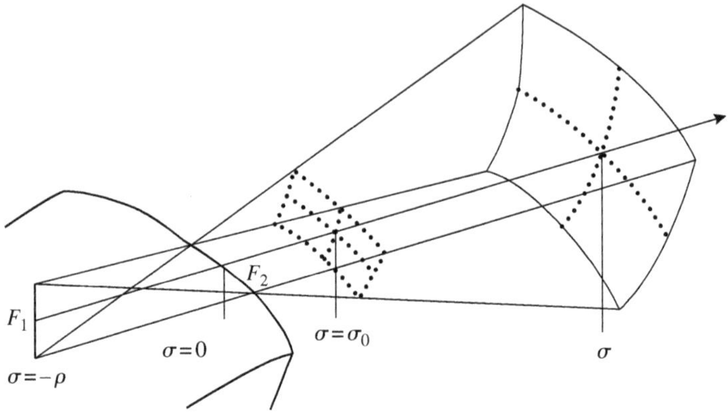

Figure 1.15 Astigmatic pencil of diffracted rays

the corresponding model or canonical problem. This form of the solution is not valid close to the edge and to the shadow boundaries of the reflected field where the field of rays is singular.

A similar form is introduced for the magnetic field $\mathbf{H}^d(\mathbf{r})$.

If we incorporate the factor $k^{-1/2}$ into the asymptotic sequence, we get an asymptotic expansion with respect to the asymptotic sequence $k^{-(n+1/2)}$.

If we insert (1.71) into Maxwell's equations (1.17), we see that $S^d(\mathbf{r})$ and $\mathbf{e}_n^d(\mathbf{r})$ verify the eikonal and transport equations (1.19) and (1.20) together with the Gauss law (1.21), the solutions of which have already been established. The diffracted field is therefore a ray field. Moreover, the diffracted rays emanate from the wedge since those emanating from a regular point of the surface are already accounted for in the expression of the general reflected field. Figure 1.15 shows an astigmatic pencil of diffracted rays. Since all diffracted rays pass through the edge, the edge is one of the caustic surfaces (here degenerated into a line) of the family of edge diffracted rays.

The family of diffracted rays with the properties shown in Figure 1.15 does not define completely the phase function $S^d(\mathbf{r})$. Indeed, we know that the eikonal equation is a first-order partial differential equation the solution of which is completely determined if we impose some complementary conditions. For example, we can impose the values of the unknown function on a regular surface which may be an initial wave front or a regular part of the illuminated surface of the body. This procedure is adequate for the reflected field since the values of the phase function $S^R(\mathbf{r})$ of the family of reflected rays, at regular points of the illuminated surface of the body, are related to those of the phase function $S^i(\mathbf{r})$ of the incident ray family by the boundary conditions and are therefore known. However, this procedure cannot be applied to the present problem since the boundary conditions are not defined on an edge. But, for the reflected field, if instead of the boundary conditions, we impose the continuity of phase on the surface of the body, we obtain the same result for $S^R(\mathbf{r})$. We extend this procedure to the edge diffracted field which consists in imposing the continuity of the phase on the edge between the incident field and the edge diffracted field:

$$S^i(\mathbf{r}) = S^d(\mathbf{r}), \qquad \mathbf{r} \in C. \tag{1.72}$$

This condition defines completely the solution of the eikonal equation.

According to (1.72) the projection of the gradient ∇S^i of the eikonal on the tangent $\mathbf{t}$ to the edge must be equal to the projection of the gradient ∇S^d on the same tangent. Hence, since $\nabla S^i = \mathbf{s}^i$, $\nabla S^d = \mathbf{s}^d$, we have (see Figure 1.16)

$$\mathbf{s}^i \cdot \mathbf{t} = \mathbf{s}^d \cdot \mathbf{t} = \cos\beta,$$

which is the law of edge diffraction already established in Section 1.1.2 by applying the generalised Fermat's principle: the diffracted rays at a given point on the edge make the same angle β with the tangent to the edge as the incident ray. The diffracted rays form therefore a cone which is known as the Keller's cone (see Figure 1.16).

In order to determine completely the diffracted field, it is necessary to find a way to relate the amplitudes $\mathbf{e}_n^d$ of the diffracted field to the amplitudes $\mathbf{e}_n^i$ of the incident field, which are known. Since the edge C is a caustic of the diffracted field, the field predicted by the solution (1.35) of the transport equation is infinite on C and it is therefore not possible to match the amplitude of the diffracted field with that of the incident field on C. However, if we relate $\mathbf{e}_n^d(\sigma)$ to its value $\mathbf{e}_n^d(\sigma_0)$ at a reference wave front $S(\sigma_0)$ and if we decompose the integral along the ray in the form:

$$\int_{\sigma_0}^{\sigma} = \int_{0}^{\sigma} - \int_{0}^{\sigma_0},$$

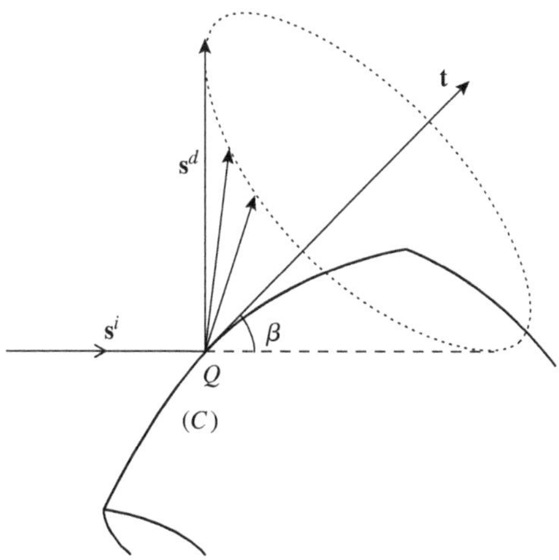

Figure 1.16 Keller's cone and unit vectors $\mathbf{s}^i, \mathbf{s}^d, \mathbf{t}$

where the finite parts of the divergent integrals are taken on the right-hand side, we get an alternative form for (1.35):

$$\sqrt{|J(\sigma)|}\mathbf{e}_n^d(\sigma) + \frac{1}{2}\int_0^{\sigma} \sqrt{|J(\sigma')|}\Delta\mathbf{e}_{n-1}^d(\sigma')\,d\sigma'$$
$$= \sqrt{|J(\sigma_0)|}\mathbf{e}_n^d(\sigma_0) + \frac{1}{2}\int_0^{\sigma_0} \sqrt{|J(\sigma')|}\Delta\mathbf{e}_{n-1}^d(\sigma')\,d\sigma'. \tag{1.73}$$

The right-hand side of (1.73) is independent of σ. If we denote its value by $\boldsymbol{\delta}_n$, we obtain:

$$\mathbf{e}_n^d(\sigma) = \frac{\boldsymbol{\delta}_n}{\sqrt{|J(\sigma)|}} - \frac{1}{2}\int_0^{\sigma} \sqrt{\left|\frac{J(\sigma')}{J(\sigma)}\right|}\Delta\mathbf{e}_{n-1}^d(\sigma')\,d\sigma'.$$

Taking the limit $\sigma \to 0$, we see that $\boldsymbol{\delta}_n$ can be written as:

$$\boldsymbol{\delta}_n = \lim_{\sigma\to 0}(\mathbf{e}_n^d(\sigma)\sqrt{|J(\sigma)|}) = a\lim_{\sigma\to 0}(\mathbf{e}_n^d(\sigma)\sqrt{\sigma}), \tag{1.74}$$

where a is a constant since $\sqrt{|J(\sigma)|}$ behaves like $\sqrt{\sigma}$ when σ tends to zero.

For $n = 0$ this limit must be the same as for a straight wedge. This is somewhat surprising since the expansion (1.71) is not valid in the vicinity of the edge, neither for the model problem (straight wedge) nor for the original problem (curved wedge). But since the inner expansions valid close to the edge for the model and original problems have the same asymptotic behaviour for the dominant term, it must be the same for the dominant term ($n = 0$) of the outer expansion (1.74) when σ tends to zero in order that both outer expansions match to the corresponding inner expansions. If $\mathbf{e}_0'^d(\sigma)$ is the zeroth-order term of the asymptotic expansion for a straight wedge, we have therefore:

$$\boldsymbol{\delta}_0 = a\lim_{\sigma\to 0}(\mathbf{e}_0'^d(\sigma)\sqrt{\sigma})$$

and since:

$$\mathbf{e}_0'^d(\sigma) = \frac{1}{\sqrt{\sigma}}\mathbf{e}_0^i(Q)\cdot D_e$$

(D_e is dyadic diffraction coefficient given in Chapter 3) we obtain by taking into account that $|J(\sigma)|$ is proportional to $(\rho + \sigma)\sigma$:

$$\mathbf{e}_0^d(\sigma) = \mathbf{e}_0^i(Q)\cdot D_e\sqrt{\frac{\rho}{\sigma(\rho+\sigma)}},$$

where ρ is the radius of curvature of the diffracted wave front in the plane of diffraction $(\mathbf{t}, \mathbf{s}^d)$. Finally the field at an observation point P is given by:

$$\mathbf{E}^d(P) = \mathbf{E}^i(Q)\cdot D_e\sqrt{\frac{\rho}{\sigma(\rho+\sigma)}}\,e^{ik\sigma},$$

where Q is the point of diffraction on the edge.

For higher-order terms ($n > 0$) the tangent plane approximation is no longer valid and the solution to the problem can only be established by using the boundary-layer method close to the edge and a matching procedure with the solution away from the edge given by the present approach.

1.1.8 Field in the shadow zone of a smooth convex object (creeping rays)

The corresponding model problem is the diffraction of a plane wave by a circular cylinder. We saw that the GO predicts a vanishing field in the shadow zone of a smooth convex object. In effect, the asymptotic expansion of Luneberg–Kline type given by (1.38), is not adapted to this zone where the asymptotic solution of the model problem shows that the field decreases as $\exp(-\alpha k^{1/3})$, that is, the decay is faster than algebraic. The field has locally the character of an inhomogeneous plane wave. To represent such a wave we can either extend the Luneberg–Kline expansion so that it represents a complex wave with phase and amplitude functions that are both complex or retain the real representation and simply augment the phase factor $\exp(ikS^d(\mathbf{r}))$ by an exponential attenuation factor.

When adopting the latter point of view, the solution of the circular cylinder problem in the deep shadow zone, far from the surface, suggests the following form for the total field known as the Friedlander and Keller expansion:

$$\mathbf{E}(\mathbf{r}) = e^{ikS(\mathbf{r})+ik^{1/3}p(\mathbf{r})} \sum_{n=0}^{N} (ik)^{-n/3} \mathbf{e}_n(\mathbf{r}), \tag{1.75}$$

with a similar expression for $\mathbf{H}(\mathbf{r})$.

Inserting this expansion into Maxwell's equations and ordering the terms, we obtain at order $O(k^2)$ the result:

$$|\nabla S|^2 = 1,$$

which is the eikonal equation of GO. Hence, far from the object, the eikonal $S(\mathbf{r})$ satisfies the eikonal equation of GO. In addition, in a homogeneous medium, the equiphase surfaces of (1.75) are orthogonal to the surfaces of equal amplitudes given by $\operatorname{Im} p(\mathbf{r}) = \text{const}$. The latter are consequently generated by the rays which are straight lines and orthogonal to the surface $S(\mathbf{r}) = \text{const}$. Hence, the value of $p(\mathbf{r})$ is constant on each ray.

It can also be shown that the amplitude vectors $\mathbf{e}_n$ verify linear ordinary differential equations along the rays similar to (1.20) and that $\mathbf{e}_0$ verifies the transport equation of GO. However, the right-hand side of the equations verified by the higher-order terms are different and depend on the derivatives of the function $p(\mathbf{r})$. Finally, the arbitrary elements in the construction of the solution are the value of $S(\mathbf{r})$ and $\mathbf{e}_n(\mathbf{r})$ on some surface and the value of $p(\mathbf{r})$ on each ray. These quantities may be adjusted in order that the expansion corresponds to the solution of a particular boundary value problem.

For $n = 0$, we know from the solution of the cylinder and the sphere that creeping waves originate at the shadow boundary, propagate along geodesics of the surface

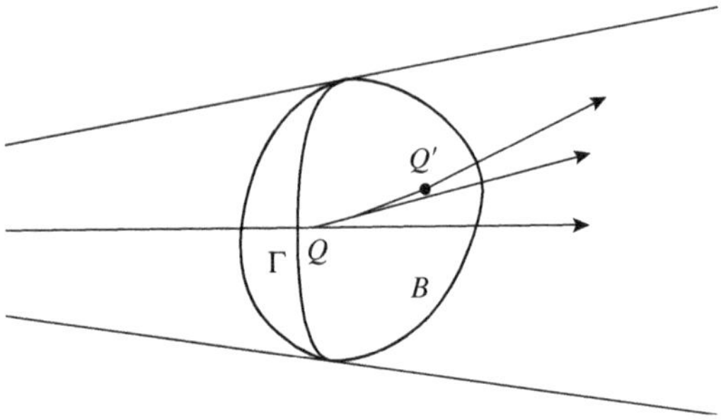

Figure 1.17 Congruence of surface diffracted rays

and continually shed diffracted rays which irradiate the shadow and also enter in the illuminated region. Hence, in a homogeneous medium, each incident ray tangential to the surface B of a smooth object at Q gives rise to a one parameter family of diffracted rays tangential to the surface B along that geodesic of B which is tangential to the incident ray at Q and starts from Q in the direction of the shadow side of B. Since Q is located on the shadow boundary curve Γ, the diffracted rays form a two parameter family or congruence of rays (see Figure 1.17).

We choose this congruence of rays as the solution of the eikonal equation. In order to define completely the phase function $S(\mathbf{r})$, the continuity of the phase is imposed along the shadow curve Γ:

$$S(Q') = S^i(Q) + kt,$$

where t is the length of the arc of geodesic between Q on Γ and the point Q', where the ray leaves the surface B.

The diffracted rays shed from the surface are tangent to B, which is therefore a caustic surface for these rays. Hence, the boundary conditions on the surface cannot be applied to the diffracted field which is infinite there. In order to calculate $\mathbf{e}_0(\sigma)$, a similar procedure to that for the edge diffracted field has to be used. It consists in taking the limit when σ tends to zero of the term $\sqrt{|J|}\mathbf{e}_0(\sigma)$, which enables one to relate $\mathbf{e}_0(\sigma)$ to the creeping wave field at Q'. The creeping wave field at Q' can be related to the incident field by an heuristic extension of the exact solution obtained for the circular cylinder or the sphere. A more satisfactory procedure consists in calculating this field by the boundary-layer method and using a matching procedure with the asymptotic expansion away from the boundary given by (1.75). Both procedures give the same result for the leading term of the asymptotic expansion which can be written in the form:

$$\mathbf{E}(P) = \mathbf{E}^i(Q) D_c e^{ikt + ik^{1/3} p(Q')} \sqrt{\frac{d\eta(Q)}{d\eta(Q')}} \sqrt{\frac{\rho}{\sigma(\rho+\sigma)}} e^{ik\sigma}, \tag{1.76}$$

where ρ is the radius of curvature at Q' of the field diffracted away from the body, in the plane of diffraction defined by the binormal at Q' to the geodesic and the direction of the ray $Q'P$. The factor $\sqrt{d\eta(Q)/d\eta(Q')}$ describes the divergence of the creeping wave along the surface. The diffraction coefficient for creeping waves D_c is given in the literature (see [19]). It depends on the polarisation of the electromagnetic field, moreover function $p(Q')$ in the argument of the exponential also depends on polarisation. Details can be found in Section 1.2.6 where the boundary-layer method is applied to the problem of diffraction by a smooth convex cylindrical surface.

1.1.9 Conclusion

The technique of GTD which consists in starting with an Ansatz suggested by the asymptotic solution of a canonical (or model) problem has been applied to the following scattering problems: reflection and refraction by a smooth object, diffraction by the edge of a curved wedge, diffraction by a smooth convex surface.

Our analysis is limited by the domain of validity of the asymptotic expansions used for solving these problems which correspond to regions in space where the ray field is regular. The domains in space where the regularity condition is not satisfied are the shadow boundaries and the caustics of the characteristic curves (or rays) associated with each diffraction phenomena. We have seen that, in these regions, the GTD generates infinite or discontinuous results, which are obviously not physical. Moreover, in these domains, the asymptotic expression of the exact solution of the model problem is no longer given by the Luneberg–Kline or the Friedlander–Keller expansions. It is usually a product of special functions or a contour integral of special functions with arguments that are asymptotic series with respect to fractional negative powers of the large parameter k. In conformity with the model problem technique, the field in the original problem must be thought to be in the same analytical form, but with different coefficients in the asymptotic series appearing in the arguments of the special functions. This procedure will be used in Chapter 3 for constructing a uniform solution for a curved wedge, valid in the transition regions close to the shadow boundaries of the direct and reflected rays.

In the diffraction by a smooth convex surface, the same technique can be applied in the transition region close to the shadow boundary of the direct rays which in this case superposes exactly that of the reflected rays. On the caustic formed by the surface of the body where the field is described by creeping ray modes, the technique of the model problem is more involved and constitutes a discipline known as the boundary-layer method which is presented in the next section.

1.2 Boundary-layer method

1.2.1 Introduction

The ray method allows wave fields to be described in domains where rays form a regular set of lines. It fails near the shadowed part of the surface of a convex body in penumbra, near caustics or focal points, etc. Asymptotic description of wave fields in all these regions may be obtained by different methods. For example, in Reference 20, a creeping waves field was obtained from the analysis of boundary integral equations.

Here we explain the general approach of the boundary-layer method. The boundary-layer method in diffraction theory has much in common with the boundary-layer method as applied in other areas of mathematical physics. It should be noted that the diffraction variant of the boundary-layer method runs into a serious difficulty: the local expansions must be compatible with one another. The matching of the local expansions is a non-trivial problem, but a solvable one as a rule.

We shall not present any rigorous mathematical justification of the local asymptotic expansions derived by the boundary-layer method and concentrate on interpretation and explanation of the formulae. In other words, we shall work on the level of rigorousness accepted usually in physics.

The basic ideas of the boundary-layer method in diffraction can be already found in the works of Fock and Leontovich in the 1940s (see, e.g. References 21–26) and in the paper of Buchal and Keller [27].

The domains where ray expansion appears not valid are usually small (in one, two or all three directions) therefore the basic idea of the method is in stretching coordinates by some powers (usually fractional) of large parameter k. Then the analytic form of the solution is prescribed, this is the so-called 'Ansatz'. Compared to quasi-optics Ansatz (1.16) boundary-layer solutions are written in stretched coordinates and asymptotic expansions are carried usually by fractional orders of $1/k$. When the Ansatz is substituted into the equation (Helmholtz or Maxwell) and into boundary and radiation conditions a recurrent sequence of simpler problems appears by equating terms of similar orders in k. If the Ansatz is correct, all these problems can be solved step-by-step up to any chosen order. There should be a sufficient number of constants (or functions) that remain undetermined. This arbitrariness is eliminated when local asymptotics are matched.

Inventing an Ansatz for a particular diffraction problem may be a complicated problem to solve for which one needs good knowledge of asymptotic technique and mathematical intuition. Physical understanding of wave phenomena usually gives a guideline to correct field representation.

We shall illustrate the way in which Fermat's principle (geometry of the field of rays) and local nature of wave phenomena allow construction of an Ansatz in some particular problems of diffraction. We start with the analysis of the wave field in the near vicinity of a light–shadow boundary on the surface of a convex body. This so-called Fock domain appears to be a cradle of creeping waves that run to the shadowed part of the boundary and of Fresnel transition field in penumbra. Analysis of the solution in the Fock domain allows the Ansatz for creeping waves to be prescribed. We consider creeping wave asymptotics and determine the amplitudes of each creeping wave by matching with the solution in the Fock domain. Then we show how the matching procedure determines the function p in the asymptotics (1.76) of a diffracted wave in deep shadow. Further, we construct the field in penumbra which is expressed via the Fresnel integral. Again, matching this solution to the asymptotics of the field in the Fock domain fixes the arbitrariness of the amplitude.

Then we study the wave field in the vicinity of a concave boundary. The Ansatz for this field of whispering gallery waves is exactly the same as in the case of creeping waves, but the radius of curvature of the surface is negative. Changing the direction of the normal allows the asymptotics of whispering gallery waves to be simply converted from the asymptotics of creeping waves.

Another analogy to the creeping rays geometry deals with the caustic of rays. Again, similarity of the geometry of rays causes similarity of fields.

We deal with two-dimensional problems for which Maxwell's equations are reduced to the Helmholtz equation for scalar waves. Three-dimensional electromagnetic problems are studied in more detail in the next chapter.

1.2.2 Diffraction by a smooth convex body

1.2.2.1 Analysis of the field of rays

Let S be the convex boundary of a body illuminated by a wave field given by its ray expansion. Figure 1.18 presents the two-dimensional cross section of the field of rays. At some line C on the surface S the rays of the incident field are tangential to S. Such rays are called limiting rays. On one side of C the surface is illuminated, that is, it is reached by incident rays and each incident ray is reflected from S. On the other side no geometrical rays reach the surface. The illuminated part of the surface and the limiting rays outgoing from C separate the illuminated domain and the shadow. Any point in the illuminated domain is reached by two rays, one incident and the other reflected from S. At sufficient distance from the curve C and from the limiting rays one can successfully apply quasi-optics Ansatz (1.18) as described in Section 1.1.4. The field in the shadow is only due to diffracted rays.

Thus, the analysis of the ray field depicted in Figure 1.18 allows the following domains, where geometrical optics fails, to be noted:

- Penumbra region (Fresnel field)
- Deep shadow near the surface (Creeping waves)
- Deep shadow far from the surface (Diffracted rays)
- Small vicinity of curve C – Fock domain.

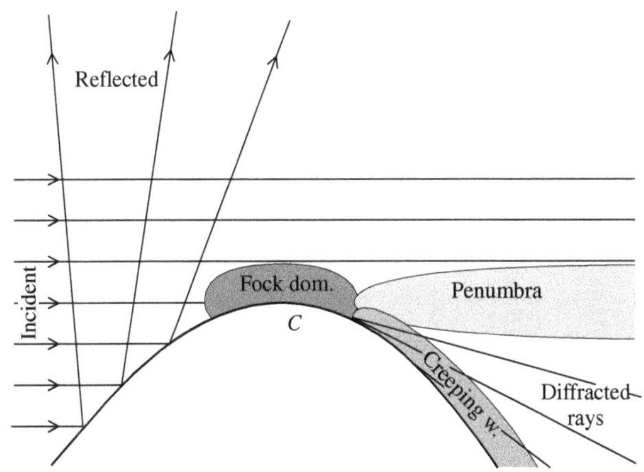

Figure 1.18 Diffraction by a convex body

Local asymptotic expansions of wave fields in all these four domains can be constructed by the boundary-layer method. Note that the wave field corresponding to diffracted (creeping waves) can be found by the ray method (see Section 1.1.8 formula (1.76)), but there remains an undetermined function p. This function is defined by matching the field to creeping waves.

1.2.3 Parabolic equation

Consider the case of acoustic waves on a cylindrical surface. This reduces the problem to a two-dimensional boundary value problem for the Helmholtz equation

$$(\Delta + k^2)U = 0.$$

We require some boundary condition on the surface S. It could be Dirichlet, Neumann or mixed type condition

$$U|_{n=0} = 0, \qquad \left.\frac{\partial U}{\partial n}\right|_{n=0} = 0, \qquad \text{or} \qquad \left.\frac{\partial U}{\partial n}\right|_{n=0} + ikZU|_{n=0} = 0.$$

Here n is the normal coordinate to the surface S and Z in the last variant of the boundary condition is the impedance.

We also require some kind of radiation conditions for large n which allow only exponentially decreasing or outgoing waves from the surface S.

We shall search for such a solution of the above problem that has the form of the wave process moving along the surface S, that is, assume that

$$U = \exp(iks)u(s,n), \tag{1.77}$$

where s is the arc-length of the surface S and new unknown function u varies with the s coordinate more slowly than the exponential multiplier.

Our first step involves substituting the above representation into the Helmholtz equation. For that, the latter should be rewritten in coordinates (s,n) as

$$\frac{1}{1+n/\rho}\left(\frac{\partial}{\partial s}\frac{1}{1+n/\rho}\frac{\partial U}{\partial s} + \frac{\partial}{\partial n}\left(1+\frac{n}{\rho}\right)\frac{\partial U}{\partial n}\right) + k^2 U = 0.$$

Here $\rho = \rho(s)$ is the radius of curvature of surface S. Multiplying the Helmholtz equation by $(1+n/\rho)^2$ and substituting expression (1.77) yields

$$2ik\frac{\partial u}{\partial s} + \frac{\partial^2 u}{\partial s^2} + \frac{n}{\rho+n}\frac{\rho'}{\rho}\left(iku + \frac{\partial u}{\partial n}\right) + \frac{\partial^2 u}{\partial n^2} + 2\frac{n}{\rho}\frac{\partial^2 u}{\partial n^2} + \frac{n^2}{\rho^2}\frac{\partial^2 u}{\partial n^2}$$
$$+ \frac{1}{\rho}\frac{\partial u}{\partial n} + \frac{n}{\rho^2}\frac{\partial u}{\partial n} + 2k^2\frac{n}{\rho}u + k^2\frac{n^2}{\rho^2}u = 0. \tag{1.78}$$

We shall satisfy this equation by assuming that u is an asymptotic series

$$u = u_0 + u_1 + u_2 + \cdots .$$

The second step of the procedure is in deciding what terms in Equation (1.78) are the principal ones, that is, how the coordinates s and n should be stretched. Let the stretched coordinates be

$$k^{\alpha}s, \qquad k^{\beta}n,$$

with yet unknown powers α and β. We assume only that $\alpha < 1$ (otherwise u varies with s quicker than the exponential factor in (1.77)). This assumption allows the term $\partial^2 u/\partial s^2$ to be excluded from principal terms and considered as a correction with respect to the term $2ik\partial u/\partial s$. Further, to be able to fix boundary condition at $n = 0$ and radiation condition at $n = +\infty$ the second-order derivative by n should be included in the principal order terms. That, by the way, means that $\beta > 0$ and the terms with $\partial u/\partial n$ are smaller than $\partial^2 u/\partial n^2$. Finally, terms with n^2 are smaller than similar terms with n in the first power. All this allows the following three terms to be considered as pretenders to be the principal terms in the equation

$$\frac{\partial^2 u}{\partial n^2}, \qquad 2ik\frac{\partial u}{\partial s} \qquad \text{and} \qquad 2k^2\frac{n}{\rho}u. \tag{1.79}$$

One can check that all the terms in (1.79) should be considered the principal order terms because otherwise boundary value problems by n have no solution. Indeed, excluding the first term yields ordinary differential equation by s with two additional conditions by n. Excluding the second term yields the Airy equation with a lack of freedom for satisfying the boundary condition at $n = 0$. Finally, if the last term is excluded, then dependence on n is given by an exponential and the boundary condition yields a trivial solution (except for the case of special impedance Z).

Preserving all three terms yields the well-known Leontovich parabolic equation

$$2ik\frac{\partial u}{\partial s} + \frac{\partial^2 u}{\partial n^2} + 2k^2\frac{n}{\rho}u = 0. \tag{1.80}$$

1.2.4 *Asymptotics of the field in the Fock domain*

Parabolic equation (1.80) fixes the scales $\alpha = \frac{1}{3}$ and $\beta = \frac{2}{3}$. That defines the size of the Fock domain. It is as small as $k^{-1/3}$ along the surface and as small as $k^{-2/3}$ in the direction of the normal. In such a small domain, one can decompose the radius of curvature $\rho(s)$ in the Taylor series

$$\rho(s) = \rho_0 + \rho_1 s + \cdots,$$

and replace ρ in the parabolic equation by the zeroth-order term. That is, the radius of curvature can be considered constant in the leading approximation. Simultaneously we define stretched coordinates to be

$$\varsigma = m_0\frac{s}{\rho_0}, \qquad \nu = 2m_0^2\frac{n}{\rho_0}, \qquad m_0 = \left(\frac{k\rho_0}{2}\right)^{1/3}.$$

We note that the large parameter k is involved in the powers $\frac{1}{3}$ and $\frac{2}{3}$, therefore it is natural to expect the asymptotic sequence u_j, $j = 0, 1, \ldots,$ to have orders $O(k^{-j/3})$.

That is, the Ansatz in the Fock domain is

$$U = \exp(iks) \sum_{j=0}^{N} U_j(\varsigma, \nu) k^{-j/3}.$$

Inserting this expression into the Helmholtz equation and ranging terms in accordance with powers of k yields the recurrent system of equations

$$L_0 U_0 = 0, \qquad L_0 U_1 + L_1 U_0 = 0, \qquad L_0 U_2 + L_1 U_1 + L_2 U_0 = 0, \ldots .$$

Here

$$L_0 = i \frac{\partial}{\partial \varsigma} + \frac{\partial^2}{\partial \nu^2} + \nu,$$

and other L_j are differential operators no higher than that of second order by ς and ν.

Let U be searched as the sum of incident $U^{\mathrm{inc}} = U^{\mathrm{inc}}(C) e^{ikx}$ and reflected U^{refl} waves. Correspondingly, all U_j can also be represented as $U_j^{\mathrm{inc}} + U_j^{\mathrm{refl}}$. Our nearest goal is to find the representation for U_0^{inc} in the coordinates (ς, ν). Hence, we represent the eikonal of the incident wave as

$$x = s + \frac{ns}{\rho_0} - \frac{s^3}{6\rho_0^2} + \cdots .$$

(This representation can be found by simple geometrical considerations or as suggested in Reference 28 by setting

$$x = a_{01} s + a_{02} s^2 + a_{03} s^3 + a_{11} sn + a_{20} n^2 + \cdots ,$$

and defining the coefficients from the eikonal equation.) Passing to variables (ς, ν) one finds

$$U_0^{\mathrm{inc}} = U^{\mathrm{inc}}(C) \exp(ikx) \sim U^{\mathrm{inc}}(C) \exp\left(iks + i \left(\varsigma \nu - \frac{\varsigma^3}{3} \right) \right).$$

That is

$$U_0^{\mathrm{inc}} = U^{\mathrm{inc}}(C) \exp\left(i \left(\varsigma \nu - \frac{\varsigma^3}{3} \right) \right).$$

In order to find the reflected part of the field and in particular U_0^{refl} it is convenient to represent both U_0^{inc} and U_0^{refl} as Fourier integrals

$$U_0^{\mathrm{inc}} = \frac{U^{\mathrm{inc}}(C)}{\sqrt{\pi}} \int_{-\infty}^{+\infty} e^{i\zeta\varsigma} v(\zeta - \nu)\, d\zeta,$$

$$U_0^{\mathrm{refl}} = \int_{-\infty}^{+\infty} e^{i\zeta\varsigma} A_0(\zeta) w_1(\zeta - \nu)\, d\zeta.$$

The choice of the Airy function w_1 in the formula for the reflected field is due to the radiation condition for $\nu \to +\infty$. The function $A_0(\zeta)$ can be found from the boundary condition on the surface. For the surface described by the Dirichlet boundary condition (or in the case of the impedance boundary condition) it is

$$A_0 = -\frac{U^{\text{inc}}(C)}{\sqrt{\pi}} \frac{v(\zeta)}{w_1(\zeta)}.$$

In the case of the Neumann boundary condition the formula changes to

$$A_0 = -\frac{U^{\text{inc}}(C)}{\sqrt{\pi}} \frac{\dot{v}(\zeta)}{\dot{w}_1(\zeta)},$$

and finally, for the impedance boundary condition with $Z = O(k^{-1/3})$ one finds

$$A_0 = -\frac{U^{\text{inc}}(C)}{\sqrt{\pi}} \frac{\dot{v}(\zeta) - im_0 Z v(\zeta)}{\dot{w}_1(\zeta) - im_0 Z w_1(\zeta)}.$$

The last expression is uniform with respect to the boundary condition, that is, it allows a limit $Z \to 0$ to Neumann and a limit $Z \to \infty$ to Dirichlet problem to be taken in the final formulae.

Combining the above formulae yields the following expression for the field in the principal order by k

$$U \sim \frac{U^{\text{inc}}(C)}{\sqrt{\pi}} e^{iks} \int_{-\infty}^{+\infty} e^{i\zeta\varsigma} \left(v(\zeta - \nu) - \frac{\dot{v}(\zeta) - im_0 Z v(\zeta)}{\dot{w}_1(\zeta) - im_0 Z w_1(\zeta)} w_1(\zeta - \nu) \right) d\zeta. \tag{1.81}$$

The next order terms can be obtained by the same approach. These terms differ in polynomial multipliers that appear in front of the Airy functions.

Formula (1.81) is formal, in particular for the next order terms the integral should be understood in a special way as the integrand increases when $\zeta \to -\infty$. The regularisation of the above integral is performed as follows. For negative ζ we replace the Airy function $v(\zeta)$ in the expression for the reflected field by $(w_1(\zeta) - w_2(\zeta))/(2i)$. Then the integral is split into two. The path of integration for $e^{i\varsigma\zeta} w_1(\zeta - \nu)$ is then shifted to the ray $\ell_1 = (\infty e^{-2\pi i/3}, 0)$ and the path of integration for

$$e^{i\varsigma\zeta} \frac{\dot{w}_2(\zeta) - imZw_2(\zeta)}{\dot{w}_1(\zeta) - imZw_1(\zeta)} w_1(\zeta - \nu),$$

is shifted to the ray $\ell_2 = (\infty e^{2\pi i/3}, 0)$ where integrands exponentially decrease.

Continuing the procedure one can find smaller order terms as well. For an arbitrary (non-cylindrical) surface the derivations for the principal order term are exactly as presented above. The second surface coordinate is not stretched in the Fock domain and derivatives by this coordinate are not involved in the higher-order operator.

1.2.5 Creeping waves

Letting ς large positive in the asymptotics (1.81) allows the integral to be replaced by the sum of residue in the poles located in the upper half-plane of complex variable ζ. These poles are the zeros of the Airy function w_1 in the case of the Dirichlet boundary condition, or zeros of its derivative w_1' in the case of the Neumann boundary condition. Each residue is characterised by an attenuation factor

$$\exp\left(i\left(\frac{k\rho_0}{2}\right)^{1/3}\frac{\zeta_j s}{\rho_0}\right), \tag{1.82}$$

which allows contributions of different residues to be separated from each other. Therefore, it is natural to expect matching of each residue to some specific wave that runs along the surface of a convex body. The attenuation factors (1.82) give a hint to search solutions of parabolic equation (1.80) in the form of the following Ansatz:

$$U = \exp(iks + ik^{1/3}p(s))\sum_{j=0}^{N} U_j(s, k^{2/3}n)k^{-j/3}. \tag{1.83}$$

Solutions of that form are not restricted to a small domain by s and are called creeping waves.

Inserting (1.83) into the Helmholtz equation, one notes that terms of order k^2 and $k^{5/3}$ disappear automatically. Equating to zero terms of order $k^{4/3}$, k^1, etc., yields the system of recurrent boundary value problems

$$L_0U_0 = 0, \qquad L_0U_1 + L_1U_0 = 0, \ldots . \tag{1.84}$$

Here

$$L_0 = \frac{\partial^2}{\partial \ell^2} + 2\left(\frac{\ell}{\rho(s)} - p'(s)\right), \qquad L_1 = 2i\frac{\partial}{\partial s}, \ldots,$$

where $\ell = k^{2/3}n$.

In the principal order one has an ordinary differential equation by ℓ with s involved as a parameter. For every fixed value of s it has the solution

$$U_0 = A_0(s)w_1(\xi - \nu), \qquad \nu = \left(\frac{2}{\rho(s)}\right)^{1/3} k^{2/3}n, \tag{1.85}$$

with

$$\xi = 2^{1/3}\rho^{2/3}(s)p'(s).$$

Again, the Airy function w_1 is chosen in accordance with the assumption that for large ν the solution should behave as a wave outgoing from the surface S. Asking (1.85) to satisfy boundary conditions yields the equation for the parameter ξ. For the case of Dirichlet or the impedance boundary condition one finds

$$w_1(\xi) = 0,$$

for the case of the Neumann boundary condition one has

$$w_1'(\xi) = 0.$$

For the impedance boundary condition with Z of order $O(k^{-1/3})$ one finds a more complicated equation for ξ

$$w_1'(\xi) = imZ(s)w_1(\xi), \text{ with } m = \left(\frac{k\rho(s)}{2}\right)^{1/3}.$$

All three equations given above have an infinite number of solutions ξ_p. In the case of the Dirichlet or Neumann boundary conditions ξ_p are constants and in the last case ξ_p are functions of s.

When the parameter $\xi = \xi_p$ is found the function $p(s)$ is a simple integral

$$p(s) = 2^{-1/3} \int^{s} \frac{\xi_p(t)}{\rho^{2/3}(t)}\, dt.$$

The amplitude factor A_0 in (1.85) is not defined from the principal order equation. However, when examining the second recurrent equation

$$L_0 U_1 = -L_1 U_0, \tag{1.86}$$

one finds that together with boundary and radiation conditions this is the Sturm–Liouville problem for U_1 with the parameter ξ taken on the spectrum. In order that this problem be solvable, the right-hand side should be compatible and the compatibility condition appears to be an equation for the amplitude A_0. Following Reference 28 we multiply Equation (1.86) by $w_1(\xi_p - \nu)$ and integrate over the semi-axis $\nu \in [0, +\infty)$ (convergence is due to rapid oscillations)

$$\int_0^\infty w_1(\xi_p - \nu) L_0 U_1\, d\nu = -\int_0^\infty w_1(\xi_p - \nu) L_1 A_0(s) w_1(\xi_p - \nu)\, d\nu.$$

Integrating by parts in the first integral yields

$$\int_0^\infty w_1(\xi_p - \nu) L_0 U_1\, d\nu = -\, w_1(\xi_p)\frac{\partial U_1(s,0)}{\partial \nu} - w_1'(\xi_p) U_1(s,0)$$
$$+ \int_0^\infty U_1 L_0 w_1(\xi_p - \nu)\, d\nu.$$

We note that $L_0 w_1(\xi_p - \nu) = 0$ and the integral in the right-hand side is equal to zero. Since U_1 satisfies the same boundary condition as the principal order term, terms outside the integral cancel each other. That is,

$$\int_0^\infty w_1(\xi_p - \nu) L_1 A_0(s) w_1(\xi_p - \nu)\, d\nu = 0,$$

for every s.

Now we substitute the operator L_1 and perform differentiation

$$2i\int_0^\infty \left(A_0' w_1^2(\xi_p - \nu) + \frac{\nu}{3}\frac{\rho'}{\rho} A_0 w_1'(\xi_p - \nu) w_1(\xi_p - \nu) \right.$$
$$\left. + \frac{1}{2}\xi_p' A_0 w_1'(\xi_p - \nu) w_1(\xi_p - \nu) \right) d\nu = 0.$$

All integrals containing Airy functions can be computed explicitly

$$\int_0^\infty w_1^2(\xi_p - \nu)\, d\nu = 2\int_0^\infty w_1'(\xi_p - \nu) w_1(\xi_p - \nu)\nu\, d\nu = \xi_p w_1^2(\xi_p) - (w_1'(\xi_p))^2,$$

$$\int_0^\infty w_1'(\xi_p - \nu) w_1(\xi_p - \nu)\, d\nu = \frac{1}{2} w_1^2(\xi_p),$$

which yields

$$\frac{A_0'}{A_0} + \frac{1}{6}\frac{\rho'}{\rho} + \frac{\xi_p' w_1^2(\xi_p)}{\xi_p w_1^2(\xi_p) - (w_1'(\xi_p))^2} = 0. \tag{1.87}$$

Introducing

$$Q = \xi_p (w_1(\xi_p))^2 - (w_1'(\xi_p))^2 = (\xi_p + m^2 Z^2) w_1^2(\xi_p), \tag{1.88}$$

we rewrite the transport equation (1.87) as

$$\frac{A_0'}{A_0} + \frac{1}{6}\frac{\rho'}{\rho} + \frac{1}{2}\frac{Q'}{Q} = 0.$$

Note that Q is not constant only if ξ_p is not a constant, that is, in the case of the impedance boundary condition with Z of order $O(k^{-1/3})$.

Solving the transport equation by variables separation allows the principal order term of the creeping wave asymptotics to be found

$$U^{(p)} = A_0(0) \exp\left(iks + i\left(\frac{k}{2}\right)^{1/3} \int_0^s \frac{\xi_p}{\rho^{2/3}}\, ds \right)$$
$$\times \left(\frac{\rho(0)}{\rho(s)}\right)^{1/6} \sqrt{\frac{Q(0)}{Q(s)}}\, w_1(\xi_p - \nu). \tag{1.89}$$

The smaller-order terms $U_1, U_2, \ldots,$ can be found by considering next order recurrent equations. At every jth step, the problem for U_j is the Sturm–Liouville problem on the spectrum. Its solvability condition allows the amplitude factor A_{j-1} of the term U_{j-1} to be determined. Then U_j is the sum of a particular solution of inhomogeneous problem $U_j^{(1)}$ and the general solution $U_j^{(0)} = A_j w_1(\xi_p - \nu)$ of the homogeneous problem. The amplitude A_j is not fixed at the jth step of the procedure, but can be determined from the compatibility condition for the inhomogeneous Sturm–Liouville problem of the next $j+1$ step.

Similar considerations can be performed in the case of an arbitrary (non-cylindrical) surface which yields a three-dimensional problem for the Helmholtz equation. Derivatives by second surface coordinate are not involved in the leading order operator and appear only in L_1. This changes only the transport equation for A_0 and results in a factor $\sqrt{d\eta(0)/d\eta(s)}$ related to divergence $d\eta$ of the geodesics on the surface S.

The amplitudes $A_j(0)$ remain undetermined in the asymptotics of creeping waves. This arbitrariness is eliminated when the field is matched with the solution (1.81) for $\varsigma \to +\infty$. Replacing the integral in (1.81) by the sum of residues yields

$$U \sim 2\sqrt{\pi} i U^{\text{inc}}(C) e^{iks} \sum_{p=0}^{\infty} e^{i\xi_p \varsigma} \frac{w_1(\xi_p - \nu)}{(\xi_p - m^2 Z^2) w_1^2(\xi_p)}.$$

Comparing this expression with the asymptotics of creeping waves at $s \approx 0$ allows the undetermined amplitude in (1.89), $A_0(0)$, to be found

$$A_0(0) = 2\sqrt{\pi} i \frac{U^{\text{inc}}(C)}{(\xi_p + m^2 Z^2) w_1^2(\xi_p)}. \tag{1.90}$$

Here we assumed that $s = 0$ at C. For the case of the Dirichlet boundary condition, the right-hand side in (1.90) should be understood as a limit $Z \to \infty$ which yields

$$A_0(0) = -2\sqrt{\pi} i \frac{U^{\text{inc}}(C)}{(w_1'(\xi_p))^2}.$$

1.2.6 Friedlander–Keller solution

As discussed in Section 1.1.8, the field in the deep shadow of a smooth obstacle is associated with surface diffracted rays. We saw in the previous section that creeping waves that propagate along the surface of a convex body are strongly attenuated. This attenuation can be interpreted as leakage of creeping wave energy carried away by waves that run from the surface along diffracted rays. A formal solution to the Helmholtz equation corresponding to diffracted (creeping) rays was found by Friedlander and Keller [29]. One can use the coordinates (s_c, l) where s_c is the surface coordinate of the creeping point Cr and l is defined as the sum of s_c and the length $\sigma = |\text{Cr M}|$ of the creeping ray (see Figure 1.19). The quadratic form matrix of this system is $\mathbf{g} = \text{diag}(\sigma^2/\rho^2, 1)$. Let the solution be searched in the form

$$U = \exp\left(ikl + ik^{1/3} p(s_c)\right) \sum_{j=0}^{N} k^{-j/3} U_j(s_c, l).$$

Substituting this Ansatz into the Helmholtz equation yields

$$L_0 U_j + L_1 U_{j-1} + L_2 U_{j-2} + L_3 U_{j-3} = 0, \qquad j = 0, 1, 2, \dots. \tag{1.91}$$

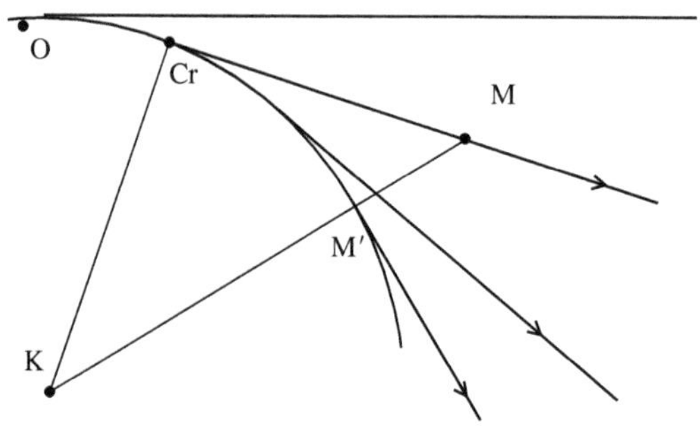

Figure 1.19 Diffracted (creeping) rays

Here

$$L_0 = 2\frac{\partial}{\partial l} + \frac{1}{\sigma}, \qquad L_1 = i\left(\frac{p'\rho}{\sigma}\right)^2,$$

$$L_2 = 2\frac{p'\rho^2}{\sigma^2}\frac{\partial}{\partial s_c} + \frac{\rho}{\sigma}\frac{\partial}{\partial s_c}\left(\frac{\rho}{\sigma}p'\right), \qquad L_3 = -i\Delta,$$

and we assume $U_{-1} \equiv U_{-2} \equiv U_{-3} \equiv 0$.

Recurrent relations (1.91) are completely analogous to transport equations of GTD. In particular one finds

$$U_0 = \frac{C(s_c)}{\sqrt{\sigma}}. \tag{1.92}$$

Here $C(s_c)$ is arbitrary and will be determined when the solution (1.92) is matched to creeping waves field (1.89). For that, one rewrites the asymptotics (1.92) near the surface in coordinates (s, ν) related to creeping waves. Geometrical considerations (see Figure 1.19, $n = |\text{M}'\,\text{M}|$, $s = s(\text{M}')$, $|\text{Cr K}| = \rho$, $|\text{M}'\,\text{K}| \approx \rho$) show that

$$l = s + \frac{2}{3k}\nu^{3/2} + O(k^{-4/3}), \qquad s_c = s - 2k^{-1/3}\left(\frac{\rho}{2}\right)^{2/3}\nu^{1/2} + O(k^{-2/3}).$$

This allows the principal order term to be written as

$$U \sim \frac{C(s)k^{1/6}}{2^{1/6}\rho^{1/3}}\exp\left(iks + ik^{1/3}p(s) + \frac{2}{3}i\nu^{3/2} - 2i\left(\frac{\rho}{2}\right)^{2/3}\nu^{1/2}\right)\nu^{-1/4}. \tag{1.93}$$

This asymptotics is valid in an intermediate domain where σ is small, but ν is large. In that domain one can replace Airy function w_1 in (1.89) by its asymptotics

$$w_1(\xi - \nu) \sim \nu^{-1/4}\exp\left(\frac{2}{3}i\nu^{3/2} - i\xi\nu^{1/2} + i\frac{\pi}{4}\right).$$

Then comparing the creeping waves asymptotics to (1.93) one finds that $p(s_c)$ is the same function as in (1.89) and

$$C(s_c) = k^{-1/6} 2^{1/6} e^{i\pi/4} \rho^{1/3} A_0(s_c).$$

Matching smaller-order terms allows all the undetermined coefficients in the Friedlander–Keller solution to be found. For the electromagnetic case, see Reference 30.

1.2.7 Boundary layer in penumbra

Let us now discuss the penumbra. Physically it is clear that the main oscillating factor in the solution should be $\exp(ikx)$ with the x coordinate directed along the limiting ray. We introduce the other Descartes coordinate y in such a way that $y > 0$ corresponds to the illuminated domain and $y < 0$ to the shadow.

Let the solution be searched in the form

$$U = \exp(ikx)u(x, y).$$

Substituting this expression into the Helmholtz equation and assuming that derivatives by x can produce large factors, but not as large as k, yields in the principal order the parabolic equation

$$2ik\frac{\partial u}{\partial x} + \frac{\partial^2 u}{\partial y^2} = 0. \tag{1.94}$$

Further, it is natural from the physical point of view (it can also be seen from the solution of a canonical problem of diffraction by a circular cylinder) that radius of curvature of the body does not play a crucial role in the asymptotic decomposition of the wave field in penumbra. Therefore, ρ should not be present in the stretched coordinates. From (1.94) it is clear that the y coordinate should be stretched with a factor $k^{1/2}$. One can check that only one dimensionless combination of k, x and y can be introduced as a stretched coordinate in the layer $y = O(k^{-1/2})$, it is given by

$$\phi = \frac{\sqrt{k}y}{\sqrt{2x}}. \tag{1.95}$$

Let the solution of the parabolic equation (1.94) be searched as a function $u(\phi)$ of that coordinate. One finds

$$-2i\phi u'(\phi) + u''(\phi) = 0.$$

Thus

$$u = C \int_{\phi_0}^{\phi} e^{i\psi^2} d\psi.$$

Here C and ϕ_0 are arbitrary constants.

The local asymptotics of the wave field in penumbra should somehow match the field in the illuminated domain with a small field in the shadow. It is possible to find such ϕ_0 that u behaves as a smooth step-function, namely

$$u = C\left(\sqrt{\pi}e^{i\pi/4} + \int_0^{\phi} e^{i\psi^2}\,d\psi\right). \tag{1.96}$$

One can rewrite the above expression in terms of Fresnel integrals.

In order to determine the multiplier C it is necessary to perform the matching of the above solution with the representation (1.81) of the field in the Fock domain. Consider for simplicity the case of Dirichlet boundary conditions. For the matching we represent w_1 as $w_2 - 2iv$ on the ray $(e^{-2\pi i/3}\infty, 0)$ of the integration path. This yields

$$U = U^{\mathrm{inc}}(C)(\Phi + \Psi),$$

$$\Phi = \frac{e^{iks}}{\sqrt{\pi}}\left(\frac{i}{2}\int_{e^{2\pi i/3}\infty}^{0} e^{i\sigma\zeta} w_2(\zeta - \nu)\,d\zeta + \int_0^{+\infty} e^{i\sigma\zeta} v(\zeta - \nu)\,d\zeta\right),$$

$$\Psi = -\frac{e^{iks}}{\sqrt{\pi}}\left(\frac{i}{2}\int_{e^{2\pi i/3}\infty}^{0} e^{i\sigma\zeta}\frac{w_2(\zeta)}{w_1(\zeta)} w_1(\zeta - \nu)\,d\zeta + \int_0^{+\infty} e^{i\sigma\zeta}\frac{v(\zeta)}{w_1(\zeta)} w_1(\zeta - \nu)\,d\zeta\right).$$

The integrals involved in Φ do not depend on the boundary condition. That means that Φ does not depend on the physical characteristics of the body. We shall show that Φ corresponds to the Fresnel part of the diffracted field in penumbra. The part Ψ of the field can be shown [26, 31] to be a slowly varying function that gives background for the rapidly varying Fresnel part. We shall consider only the Fresnel part Φ.

Let us again use the relation between Airy functions and represent Φ in the form

$$\Phi = \frac{ie^{iks}}{2\sqrt{\pi}}\int_{-\infty}^{0} e^{i\sigma\zeta} w_1(\zeta - \nu)\,d\zeta + \frac{e^{iks}}{\sqrt{\pi}}\int_{-\infty}^{+\infty} e^{i\sigma\zeta} v(\zeta - \nu)\,d\zeta.$$

As we already know, the second term is the incident field e^{ikx}. In penumbra we have $\nu \to \infty$ and in the first integral we can replace the Airy function w_1 by its asymptotics. Changing the integration variable to $t = \nu - \zeta$ yields

$$\Phi \sim \frac{e^{iks + i\sigma\nu + 3i\pi/4}}{2\sqrt{\pi}}\int_{\nu}^{+\infty} \exp\left(-i\sigma t + \frac{2i}{3}t^{3/2}\right) t^{-1/4}\,dt + e^{ikx}.$$

The main contribution to the above integral is due to vicinity of the stationary point $t_{\mathrm{st}} = \sigma^2$. Decomposing the integrated function into Taylor series by powers of

$\psi = (t - \sigma^2)/2\sqrt{\sigma}$ and neglecting higher-order terms, one gets

$$\Phi \sim e^{ikx}\left(1 + \frac{e^{3\pi i/4}}{\sqrt{\pi}} \int_{\phi}^{+\infty} e^{i\psi^2}\, d\psi\right),$$

where

$$\phi = \frac{\nu - \sigma^2}{2\sqrt{\sigma}} \approx \sqrt{\frac{k}{2s}}\left(n - \frac{s^2}{2\rho_0}\right).$$

One can check that the expression for ϕ asymptotically coincides with (1.95) for small s and n. Finally, one determines the amplitude in (1.96) as

$$C = \frac{e^{-i\pi/4}}{\sqrt{\pi}} U^{\text{inc}}(C).$$

1.2.8 *Whispering gallery waves*

In the formulae corresponding to the creeping waves asymptotics the radius of curvature was assumed positive. However, all derivations remain valid if ρ is strictly negative, that is, in the case of a concave surface. Solutions analogous to creeping waves (i.e., having the form (1.83)) on a concave surface are called whispering gallery waves. For the asymptotic analysis of whispering gallery waves it is more convenient to change the direction of the normal and by this to make ρ positive again. Then the only difference will be in considering Equations (1.84) on the semi-axis $n < 0$ rather than at $n > 0$. The condition at infinity now specifies an exponential decrease which results in replacing Airy function w_1 in all the formulae corresponding to creeping waves by another solution of Airy equation v. In the principal order one has

$$U = A_0(0) \exp\left(iks + i\left(\frac{k}{2}\right)^{1/3} \int_0^s \frac{\eta}{\rho^{2/3}}\, ds\right)\left(\frac{\rho(0)}{\rho(s)}\right)^{1/6} \sqrt{\frac{Q(0)}{Q(s)}}\, v(\eta - \nu),$$

where η is any solution of the equation

$$v'(\eta) = i\left(\frac{k\rho(s)}{2}\right)^{1/3} Z(s) v(\eta).$$

One should note, however, an important difference between creeping and whispering gallery waves. Creeping waves are characterised by the complex parameter ξ which causes exponential decay of creeping waves along the surface S. On lossless surfaces, whispering gallery waves are, on the contrary, characterised by real parameters η and therefore propagate along the surface without attenuation. This, by the way, means that the contributions of whispering gallery waves corresponding to different parameters η cannot be separated from each other.

1.2.9 Wave field near a caustic

Another case of wave energy concentration deals with the field in the vicinity of a caustic. The caustic surface is the envelope of the rays (namely the surface tangential to the rays). The caustic is also the locus of the centres of curvature of the wavefronts. On this surface, the field computed by the geometrical optics is infinite. The physical field is of course finite, but more intense than the field in a classical ray region (the caustic is the place 'where light burns', which explains the name).

The caustics occur in a number of different situations, as for instance in the study of the reflection of a plane wave by a concave surface, or, of the whispering gallery modes. In the two-dimensional case, the caustic is a curve with cusp points ('points de rebroussement') where the tangent vector to the caustic curve changes sign. The cusp of the caustic is composed of points. A familiar two-dimensional caustic (the breakfast caustic) appears on the surface of a bowl filled with liquid illuminated by the sun or by a lamp. Another familiar caustic is the focus of a converging lens.

In three dimensions, the wavefront has two principal radii of curvature. Therefore, the caustic, locus of centres of curvature of the wavefronts, comprises two sheets. These sheets are joined together along cusp lines. These cusp lines present in turn some singular points. For a 'generic' (i.e., stable under a small perturbation) caustic, these singular points can be classified: swallowtails which are cusps on the cusp lines, and umbilics, where the two sheets of the caustic meet. Caustics have been studied in higher-dimensional space and their singularities have been classified [32].

Thus, the 'generic' caustic surface presents a rather involved geometry, with cusp lines and singular points. The field in the vicinity of the regular part of the caustic surface turns out to be described in terms of Airy functions, as explained in Section 1.2.9.1, whereas the field near the cusp line and near special points on the cusp line is expressed in terms of other special functions of catastrophe theory [33]. Although these points are somewhat fascinating from the mathematical point of view, we will first consider the most frequent case, that is, a regular point of a caustic, namely a point where the caustic surface is regular, far away from the cusps.

1.2.9.1 Asymptotic decomposition near a smooth caustic

The geometry of rays near the caustic is presented in Figure 1.20. One can note similarity of rays going from the caustic with creeping rays. We refer here to the locality principle established by Fock. It says that if the rays in two problems is similar, then wave fields are also similar. Therefore, it is natural to expect the boundary layer near the caustic to have the thickness $O(k^{-2/3})$ same as for creeping waves. However, in the caustic problem there is no obstacle and as a consequence there is no shadow. That is, waves propagate along the caustic without attenuation. All this allows the following Ansatz to be suggested

$$U = e^{iks} \sum_{j=0}^{N} U_j(s, k^{2/3}n)k^{-j/3}.$$

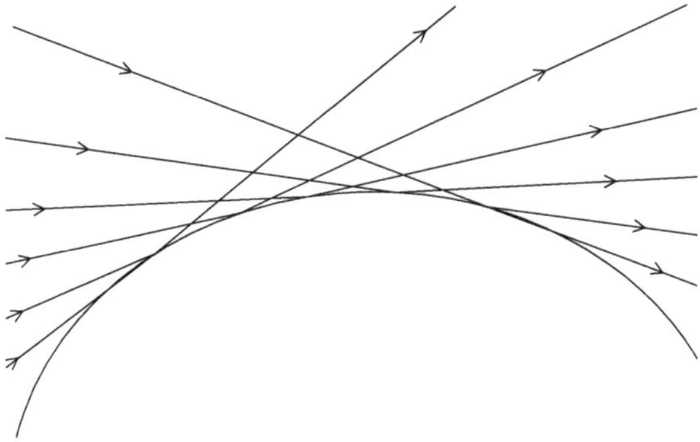

Figure 1.20 Rays near a smooth caustic

Here the coordinates (s, n) are introduced as the arc-length measured along the caustic and the normal to the caustic.

Substituting this Ansatz into the Helmholtz equation and equating terms of similar orders in k yields the recurrent system

$$L_0U_0 = 0, \qquad L_0U_1 + L_1U_0 = 0, \quad \ldots .$$

Introducing $\nu = 2^{1/3}k^{2/3}\rho^{-1/3}(s)n$, the operator L_0 can be written as

$$L_0 = \frac{\partial^2}{\partial\nu^2} + \nu.$$

Having in mind the property $U \to 0$ when $\nu \to -\infty$, we choose the solution of the Airy equation $L_0U_0 = 0$ in the form

$$U_0 = A_0(s)v(-\nu).$$

In the illuminated region $\nu \gg 1$ and the Airy function can be replaced by its asymptotics

$$U_0 \sim \frac{A_0(s)}{2\nu^{1/4}}\left(\exp\left(-\frac{2i}{3}\nu^{3/2} + \frac{i\pi}{4}\right) + \exp\left(\frac{2i}{3}\nu^{3/2} - \frac{i\pi}{4}\right)\right). \tag{1.97}$$

The first term corresponds to the wave approaching the caustic and the second term corresponds to the wave leaving it. Matching the field of the incoming wave to its ray expansion

$$U^{\text{inc}} = e^{ikS(\mathbf{r})}u(\mathbf{r}), \tag{1.98}$$

allows the amplitude A_0 to be found. For that the expression (1.98) is written in coordinates (s, n) and matched to the first term of (1.97). In the principal order, one finds

$$U^{\text{inc}} \sim \exp\left(iks - i\tfrac{2}{3}\nu^{3/2}\right)k^{1/6}\nu^{-1/4}\rho^{1/6}2^{-1/6}\psi_0^{\text{inc}}.$$

Here $\psi_0^{\text{inc}} = |J(0)|^{1/2}u_0(0)$ is the diffraction coefficient (see (1.34)). Finally, one finds

$$A_0 = 2^{5/6}(k\rho)^{1/6}e^{-i\pi/4}\psi_0^{\text{inc}}.$$

That is, the wave field near the caustic is $k^{1/6}$ times larger than far from it.

From (1.97) it follows that the wave leaving the caustic has the diffraction coefficient

$$\psi_0^{\text{out}} = e^{-i\pi/2}\psi_0^{\text{inc}}.$$

That is, on passing through a caustic the phase of the wave jumps downwards by $\pi/2$.

1.2.9.2 Wave field near a cusp of a caustic

We use the method of spectral reconstruction which is a simplification of the Maslov method, as explained in Reference 34. In practice, the application of this method involves the steps outlined below. Consider the rays in the neighbourhood of M, where we wish to calculate the diffracted field (see Figure 1.21). With each ray passing through N close to M, we associate a plane wave propagating in the direction $\mathbf{p}$ of the ray passing through N with the eikonal $S(\mathbf{p})$ of geometrical optics at point N, and with yet undetermined amplitude $A(\mathbf{p})$. We seek the solution at point M in the form

$$u(M) = \int_V A(\mathbf{p}) \exp(ik(S(\mathbf{p}) + \mathbf{p} \cdot \mathbf{t}(N, M)))\, d\mathbf{p}. \tag{1.99}$$

Here $\mathbf{t}(N, M)$ is the vector from N to M. The integral is performed in the angular range in the neighbourhood V of the direction of the ray passing through M.

The point N can be located on a plane passing through M, and one reverts to the Maslov method, or more generally, on an arbitrary surface close to M which provides more flexibility to the method. The amplitude $A(\mathbf{p})$ is then calculated by imposing the following condition: the first term of the approximation of (1.99) by the stationary phase method must be the same as that given by the geometrical optics. This condition is in fact the result of matching procedure. Justification of the method and some examples can be found in the publications by Arnold [35] and also in the book by Kravtsov and Orlov [36] (where it is called the method of interference integral).

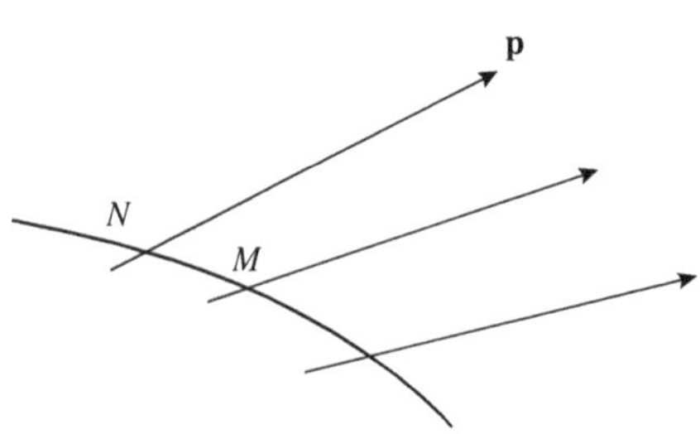

Figure 1.21 Spectral reconstruction method

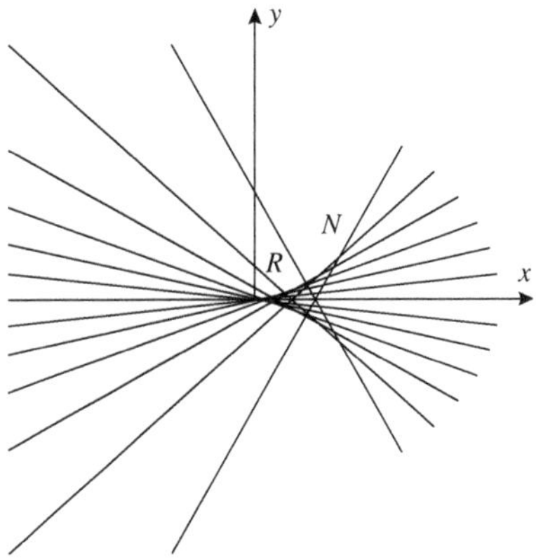

Figure 1.22 Cusp of a caustic formed by rays reflected from a concave mirror

Let R be the cusp point of the caustic line (see Figure 1.22). Introduce coordinates (x, y) with $R = (0, 0)$ and, following Pearcey [37], define the caustic C in the vicinity of point R via the equation (with x changed to $-x$ as contrasted with Reference 37)

$$y = \sqrt{\frac{8}{9\sigma}x^3} = \frac{2\sqrt{2}}{3}\frac{x^{3/2}}{\sigma^{1/2}}.$$

Where σ is a geometrical parameter.

The rays are tangent to C and the ray vector passing through point $N(x, y(x))$ is

$$\mathbf{p} = \left(\frac{1}{\sqrt{1+2x/\sigma}}, \sqrt{\frac{2x/\sigma}{1+2x/\sigma}}\right).$$

The eikonal at N equals s, where s is the curvilinear abscissa on C measured from R

$$s = \int_0^x \sqrt{1+\frac{2x}{\sigma}}\,dx.$$

The eikonal $S(\mathbf{p})$ of the plane wave corresponding to the ray passing through N at the point $M = (x_0, y_0)$ is given by

$$S(\mathbf{p}) = s + p_x(x - x_0) + p_y(y - y_0),$$

where x, y, x_0, y_0 are small quantities of the same order. Next, we perform a Taylor expansion, retaining only the terms up to $O(x^2)$ to obtain

$$S = x_0 + \sqrt{\frac{2x}{\sigma}}y_0 - \frac{x_0 x}{\sigma} + \frac{1}{6}\frac{x^2}{\sigma} + O(x^2).$$

The integral representation then takes the form

$$\exp(ikx_0)\int B(x)\exp\left(ik\left(\sqrt{\frac{2x}{\sigma}}y_0 - \frac{x_0 x}{\sigma} + \frac{1}{6}\frac{x^2}{\sigma}\right)\right)dx. \qquad (1.100)$$

We now introduce a change of the variables $t^4 = (k/6)(x^2/\sigma)$ to obtain the following expression for the phase Φ in the integrand:

$$\Phi = \left(\frac{24}{k\sigma}\right)^{1/4}(ky_0)t - \left(\frac{6}{k\sigma}\right)^{1/2}(kx_0)t^2 + t^4.$$

Setting

$$Y = \left(\frac{24}{k\sigma}\right)^{1/4} ky, \qquad X = -\left(\frac{6}{k\sigma}\right)^{1/2} kx,$$

we obtain

$$\Phi = Yt + Xt^2 + t^4,$$

which is the phase of Pearcey function. Depending upon the values of X and Y, there can be either one or three values of t that cancel $d\Phi/dt$, implying that there is either one or three stationary phase points in (1.100). In other words, the application of the method of the stationary phase to the integral representation in (1.100) will yield either one or three rays depending upon whether the point is outside or inside the beak. The difficulties in geometrical optics arise from its application in the neighbourhood of the point R, where these three points coalesce at the point $t = 0$.

The representation

$$u = e^{ikx}\left(\int \exp(ik(Yt + Xt^2 + t^4))\,dt + k^{-1/4}U_1(X, Y) + \cdots\right) \tag{1.101}$$

can be considered as the Ansatz of the boundary-layer method. Finding terms U_j of asymptotic decomposition in all orders is then done by the usual procedure of the boundary-layer method [38].

1.2.10 Diffraction by a transparent body

In the case of a dielectric body the picture of the field of rays presented in Figure 1.18 is completed by rays transmitted to the other media inside the body. Two cases can be distinguished. If the velocity of a wave in the outer space is greater than the velocity inside the body, then all transmitted rays are directed inside the body. In that case for computing the field in the outer space one can replace the body by an impedance surface with the impedance corresponding to the formulae (1.68) for the coefficients of reflection. This impedance depends on the angle of incidence, in particular for the description of diffraction phenomena in the Fock domain and in deep shadow one should take the impedance corresponding to tangential incidence. The transmitted waves field in that case is subject to ray asymptotics as described in Section 1.1.6.

A different structure of transmitted field is observed if the velocity in the outer space is less than the velocity inside the body. In that case there exists a critical direction of incidence (see Figure 1.23). The transmitted wave runs parallel to the interface and formula (1.69) of geometrical optics becomes inapplicable. It is natural to expect whispering gallery wave generation by an incident wave running at critical direction. Excitation of the whispering gallery wave was studied in Reference 39 by

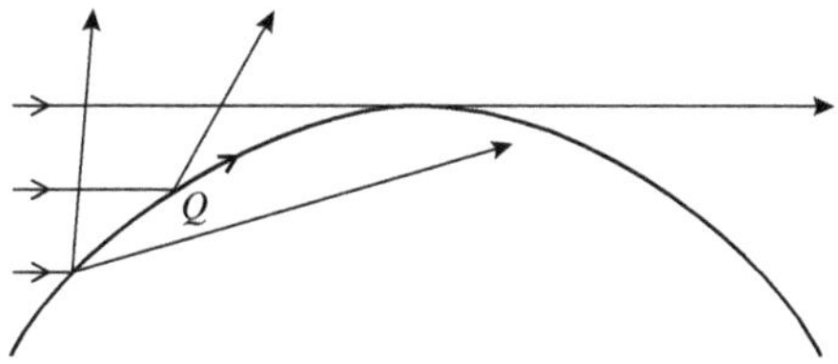

Figure 1.23 Critical direction

using reciprocity principle. Direct analysis of the whispering gallery waves excitation requires asymptotics of the field in the vicinity of point Q at which the critical ray reflects.

Let the incident wave be given by its ray expansion

$$u^i = e^{ikS^i(\mathbf{r})} \sum_{j=0}^{N} \frac{U_j^i(\mathbf{r})}{(-ik_1)^j}.$$

We assume that the total field $u(\mathbf{r})$ and its normal derivatives are continuous at the interface, that is,

$$u^i|_S + u^r|_S = u^t|_S, \qquad \left.\frac{\partial u^i}{\partial n}\right|_S + \left.\frac{\partial u^r}{\partial n}\right|_S = \left.\frac{\partial u^t}{\partial n}\right|_S.$$

Consider the near vicinity of the point Q. Introduce stretched coordinates (we assume $s = 0$ at point Q)

$$\nu = k^{1/2} n, \qquad \eta = k^{3/4} n, \qquad \sigma = k^{1/2} s. \tag{1.102}$$

The coordinates (σ, ν) will be used for the reflected wave and the coordinates (σ, η) will be used for transmitted wave in the domain $n < 0$ inside the body where the wave number is k_2. These scales can be derived by analysing the principal order terms in the equations satisfied by the field and will be justified below.

The reflected field is searched in the form

$$u^r = e^{ikS^r(s,n)} \sum_{j=0}^{\infty} k^{-j/4} U_j^r(\sigma, \nu). \tag{1.103}$$

For the eikonal S^r we assume equations of geometrical optics to be satisfied, that is,

$$|\nabla S^r| = 1, \qquad S^r(s,0) = S^i(s,0), \qquad \frac{\partial S^r(s,0)}{\partial n} = -\frac{\partial S^i(s,0)}{\partial n}.$$

Compared to the general case studied in Section 1.1.6 we add in (1.103) terms with $j = 4l + 1, 4l + 2$ and $4l + 3$.

For the transmitted wave, we accept the Ansatz

$$U^t = e^{ikS_0(s)} \sum_{j=0}^{\infty} k^{-j/4} U_j^t(\sigma, \eta), \tag{1.104}$$

where $S_0(s) \equiv S^i(s, 0)$.

After substituting the representation (1.103) into the Helmholtz equation, decomposing ρ in the vicinity of the point Q into series by s and switching to stretched coordinates (σ, ν) we equate terms at powers $k^{j/4}$. In the leading order, it gives

$$\nabla S^r(0,0) \cdot \nabla U_j^r = 0.$$

Analogous procedure for the transmitted wave yields

$$\frac{\partial^2 U_0^t}{\partial \eta^2} + 2ia\frac{\partial U_0^t}{\partial \sigma} - 2ab^2\sigma U_0^t = 0. \tag{1.105}$$

It is easy to find that $a \equiv S_0'(0) = k_2/k$ and $b^2 \equiv S_0''(0) > 0$. The scales in (1.102) are introduced in such a way that the principal order terms in the Helmholtz equation inside the body are

$$\frac{\partial^2 U}{\partial n^2}, \qquad \frac{\partial U}{\partial s} \qquad \text{and} \qquad sU,$$

and all these terms have the same order in large parameter k.

Substituting the representations (1.103), (1.104) and the ray expansion of the incident wave into continuity conditions at the interface yields

$$U_0^r(\sigma, 0) = U_0^i(\sigma_0), \qquad U_0^t(\sigma, 0) = 2U_0^i(\sigma, 0).$$

For the reflected wave, one finds unique solution $U_0^r = U_0^i(0,0)$. The problem for the transmitted wave requires additional conditions for the uniqueness. We consider the transmitted field to represent a wave running from the interface. Besides it should match to ray expansion of transmitted wave for large negative σ.

Introducing $\mathcal{U} = \exp(ib^2\sigma^2/2)U_0^t(\sigma, \eta)$ allows Equation (1.105) to be rewritten as the usual parabolic equation

$$\frac{\partial^t \mathcal{U}}{\partial \eta^2} + 2ia\frac{\partial \mathcal{U}}{\partial \sigma} = 0.$$

Its solution corresponding to the outgoing wave is given by the Fourier integral

$$U_0^t = e^{-ib^2\sigma^2/2} \int e^{i\xi\sigma} A(\xi) e^{\sqrt{2a\xi}\eta}\, d\xi. \tag{1.106}$$

The factor $A(\xi)$ is found from the boundary condition $U^t(\sigma, 0) = 2U_0^i(0,0)$ which gives

$$A(\xi) = \frac{U_0^i(0,0)}{\pi} \int e^{-i\xi\varsigma} e^{ib^2\varsigma^2/2}\, d\varsigma. \tag{1.107}$$

Further, we perform a series of formal transformations. Represent the exponent in (1.107) in the form

$$i\frac{b^2}{2}\left(\varsigma - \frac{\xi}{b^2}\right)^2 - i\frac{\xi^2}{2b^2},$$

change the variable of integration to $t = e^{-i\pi/4}\varsigma$ and recognise the canonical integral

$$\int_{-\infty}^{+\infty} e^{-b^2t^2/2}\,dt = \frac{\sqrt{2\pi}}{b}.$$

Substituting the result into (1.106) yields

$$U_0^t = U_0^i(0,0)e^{i\pi/4}\sqrt{\frac{2}{\pi}}\frac{1}{b}e^{-ib^2\sigma^2/2}\int \exp\left(i\xi\sigma + \sqrt{2a\xi}\eta - i\frac{\xi^2}{2b^2}\right)d\xi.$$

Changing the integration variable to z such that $\xi = bz^2$ and combining the exponential factors we get the final formula [40]

$$U_0^t = \frac{4U_0^i(0,0)}{\sqrt{2\pi}}e^{i\pi/4}\int_\gamma \exp\left(-\frac{i}{2}(b\sigma - z^2)^2 + \sqrt{2ab}\eta z\right)z\,dz. \tag{1.108}$$

The path of integration γ comes from infinity along the ray $\arg(z) = -5\pi/8$ and goes to infinity along the ray $\arg(z) = -\pi/8$.

The integral in (1.108) is a modified Pearcey integral, it differs from the Pearcey integral in (1.101) by the path of integration. The phase function has three saddle points which for $\eta = 0$ are at zero and at $\pm\sqrt{b\sigma}$. For $\sigma \to -\infty$ the saddle point $z = -i\sqrt{-b\sigma}$ approaches the integration path. Its contribution is

$$U_0^t \sim 2U_0^i e^{-i\sqrt{-ab^2\sigma}\eta} + O\left(\frac{1}{\sqrt{-\sigma}}\right).$$

This expression matches the ray expansion (1.69) in the domain $\sqrt{k}s = O(1)$.

For large positive σ there are two saddle points z_1 and z_2 on the integration path. When η is small the saddle points are given by the expressions

$$z_1 = \sqrt{b\sigma} - \frac{i}{2}\sqrt{\frac{a}{2b}\frac{\eta}{\sigma}} + \cdots, \qquad z_2 = i\sqrt{\frac{a}{2b}\frac{\eta}{\sigma}} + \cdots.$$

Contribution of the saddle point z_1 corresponds to a wave that exponentially decreases with $\eta \to -\infty$. On the interface this wave is equal to $2U_0^i(0,0)$. The other saddle point corresponds to a non-decreasing wave which corresponds to transmitted rays originated on the part $s < 0$ of the interface.

1.2.11 Conclusion

We have discussed the boundary-layer method in diffraction problems for the Helmholtz equation, however it can be effectively applied to problems of electromagnetic and elastic waves diffraction. It can also be generalised to the case of anisotropic materials, to wave phenomena in piezoelectrics, etc. More complicated geometries that involve additional large or small parameters can also be examined with the use of the boundary-layer method.

If heuristic ideas and physical understanding of the wave phenomena do not allow the Ansatz to be suggested one can first invent a canonical problem which allows explicit solution. Then asymptotic analysis of this solution can give the idea of correct wave field representation.

We have not justified the asymptotics derived by the boundary-layer method and only checked that the recurrent procedure allows any desired number of terms to be constructed and that the local asymptotics can be matched to each other and to ray expansions where the latter are available. If this is so, it usually means that the solution is correct, however rigorous mathematical justification remains a separate serious problem. Justification of the field asymptotics in deep shadow, for example, was done by Philippov [41, 42] by means of reducing the problem to boundary integrals. Ideas of microlocal analysis give another way of justifying asymptotic expansions [43].

1.3 Numerical examples

We present numerical results for surface currents on circular and elliptic cylinders. Currents are defined for TM and TE waves correspondingly by the formulae

$$\mathbf{J}_s = H_z|_{n=0}\mathbf{s}, \qquad \mathbf{J}_s = \frac{i}{k}\frac{\partial E_z}{\partial n}\bigg|_{n=0}\mathbf{z}.$$

Here **s** and **z** are unit vectors tangential to the surface, **s** is along the direction of creeping waves propagation and **z** is along the generatrix of the cylinder.

The cylinders are illuminated by plane waves. The geometrical parameters are defined in Figure 1.24. The direction of propagation of the incident plane wave is parallel to axis Ox and is fixed whereas the surface current and the radiated field are calculated as functions of the polar angle θ which varies from 0 to 360°. In a point M on the illuminated side, the contributions of three different rays are added: the direct ray and two creeping rays excited at Q' and Q and propagating along the surface from Q' to M via Q and from Q to M via Q'. Creeping waves that make more than a tour around the cylinder are neglected. At a point M' on the shadowed side,

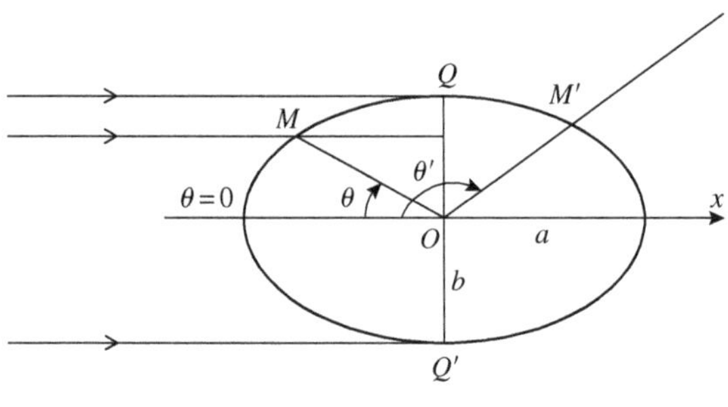

Figure 1.24 Geometry of the problem

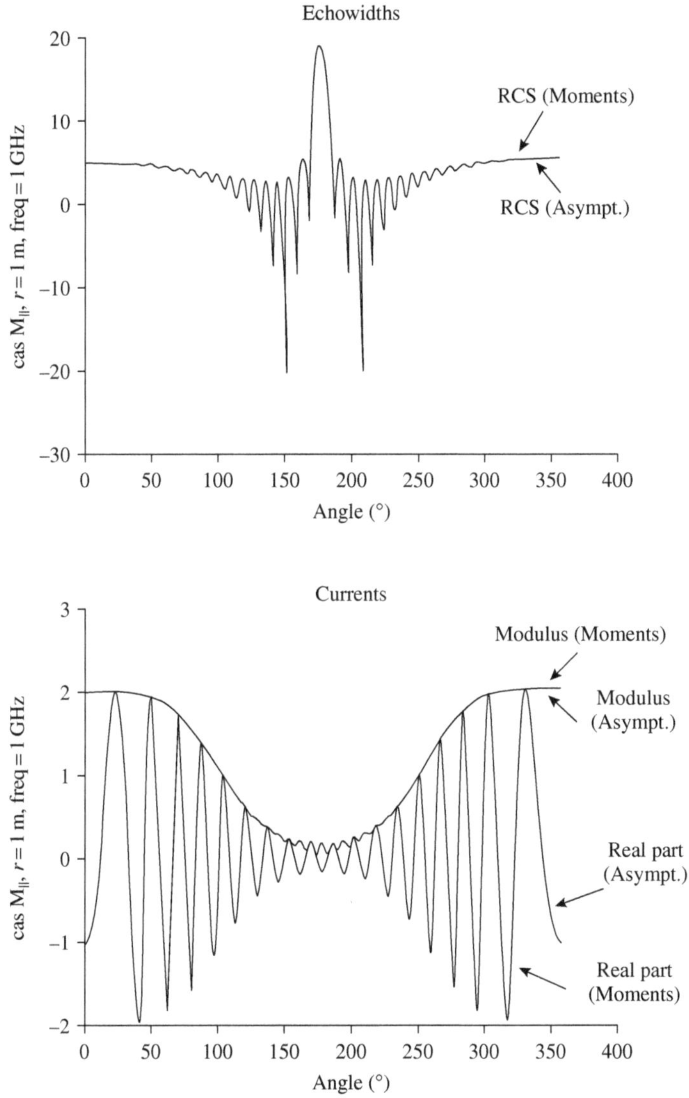

Figure 1.25 Perfectly conducting circular cylinder, $a = 1\,m$, $f = 1\,GHz$, TM polarisation

the contributions of two rays are added: the creeping ray QM' excited at Q and the creeping ray $Q'M'$ excited at Q'.

The results are given at 1 GHz for a perfectly conducting elliptic cylinder and a coated cylinder with the coating characterised by $\varepsilon_r = 3, \mu_r = 1$ and thickness 0.5 cm. The asymptotic results are compared to that obtained by the method of moments.

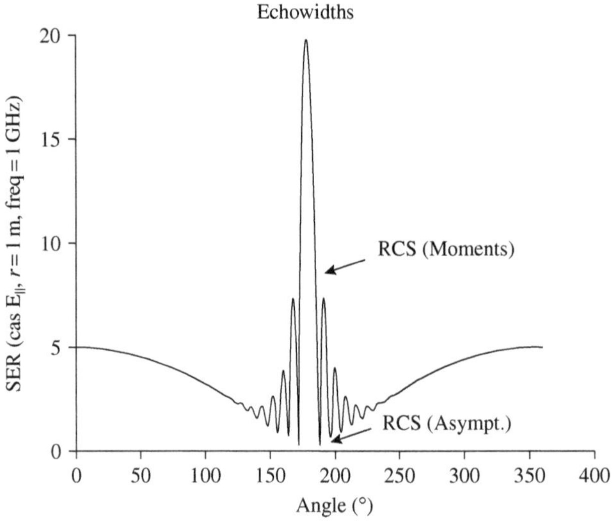

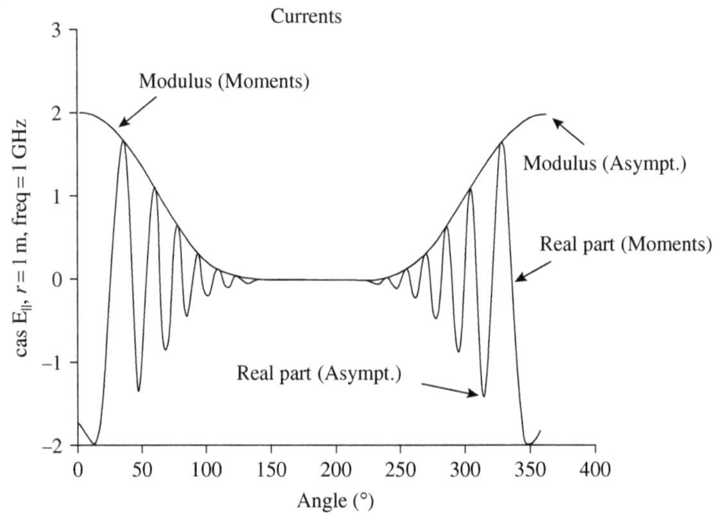

Figure 1.26 Perfectly conducting circular cylinder, $a = 1\,m$, $f = 1\,GHz$, TE polarisation

The curves in Figures 1.25 and 1.26 show the variation of the radiated field, of the modulus and the real part of the surface current, for TM and TE waves, respectively. In these figures the cylinder is a perfectly conducting circular of radius $a = 1$ m.

In Figures 1.27 and 1.28 the radiated fields for a perfectly conducting elliptic cylinder with $a = 1$ m and $b = 0.5$ m for TM and TE incident waves are shown. In this case the smallest radius of curvature is 0.25 m, which is less than the wavelength at a frequency of 1 GHz.

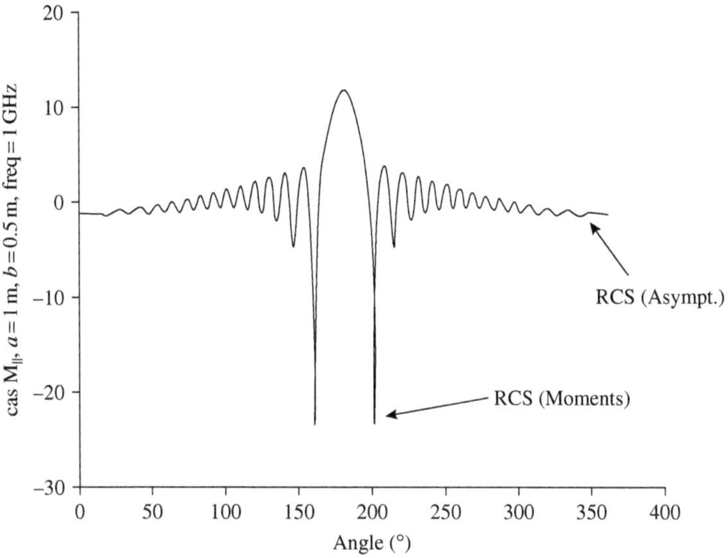

Figure 1.27 *Perfectly conducting elliptic cylinder, $a = 1\,m$, $b = 0.5\,m$, $f = 1\,GHz$, TM polarisation*

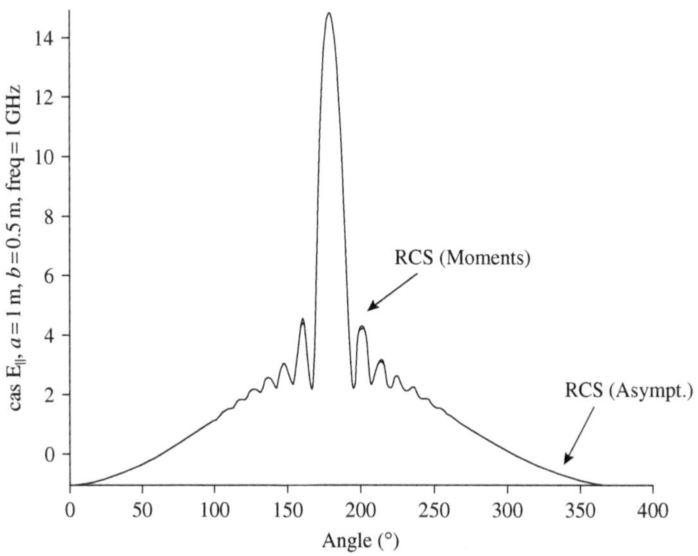

Figure 1.28 *Perfectly conducting elliptic cylinder, $a = 1\,m$, $b = 0.5\,m$, $f = 1\,GHz$, TE polarisation*

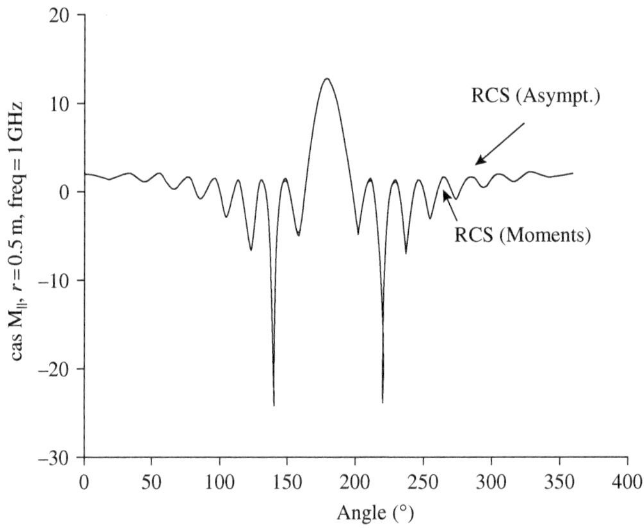

Figure 1.29 Coated circular cylinder, $a = 0.5\,m$, $f = 1\,GHz$, TM polarisation

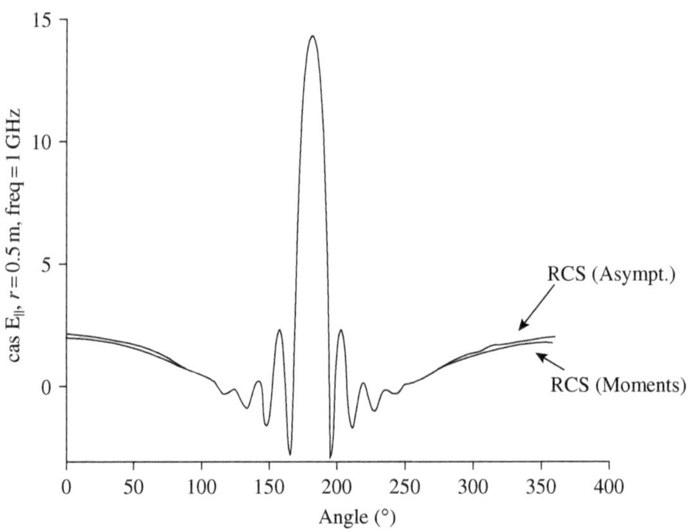

Figure 1.30 Coated circular cylinder, $a = 0.5\,m$, $f = 1\,GHz$, TE polarisation

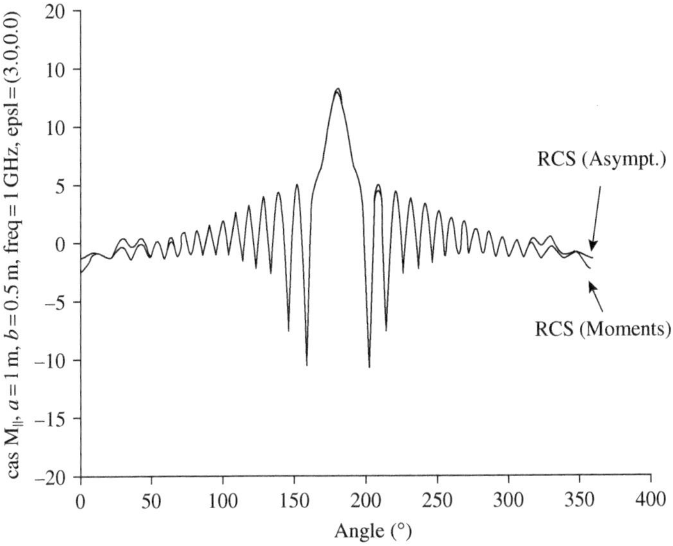

Figure 1.31 *Coated elliptic cylinder, a = 1 m, b = 0.5 m, f = 1 GHz, TM polarisation*

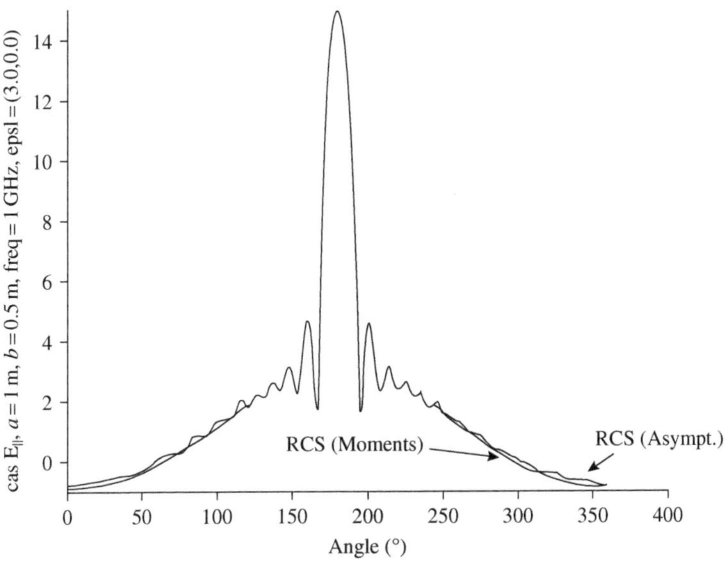

Figure 1.32 *Coated elliptic cylinder, a = 1 m, b = 0.5 m, f = 1 GHz, TE polarisation*

For the coated circular cylinder of radius $a = 0.5$ m, the results are presented in Figures 1.29 and 1.30. For the coated elliptic cylinder with $a = 0.5$ m and $b = 1$ m, the results are given in Figures 1.31 and 1.32.

In all the cases, the correspondence with the results of the method of moments is excellent.

1.4 References

1 KELLER, J. B.: 'Geometrical theory of diffraction', *Journal of Optical Society of America*, 1962, **52**, pp. 116–30

2 KELLER, J. B.: 'Diffraction by an aperture', *Journal of Applied Physics*, 1957, **28**, pp. 426–44

3 KELLER, J. B.: 'Diffraction by a convex cylinder', *IRE Transactions on Antennas and Propagation*, 1956, **AP-24**, pp. 312–21

4 KELLER, J. B.: 'One hundred years of diffraction theory', *IEEE Transactions on Antennas and Propagation*, 1985, **AP-33** (2), pp. 123–6

5 SOMMERFELD, A., and RUNGE, J.: 'Anwendung der Vektorrechnung auf die Grundlagen der geometrischen Optik', *Annalen der Physik*, 1911, **35**, pp. 277–98

6 KLINE, M.: 'An asymptotic solution of Maxwell's equations', *Communications on Pure & Applied Mathematics*, 1951, **4**, pp. 225–62

7 BABICH, V. M., and BULDYREV, V. S.: 'Short-wavelength diffraction theory' (Springer-Verlag, Berlin, 1991)

8 BOROVIKOV, V. A., and KINBER, B. YE.: 'Geometrical theory of diffraction' (IEE Electromagnetic Waves Series, vol. 37, 1994)

9 KOUYOUMJIAN, R. G.: 'Asymptotic high-frequency methods', *Proceedings of the IEEE*, 1965, **53**, pp. 864–76

10 POINCARÉ, H.: 'Les méthodes nouvelles de la mécanique, vol. II, 1892' (reedited by Dover Publications, New York, 1957)

11 ERDELYI, A.: 'Asymptotic expansions' (Dover Publications, New York, 1957)

12 VAN DYKE, M.: 'Perturbation methods in fluid mechanics' (The Parabolic Press, Stanford, CA, 1975)

13 KEVORKIAN, J., and COLE, J. D.: 'Perturbation methods in applied mathematics' (Springer-Verlag, New York, 1981)

14 FRANCOIS, C.: 'Les méthodes de perturbation en mécanique' (Course no 252 edited by ENSTA, 32 Bd Victor, Paris, 1973)

15 DESCHAMPS, G. A.: 'Ray techniques in electromagnetics', *Proceedings of the IEEE*, 1972, **60** (9), pp. 1022–35

16 KOUYOUMJIAN, R. G., and PATHAK, P. H.: 'A uniform geometrical theory of diffraction for an edge in a perfectly conducting surface', *Proceedings of the IEEE*, 1974, **62**, pp. 1448–61

17 BREMMER, H., and LEE, S. W.: 'Geometrical optics solution of reflection from arbitrarily curved surface', *Radio Science*, 1982, **17** (5), pp. 1117–31

18 CHALOUPKA, H., and MECKELBURG, H. J.: 'Improved high frequency approximation for curved conducting surfaces', *AEU*, 1985, *Band* **39**, pp. 245–50
19 KOUYOUMJIAN, R. G.: 'The geometrical theory of diffraction and its application', in MITTRA, R. (Ed.): 'Numerical and asymptotic techniques in electromagnetics' (Springer-Verlag, Berlin, 1975)
20 HONG, S.: 'Asymptotic theory of electromagnetic and acoustic diffraction by smooth convex surfaces of variable curvature', *Journal of Mathematical Physics*, 1967, **8** (6), p. 1223
21 LEONTOVICH, M. A.: 'On one method for the solution of the problem on electromagnetic waves propagation along the surface of the Earth', *Izvestiya AN SSSR Physics*, 1944, **8** (1), p. 16 (in Russian)
22 FOCK, V. A.: 'Diffraction of radio-waves around the earth's surface', *Journal of Physics of the USSR*, 1945, **9** (4), p. 255
23 LEONTOVICH, M. A., and FOCK, V. A.: 'Solution of the problem of propagation of electromagnetic waves along the earth's surface by the method of parabolic equation', *Journal of Physics of the USSR*, 1946, **10** (1), p. 13
24 FOCK, V. A.: 'The distribution of currents induced by a plane wave on the surface of a conductor', *Journal of Physics of the USSR*, 1946, **10** (2), p. 130
25 FOCK, V. A.: 'New methods in diffraction theory', *Philosophical Magazine Series* 7, 1948, **39**, p. 149
26 FOCK, V. A.: 'Fresnel diffraction from convex bodies', *Uspekhi Fizicheskikh Nauk*, 1950, **43** (4), p. 587 (in Russian)
27 BUCHAL, R. N., and KELLER, J. B.: 'Boundary layer problems in diffraction theory', *Communications on Pure & Applied Mathematics*, 1960, **13**, p. 85
28 BABICH, V. M., and KIRPICHNIKOVA, N. YA.: 'The boundary-layer method in diffraction problems' (Springer-Verlag, Berlin, 1979)
29 FRIEDLANDER, F. G., and KELLER, J. B.: 'Asymptotic expansions of solutions of $(\nabla^2 + k^2)u = 0$', *Communications on Pure & Applied Mathematics*, 1955, **8** (3), p. 378
30 ANDRONOV, I. V., and BOUCHE, D.: 'Friedlander–Keller solution for the 3D Maxwell case', in KONG, J. (Ed.): 'Electromagnetic waves PIER 15' (EMW Publishing, Cambridge, USA, 1997) p. 87
31 FOCK, V. A.: 'Electromagnetic diffraction and propagation problems' (Pergamon Press, Oxford, 1965)
32 ARNOLD, V. I.: 'The theory of singularities and its applications' (Cambridge University Press, 1991)
33 POSTON, T., and STEWART, I.: 'Catastrophe theory and its applications' (Dover Publications, New York, 1996)
34 BOUCHE, D., MOLINET, F., and MITTRA, R.: 'Asymptotic methods in electromagnetics' (Springer-Verlag, Berlin, 1997)
35 ARNOLD, J. M.: 'Spectral synthesis of uniform wave functions', *Wave Motion*, 1986, **8**, pp. 135–50
36 KRAVTSOV, Y., and ORLOV, Y.: 'Caustics, catastrophes and wave fields' (Springer-Verlag, Berlin, 1999)

37 PEARCEY, T.: 'The structure of an electromagnetic field in a neighborhood of a cusp of a caustic', *Philosophical Magazine*, 1946, **37**, pp. 311–27
38 STARKOV, A. S.: 'Wave field asymptotics in a neighborhood of dicaustic'. Proceedings of the 9th USSR symposium on *Diffraction and wave propagation*, Moscow-Tbilisi, 1985, vol. 1, pp. 82–85 (in Russian)
39 BULDYREV, V. S.: 'Short wave inteference in the problem of diffraction on inhomogeneous arbitrary-section cylinder', *Radiofizika USSR*, 1964, **10** (5), pp. 699–711
40 BABICH, V. M., and KAMOTSKII, V. M.: 'On the wave field in the small neighborhood of the point of incidence of the limiting ray', *Zapiski Nauchnyk Seminarov Leningradskogo Otdeleniya Matemtidneskogo Institute*, 2000, **264**, pp. 22–32
41 PHILIPPOV, V. B.: 'On exact justification of shortwave asymptotics of diffraction problem in the shadow', *Zapiski Nauchnyk Seminarov Leningradskogo Otdeleniya Matemtidneskogo Institute*, 1973, **34**, p. 142
42 ZAYAEV, A. B., and PHILIPPOV, V. B.: 'On exact justification of creeping wave solutions', *Zapiski Nauchnyk Seminarov Leningradskogo Otdeleniya Matemtidneskogo Institute*, 1983, **128**, p. 48
43 EGOROV, Y. V., and SHUBIN, M. A.: 'Partial Differential Equations IV, Microlocal Analysis and Hyperbolic Equations' (Springer-Verlag, Berlin, 1993)

Chapter 2
Electromagnetic creeping waves

2.1 Creeping waves on a general surface

2.1.1 Introduction

In this section, we turn our attention to the boundary-layer problem associated with creeping waves on a general surface. Creeping waves can be launched by a distant incident wave or by a source located on the surface of the object.

To address the boundary-layer problem, we follow exactly the same steps as in Chapter 1 for the case of the cylinder. These are:

- choosing an Ansatz
- expressing the equations in the surface coordinate system
- introducing the Ansatz in the above equations
- solving the equations for each order.

Before going into the details of the procedure, we make some preliminary observations pertinent to the method and describe the structure of our presentation.

As shown in Section 1.1.2 of Chapter 1, the surface rays follow the geodesics of the surface. If the creeping waves are generated by a distant incident wave, these geodesics start at the light–shadow boundary on the surface and are directed at that boundary along the rays of the incident field. If creeping waves are generated by a point source on the surface, all geodesics start in the point of the source. On a general surface geodesics form a system of curves with one or several caustics that have cusps. Away from caustics the system of geodesics can be chosen as the coordinate system. If s is the curvilinear abscissa along the geodesics, then the main phase factor $\exp(ikS(\mathbf{r}))$ is simply $\exp(iks)$.

As is well known, on a cylindrical surface, the electromagnetic field is decomposed into transverse electric (TE) and transverse magnetic (TM) polarisations and therefore the problems are scalar. The only characteristic of the surface that influences the asymptotics of creeping waves in that case is radius of curvature ρ. On a general surface, the problems are not reduced to scalar ones and additionally to ρ other characteristics of the surface (torsion τ and radius of transverse curvature ρ_t) appear in the asymptotic expansion of creeping waves, however only in the correction terms at order $k^{-1/3}$. We mainly concentrate on deriving the principal order terms of the creeping waves expansion and give smaller order correction only for the amplitude factor in Section 2.1.8.

The attenuation parameter ξ depends on the boundary condition satisfied by the electromagnetic field on the surface. For the same reason of uniformity as in Chapter 1, we assume an impedance boundary condition and accept quantities mZ and m/Z to be of order 1 when applying the asymptotic procedure. This makes it possible to follow the transformation of our formulae from the case of an impedance boundary condition to the case of a perfect conductor.

We now proceed with the task of deriving the solution.

2.1.2 Equations and boundary conditions

We seek a solution of Maxwell's equations, satisfying the impedance condition $\mathbf{E} - (\mathbf{n} \cdot \mathbf{E})\mathbf{n} = Z_a \mathbf{n} \times \mathbf{H}$ on the surface, where $\mathbf{n}$ is the unit normal vector to the surface and the complex number Z_a is the absolute impedance. (Taking $Z_a = 0$ gives the case of a perfect conductor.) We require the solution to satisfy the Silver–Müller radiation condition at infinity. With the $e^{-i\omega t}$ time convention, Maxwell's equations read

$$\operatorname{curl} \mathbf{E} = i\omega\mu\mathbf{H},$$

$$\operatorname{curl} \mathbf{H} = -i\omega\varepsilon\mathbf{E},$$

or, alternatively,

$$\operatorname{curl} \sqrt{\varepsilon}\,\mathbf{E} = i\omega\sqrt{\varepsilon\mu}\sqrt{\mu}\,\mathbf{H},$$

$$\operatorname{curl} \sqrt{\mu}\,\mathbf{H} = -i\omega\sqrt{\varepsilon\mu}\sqrt{\varepsilon}\mathbf{E}.$$

The impedance condition can be written in the form

$$\sqrt{\varepsilon}\mathbf{E} - (\mathbf{n} \cdot \sqrt{\varepsilon}\mathbf{E})\mathbf{n} = Z\mathbf{n} \times \sqrt{\mu}\,\mathbf{H},$$

where Z is the relative impedance, that is, the absolute impedance normalised to the impedance of free space. Next, following Fock [1], we introduce scaled field quantities $\hat{\mathbf{E}} = \sqrt{\varepsilon}\,\mathbf{E}$ and $\hat{\mathbf{H}} = \sqrt{\mu}\,\mathbf{H}$ and rewrite Maxwell's equations as (hats here and below are discarded)

$$\begin{aligned} &\operatorname{curl} \mathbf{E} = ik\mathbf{H}, && \mathbf{r} \in \mathbb{R}^3 \backslash \Omega, \\ &\operatorname{curl} \mathbf{H} = -ik\mathbf{E}, && \mathbf{r} \in \mathbb{R}^3 \backslash \Omega, \\ &\mathbf{E} - (\mathbf{n} \cdot \mathbf{E})\mathbf{n} = Z\mathbf{n} \times \mathbf{H}, && \mathbf{r} \in \partial\Omega, \end{aligned} \tag{2.1}$$

where $\omega\sqrt{\varepsilon\mu} = \omega/c = k$, $Z_a = Z\sqrt{\mu/\varepsilon}$. The fields $\mathbf{E}$ and $\mathbf{H}$ must satisfy the Silver–Müller radiation condition, which reads

$$\mathbf{E} = O(r^{-1}), \quad \mathbf{H} = O(r^{-1}), \qquad r = |\mathbf{r}| \to \infty,$$
$$\mathbf{E} + r^{-1}\mathbf{r} \times \mathbf{H} = o(r^{-1}), \quad \mathbf{H} - r^{-1}\mathbf{r} \times \mathbf{E} = o(r^{-1}), \qquad r \to \infty.$$

2.1.3 Form of the asymptotic expansion

Let us now turn to the derivation of the asymptotic representation. Introduce a system of geodesic coordinates (s, a, n), where s denotes the curvilinear abscissa along an arbitrary geodesics; a denotes the curvilinear abscissa along the orthogonal curve on $\partial\Omega$, which is the coordinate axis of the geodesic coordinates considered and n denotes the distance to the surface along the normal. The asymptotic expansion of the solution is written in the form similar to (1.83) of Chapter 1

$$\begin{aligned} \mathbf{E} &= \exp(iks + ik^{1/3}p(s,a)) \sum_{j=0}^{N} \mathbf{E}_j k^{-j/3}, \\ \mathbf{H} &= \exp(iks + ik^{1/3}p(s,a)) \sum_{j=0}^{N} \mathbf{H}_j k^{-j/3}. \end{aligned} \tag{2.2}$$

Here the first exponential multiplier represents the quick phase factor, while the second accounts for slower amplitude behaviour. Phase function $p(s,a)$ is complex, with its imaginary part describing the attenuation of the creeping waves, and $\mathbf{E}_j$ and $\mathbf{H}_j$ are the coefficients of the amplitude expansions with respect to the powers $k^{-j/3}$ of the wave number k. The coefficients $\mathbf{E}_j$ and $\mathbf{H}_j$ are complex-valued functions of the stretched coordinate $\ell = k^{2/3}n$ and of the coordinates s and a. In the boundary layer near the surface we assume ℓ to be of order $O(1)$, that is, $n = O(k^{-2/3})$.

The calculation of $p(s,a)$ and of the fields $\mathbf{E}_j$ and $\mathbf{H}_j$, is performed by inserting the expansions (2.2) in Maxwell's equations and by taking into account the impedance condition. Following this procedure, we can derive, for each order, the relevant equations and boundary conditions which, when combined with the radiation condition, enable us to determine the fields $\mathbf{E}_j$ and $\mathbf{H}_j$. In practice, we typically restrict ourselves to deriving the first-order terms, that is to say to the calculation of $\mathbf{E}_0$ and $\mathbf{H}_0$. Using this procedure, Maxwell's equations and the boundary conditions will be satisfied only up to a certain order and, consequently, our solutions will be an approximation to the scattering problem at high frequencies. We should point out that we have not provided an upper bound for the terms we have neglected.

2.1.4 Derivation of the solution of Maxwell's equations in the coordinate system (s, a, n)

The coordinates (s, a, n) are neither orthogonal nor are they normalised. In order to express the differential operator curl, with these coordinates, we have to introduce (see, e.g., Reference 2) the covariant and contravariant components of the vectors $\mathbf{E}_j$ and $\mathbf{H}_j$ that are under consideration. Denoting as (E^s, E^a, E^n) and (H^s, H^a, H^n) the contravariant components of $\mathbf{E}$ and $\mathbf{H}$ and as (E_a, E_s, E_n) and (H_s, H_a, H_n) those of the associated linear form (covariant), the expression for the curl operator in the coordinate system (s, a, n) becomes

$$
\begin{aligned}
(\text{curl }\mathbf{E})^s &= \frac{1}{\sqrt{g}}\left(\frac{\partial E_n}{\partial a} - \frac{\partial E_a}{\partial n}\right),\\
(\text{curl }\mathbf{E})^a &= \frac{1}{\sqrt{g}}\left(\frac{\partial E_s}{\partial n} - \frac{\partial E_n}{\partial s}\right),\\
(\text{curl }\mathbf{E})^n &= \frac{1}{\sqrt{g}}\left(\frac{\partial E_a}{\partial s} - \frac{\partial E_s}{\partial a}\right),
\end{aligned}
$$

where

$$
\begin{pmatrix} E_s \\ E_a \\ E_n \end{pmatrix} = g_{ij} \begin{pmatrix} E^s \\ E^a \\ E^n \end{pmatrix}.
$$

The matrix g_{ij} characterises the metric of the coordinate system and g is its determinant. The matrix is composed of the coefficients of the second quadratic form of the coordinate system and takes the form

$$
g_{ij} = \begin{pmatrix}
\left(1 + \dfrac{n}{\rho}\right)^2 + \tau^2 n^2 & -h\tau\left(2n + \dfrac{n^2}{\rho} + \dfrac{n^2}{\rho_t}\right) & 0 \\
-h\tau\left(2n + \dfrac{n^2}{\rho} + \dfrac{n^2}{\rho_t}\right) & h^2\left(\left(1 + \dfrac{n}{\rho_t}\right)^2 + \tau^2 n^2\right) & 0 \\
0 & 0 & 1
\end{pmatrix}, \tag{2.3}
$$

where $\rho = \rho(s, a)$ and $\tau = \tau(s, a)$ are the radii of curvature of the geodesics $a = \text{const}$ and its torsion, $\rho_t = \rho_t(s, a)$ is the normal radius of curvature of the surface in the direction orthogonal to the geodesics, the parameter $h = h(s, a)$ measures the narrowing (or broadening) of an infinitesimal geodesics pencil between a point with coordinates (s, a) and the axis of a, that is, the line $s = 0$.

We restrict our consideration to the terms of $O(n)$ and neglect those that are $O(n^2)$ because of the following reason. After stretching, the latter terms are found to be of order $O(k^{-4/3})$, since n is of order $O(k^{-2/3})$. When we multiply these terms by k, in accordance with Maxwell's equations, they become of order $k^{-1/3}$ and as can be seen below, they do not influence the leading approximation for the electric and magnetic vectors. Derivation of next order terms can be found in Reference 3 and in Section 2.1.8. We find

$$
\begin{aligned}
E_s &= \left(1 + \frac{2n}{\rho}\right) E^s - h\tau n E^a + O(n^2),\\
E_a &= -h\tau n E^s + h^2\left(1 + \frac{2n}{\rho_t}\right) E^a + O(n^2),\\
E_n &= E^n + O(n^2).
\end{aligned}
$$

The first Maxwell equation (2.1) can be written as three scalar equations, each of which yields four equations, if we sort out the terms that are of order k, $k^{2/3}$, $k^{1/3}$ and k^0. Thus, all together a total of 12 equations are derived from the first Maxwell equation. Likewise, the second Maxwell equation provides 12 more equations – the dual set, which we could obtain from the first set of 12 equations by replacing **E** with **H**, and **H** with −**E**. The following is a catalogue of the first set of equations:

Order k

$$H_0^s = 0, \tag{2.4}$$

$$E_0^n = -hH_0^a, \tag{2.5}$$

$$H_0^n = hE_0^a; \tag{2.6}$$

Order $k^{2/3}$

$$H_1^s = ih\frac{\partial E_0^a}{\partial \ell}, \tag{2.7}$$

$$E_1^n = -hH_1^a, \tag{2.8}$$

$$H_1^n = hE_1^a; \tag{2.9}$$

Order $k^{1/3}$

$$H_2^s = ih\frac{\partial E_1^a}{\partial \ell} - \frac{\partial p}{\partial a}H_0^a, \tag{2.10}$$

$$hH_2^a + E_2^n = h\left(\frac{\partial p}{\partial s}H_0^a - \ell\left(\frac{1}{\rho} + \frac{1}{\rho_t}\right)H_0^a - \frac{\partial^2 H_0^a}{\partial \ell^2}\right), \tag{2.11}$$

$$hE_2^a - H_2^n = h\left(-\frac{\partial p}{\partial s}E_0^a + \ell\left(\frac{1}{\rho} - \frac{1}{\rho_t}\right)E_0^a\right); \tag{2.12}$$

Order k^0

$$H_3^s = ih\frac{\partial E_2^a}{\partial \ell} - \frac{\partial p}{\partial a}H_1^a + 2i\frac{h}{\rho_t}E_0^a - ih\ell\left(\frac{1}{\rho} - \frac{1}{\rho_t}\right)\frac{\partial E_0^a}{\partial \ell}, \tag{2.13}$$

$$\begin{aligned} hH_3^a + E_3^n = {}& h\left(\frac{\partial p}{\partial s}H_1^a - \ell\left(\frac{1}{\rho} + \frac{1}{\rho_t}\right)H_1^a - \frac{\partial^2 H_1^a}{\partial \ell^2}\right) \\ & + 2ih\tau E_0^a + 2ih\tau\ell\frac{\partial E_0^a}{\partial \ell} - \frac{\partial}{\partial s}(hH_0^a) + i\frac{\partial p}{\partial a}\frac{\partial E_0^a}{\partial \ell}, \end{aligned} \tag{2.14}$$

$$\begin{aligned} hE_3^a - H_3^n = {}& h\left(-\frac{\partial p}{\partial s}E_1^a + \ell\left(\frac{1}{\rho} - \frac{1}{\rho_t}\right)E_1^a\right) - 2ih\tau\ell\frac{\partial H_0^a}{\partial \ell} \\ & + \frac{i}{h}\frac{\partial}{\partial s}(h^2E_0^a) - i\frac{\partial p}{\partial a}\frac{\partial H_0^a}{\partial \ell}. \end{aligned} \tag{2.15}$$

This set of equations must be complemented by its dual, that is, the equations obtained from the equations above, by replacing E with H and H with $-E$.

We will now interpret the equations associated with the different orders and will use them to calculate the functions $\mathbf{E}_0$ and $\mathbf{H}_0$, by taking into account the boundary and radiation conditions.

2.1.5 *Interpretation of the equations associated with the first three orders*

The equations of order k, that is (2.4)–(2.6) and their dual equations, show that H_0^n is a function of E_0^a, and E_0^n is a function of H_0^a. They also show that the components with the superscript s, namely those in the direction of propagation of the creeping mode we are looking for, vanish for the zeroth order.

One can provide the following interpretation of the above result which is consistent with the geometrical theory of diffraction (GTD). There are two types of creeping waves that are solutions of Maxwell's equations. The first of these is called the electric creeping wave, because, for the zero order, the electric and magnetic fields, $\mathbf{E}$ and $\mathbf{H}$, are expressed as functions of the a-component of the electric field E_0^a. For the zero order, the components of this electric creeping wave are $E_0^a \neq 0$, $E_0^n = 0$, $E_0^s = 0$, $H_0^a = 0$, $H_0^n = hE_0^a$, $H_0^s = 0$. At this point, it is useful to remember that the frame $(\mathbf{e}_s, \mathbf{e}_a, \mathbf{e}_n)$ is not normalised. Thus, the equation dual to (2.5) simply implies that H_0^n and E_0^a have the same magnitude. The two dominant components (i.e., the highest order with respect to k) of the electric creeping wave are E_0^a and H_0^n; hence, at the dominant order the electric creeping waves propagate as plane waves. Similarly, for the zeroth- and first-order components of the magnetic creeping wave, the $\mathbf{E}$ and $\mathbf{H}$ fields are expressible in terms of H_0^a and H_1^a. For the zero order, the components of such a creeping wave are $H_0^a \neq 0$, $H_0^n = 0$, $H_0^s = 0$, $E_0^a = 0$, $E_0^n = -hH_0^a$, $E_0^s = 0$. Once again, the equation dual to (2.6) simply means that E_0^n and H_0^a have equal magnitudes, because h in the above equations is the norm of the vector $\mathbf{e}_a$. The two dominant components of the magnetic creeping wave are H_0^a and E_0^n, and, to a first approximation, the magnetic creeping wave propagates like a plane wave.

We see from the above that the dominant order components of both the electric and magnetic creeping waves propagate as plane waves. The electric creeping wave is polarised with its electric field vector parallel to the surface, while correspondingly, the magnetic creeping wave has its magnetic field vector parallel to the surface.

Equations of order $k^{2/3}$, (2.7)–(2.9) and their dual equations, show that in contrast to plane wave fields, there are now non-zero components of electric and magnetic fields in the direction of propagation. More precisely, the electric (magnetic) creeping wave has a first-order component of its magnetic (electric) field, which does not vanish in the direction of propagation. In fact, for the electric creeping wave

$$H_1^s = ih\frac{\partial E_0^a}{\partial \ell} \neq 0,$$

and for the magnetic creeping wave

$$E_1^s = -ih\frac{\partial H_0^a}{\partial \ell} \neq 0.$$

Once again, the coefficient h in the above equations is nothing, but the norm of the vector $\mathbf{e}_a$.

Equations (2.8), (2.9) and their dual equations express H_1^n and E_1^n as functions of E_1^a and H_1^a, respectively. To zeroth and first orders, all the components of **E** and **H** are expressible in terms of E_0^a, H_0^a, E_1^a and H_1^a. It is therefore natural to choose E^a and H^a as the potentials in terms of which we express the rest of the field components. We should remark that we could have just as well chosen the normal components of the fields **E** and **H** for this purpose, and the results would have been equivalent. We also observe that E_0^a and H_0^a are yet to be obtained and we accomplish this by considering the equations of the next order.

Equations of order $k^{1/3}$ are equations (2.10)–(2.12) and their dual equations. We note that Equation (2.10) and dual to it yield H_2^s and E_2^s as functions of E_0^a and H_0^a, respectively, and of the derivatives of the first-order components E_1^a and H_1^a. Equations (2.11), (2.12) and their dual are written for $hH_2^a + E_2^a$ and $hE_2^a - H_2^a$. They are the counterparts of the two sets of equations in the higher order

$$hH_0^a + E_0^n = 0, \qquad hE_0^a - H_0^n = 0,$$
$$hH_1^a + E_1^n = 0, \qquad hE_1^a - H_1^n = 0.$$

However, at order $k^{1/3}$ the equations are somewhat different, now we have the presence of a non-zero right-hand side that entails the derivatives of E_0^a or H_0^a.

Using (2.11) and equation dual to (2.12), which lead to two different formulae for $hH_2^a + E_2^n$, we derive a compatibility condition, which after some manipulation reduces to the equation

$$\frac{\partial^2 H_0^a}{\partial \ell^2} + 2\left(\frac{\ell}{\rho} - \frac{\partial p}{\partial s}\right) H_0^a = 0. \tag{2.16}$$

Likewise, using (2.12) and equation dual to (2.11), which lead to two different formulae for $hE_2^a - H_2^n$, we obtain the second compatibility condition

$$\frac{\partial^2 E_0^a}{\partial \ell^2} + 2\left(\frac{\ell}{\rho} - \frac{\partial p}{\partial s}\right) E_0^a = 0. \tag{2.17}$$

Equations (2.16) and (2.17) are similar to the leading order equation in (1.84) of Chapter 1 obtained for the case of scalar creeping waves on a cylinder. The dependence of E_0^a and H_0^a on the coordinate n is then determined, as in (1.85) of Chapter 1, by the function w_1

$$E_0^a = h^{-1}(s,a)A_0(s,a)w_1(\xi - \nu), \tag{2.18}$$
$$H_0^a = h^{-1}(s,a)B_0(s,a)w_1(\xi - \nu), \tag{2.19}$$

where

$$\nu = \left(\frac{2}{\rho}\right)^{1/3} \ell, \tag{2.20}$$

$$\xi = 2^{1/3} \rho^{2/3} \frac{\partial p}{\partial s}. \tag{2.21}$$

In order to find the parameter ξ which then gives the expression for the function p by (2.21), we have to take into account the boundary conditions on the surface $\partial\Omega$. This will be carried out in the following section. The amplitudes A_0 and B_0 remain arbitrary in that order of considered equations. The variation of the amplitudes A_0 and B_0 with respect to s will be studied when equations of order $O(k^0)$ will be considered.

Before concluding this section, we point out that Equations (2.11) and (2.12) enable us to derive expressions for E_2^n and H_2^n as functions of H_0^a, H_2^a and E_0^a, E_2^a, respectively. For the second order, as the previous ones, all the components of $\mathbf{E}$ and $\mathbf{H}$ can be expressed in terms of the a-components of $\mathbf{E}$ and $\mathbf{H}$ fields only.

2.1.6 Boundary conditions and the determination of $p(s, a)$

We will now turn to the boundary conditions to be satisfied by the field components. We begin with the case of a perfect conductor, and express the boundary conditions on the surface of the object.

2.1.6.1 Perfect conductor

The electric field tangent to the surface of a perfect conductor must vanish, consequently, we have

$$E^a = 0, \qquad E^s = 0, \qquad \ell = 0. \tag{2.22}$$

Expanding (2.22) according to the powers of $k^{1/3}$, we obtain

$$E_0^a = E_1^a = \cdots = E_j^a = 0, \qquad \ell = 0 \tag{2.23}$$

and

$$E_0^s = E_1^s = \cdots = E_j^s = 0, \qquad \ell = 0. \tag{2.24}$$

We note that $E_0^s = 0$ is satisfied for all ℓ, thus, the first condition given by (2.24) is applicable to E_1^s. After expressing E_1^s from the equation dual to (2.7) as

$$E_1^s = -ih\frac{\partial H_0^a}{\partial \ell},$$

we find that the fields E_0^a and H_0^a satisfy the following boundary conditions:

$$E_0^a = 0, \qquad \ell = 0 \tag{2.25}$$

and

$$\frac{\partial H_0^a}{\partial \ell} = 0, \qquad \ell = 0. \tag{2.26}$$

Thus, we see that for the case of a perfect conductor, the electric creeping wave satisfies Dirichlet and the magnetic wave satisfies Neumann boundary conditions, respectively.

Inserting (2.18) in the boundary condition (2.25) we introduce ξ_E as a solution of the equation

$$w_1(\xi_E) = 0.$$

In a similar manner, by using (2.19) and (2.26), we introduce ξ_H such that

$$w_1'(\xi_H) = 0.$$

Accordingly ξ_E is a zero of w_1, ξ_H is a zero of the derivative of w_1 and $\xi_{E,H}$ are related to p by (2.21). Thus, we deduce two values for $\partial p/\partial s$

$$\frac{\partial p}{\partial s} = 2^{-1/3}\rho^{-2/3}\xi_E \tag{2.27}$$

and

$$\frac{\partial p}{\partial s} = 2^{-1/3}\rho^{-2/3}\xi_H. \tag{2.28}$$

Equations (2.27) and (2.28) are not compatible because the function w_1 does not have a double zero, hence $\xi_E \neq \xi_H$. From the above discussion it follows that either $H_0^a = 0$, or $E_0^a = 0$ everywhere. In the first case, we obtain an electric creeping wave all of whose components can be calculated from E_0^a. It propagates with the phase function given by

$$\exp\left(iks + i\left(\frac{k}{2}\right)^{1/3}\xi_E\int_0^s \frac{ds'}{\rho^{2/3}(s',a)}\right).$$

In the second case, we obtain a magnetic creeping wave all of whose components can be calculated from H_0^a. It propagates with the phase function

$$\exp\left(iks + i\left(\frac{k}{2}\right)^{1/3}\xi_H\int_0^s \frac{ds'}{\rho^{2/3}(s',a)}\right). \tag{2.29}$$

The electric and magnetic creeping waves propagate with different velocities and attenuation parameters. This result is well known in GTD, and is recognised, for instance, when solving the canonical problem of the sphere. The advantage of the asymptotic expansion method is that it leads us to this result in a more deductive manner.

We have just derived functions $p(s,a)$ for the electric and magnetic creeping waves propagating on a perfect conductor. To find A_0 and B_0 from (2.18) and (2.19), we have to impose the boundary conditions on E_1^a and H_1^a, as shown below, and to work with the equation of order k^0 as discussed in the last paragraph of this section.

Let us turn now to the boundary condition on E_1^a and H_1^a. From (2.23) we have

$$E_1^a = 0 \qquad \text{for } \ell = 0.$$

The boundary condition on H_1^a is obtained from $E_2^s = 0$ and from the expression dual to (2.10) that relates E_2^s to H_1^a. Taking into account that $E_0^a = 0$ for $\ell = 0$, we obtain

$$\frac{\partial H_1^a}{\partial \ell} = 0.$$

Thus, for the case of a perfect conductor, the quantities E_1^a and H_1^a satisfy homogeneous Dirichlet and Neumann conditions, respectively. Next, we proceed to discuss the general case of a surface described by an impedance boundary condition.

2.1.6.2 Impedance conditions

The isotropic impedance boundary condition relates the tangential electric and magnetic fields as follows

$$\mathbf{E}_t = Z(\mathbf{n} \times \mathbf{H}_t), \tag{2.30}$$

where

$$\mathbf{E}_t = E^s\mathbf{e}_s + E^a\mathbf{e}_a, \qquad \mathbf{H}_t = H^s\mathbf{e}_s + H^a\mathbf{e}_a.$$

Taking into account the fact that, on the surface, the frame $(\mathbf{e}_s, \mathbf{e}_a, \mathbf{e}_n)$ is orthogonal, but not normalised, we have

$$\mathbf{n} \times \mathbf{e}_s = h^{-1}\mathbf{e}_a, \qquad \mathbf{n} \times \mathbf{e}_a = -h^{-1}\mathbf{e}_s.$$

Then we can rewrite (2.30) in the form

$$E^s = -ZhH^a$$

and

$$hE^a = ZH^s.$$

Finally, by using the fact that $E_0^s = H_0^s = 0$, we get

$$k^{-1/3}E_1^s + \cdots + k^{-j/3}E_j^s + \cdots = -Zh(H_0^a + k^{-1/3}H_1^a + \cdots + k^{-j/3}H_j^a + \cdots), \tag{2.31}$$

$$k^{-1/3}H_1^s + \cdots + k^{-j/3}H_j^s + \cdots = \frac{h}{Z}(E_0^a + k^{-1/3}E_1^a + \cdots + k^{-j/3}E_j^a + \cdots). \tag{2.32}$$

Equation (2.32) can also be obtained from (2.31) by introducing the changes E to H, H to $-E$ and Z to Z^{-1}.

Boundary conditions for E_0^a and H_0^a can be determined if we choose the order of the surface impedance Z. The assumption $Z = O(1)$ will be studied in Section 2.2.2 and it will be found that it leads to the Dirichlet condition for E_0^a and H_0^a. However, this result is not uniform with respect to Z, because, when $Z \to 0$, the component H_0^a

continues to satisfy the Dirichlet boundary condition, while when $Z = 0$, that is, for the perfectly conducting case, it satisfies the Neumann condition.

In order to avoid these problems and to derive a uniform solution with respect to Z, when $Z \to 0$, we assume, as in Chapter 1, an impedance behaviour as follows:

$$Z = k^{-1/3} Z_H, \qquad \text{in Equation (2.31)},$$

$$Z = k^{1/3} Z_E, \qquad \text{in Equation (2.32)},$$

where Z_H and Z_E are assumed to be of order $O(1)$. This, in turn, implies that the impedance is small in (2.31) and large in (2.32). These statements look somewhat contradictory and should not be taken too literally. These hypotheses are chosen so that the impedance has an effect on the boundary conditions, and in turn on the propagation constants for the creeping waves, that is, to fit the physics as closely as possible. The hypothesis $Z = O(1)$, which leads to impedance effects appearing only in order $j = 1$, that is, as perturbation of the perfectly conducting case, will be analysed in Section 2.2.2.

By equating the coefficients of the powers $k^{-1/3}$, we obtain, to the first order

$$Z_H H_0^a = i \frac{\partial H_0^a}{\partial \ell}.$$

In the variable ν introduced by (2.20) and the quantity

$$m = \left(\frac{k\rho}{2} \right)^{1/3},$$

the above equation can be written in the form

$$\frac{\partial H_0^a}{\partial \nu} + imZ H_0^a = 0, \qquad \nu = 0. \tag{2.33}$$

Similarly for (2.32) we have, to the first order

$$\frac{\partial E_0^a}{\partial \nu} + i \frac{m}{Z} E_0^a = 0, \qquad \nu = 0, \tag{2.34}$$

which can be deduced from (2.33) by substituting E in place of H and $1/Z$ in place of Z.

Substituting (2.18) and (2.19) into the boundary conditions (2.33) and (2.34), enables us to calculate ξ_E and ξ_H, and, hence, functions $p(s, a)$

$$w_1'(\xi_E) = i \frac{m}{Z} w_1(\xi_E), \tag{2.35}$$

$$w_1'(\xi_H) = imZ w_1(\xi_H). \tag{2.36}$$

As for the case of a perfect conductor, the quantities ξ_E and ξ_H cannot be equal except when $Z = 1$, because w_1 and w_1' do not vanish simultaneously.

As for the perfect conductor, we define two modes. The first of these is the electric creeping wave. For this mode, $E_0^a \neq 0$, but $H_0^a = 0$ everywhere. The phase function of this mode is

$$\exp\left(iks + i\left(\frac{k}{2}\right)^{1/3} \int_0^s \xi_E(s', a) \frac{ds'}{\rho^{2/3}(s', a)}\right).$$

This time ξ_E depends upon Z and is determined from Equation (2.35). As a result, it is not a constant as in the case of a perfect conductor, but is a function of point on the surface $\partial\Omega$.

The second mode is the magnetic creeping wave, with $H_0^a \neq 0, E_0^a = 0$. The equations are the same as those derived for a perfect conductor. The phase function is

$$\exp\left(iks + i\left(\frac{k}{2}\right)^{1/3} \int_0^s \xi_H(s', a) \frac{ds'}{\rho^{2/3}(s', a)}\right).$$

Here ξ_H is determined from Equation (2.36).

Equations (2.35) and (2.36) can be deduced from each other by replacing Z with $1/Z$ and vice versa. We can verify, *a posteriori*, that the assumptions embodied in (2.31) and (2.32) are not contradictory. Equation (2.31) assumes that the impedance is small (of order $O(m^{-1})$) for the magnetic creeping wave, whereas (2.32) implies that the impedance is large for the electric creeping wave. We observe that electric and magnetic creeping waves are independent solutions of the problem and therefore these two assumptions are indeed uncoupled. Furthermore, they must yield good results even outside their initial domain. In effect, if we let Z tend to infinity in (2.33) even though it should be small, in principle, we regain the Dirichlet boundary condition for H_0^a, which corresponds to the magnetic conducting case. We conclude, therefore, that the limit obtained is correct for $Z \to +\infty$. Similarly, if we let Z tend to 0 in (2.35), despite the fact that it is assumed to be large, we recover the Dirichlet boundary condition for E_0^a for the perfect electric conductor case. Again, the limit is correct as $Z \to 0$, and we conclude, therefore, that the solution for the impedance boundary condition case is general enough to include the conducting case.

Boundary conditions satisfied by E_1^a and H_1^a will also be needed. To this end, we return to (2.31) and (2.32) and obtain a homogeneous boundary condition on H_1^a for the magnetic creeping wave by using the second-order terms. The condition reads

$$\frac{\partial H_1^a}{\partial \nu} + imZH_1^a = 0, \qquad \nu = 0.$$

The condition on E_1^a also obtained by using the second-order terms is

$$\frac{\partial E_1^a}{\partial \nu} + i\frac{m}{Z}E_1^a = -\frac{i}{h}\frac{\partial \rho}{\partial a}\left(\frac{\rho}{2}\right)^{1/3} H_0^a, \qquad \nu = 0. \tag{2.37}$$

In the same way, we obtain for the electric creeping wave a homogeneous condition on E_1^a

$$\frac{\partial E_1^a}{\partial \nu} + i\frac{m}{Z}E_1^a = 0, \qquad \nu = 0,$$

and an inhomogeneous condition on H_1^a

$$\frac{\partial H_1^a}{\partial \nu} + imZH_1^a = \frac{i}{h}\frac{\partial p}{\partial a}\left(\frac{\rho}{2}\right)^{1/3} E_0^a, \qquad \nu = 0. \tag{2.38}$$

These boundary conditions will enable us to determine amplitudes A_0 and B_0 as shown below.

2.1.6.3 Determination of A_0 and B_0

Equations of order k^0 should be considered. These are Equations (2.13)–(2.15) and their dual equations.

We see that (2.14) and the dual equation of (2.15) provide two different expressions for $hH_3^a + E_3^n$, whereas (2.15) and the dual equation of (2.14) yield corresponding expressions for $hE_3^a - H_3^n$. From these equations, we obtain two compatibility conditions and the expressions for H_3^n and E_3^n as functions of E_j^a and H_j^a with $j = 0, 1, 2, 3$. Likewise, Equation (2.13) and its dual provide the expressions for E_3^s and H_3^s. Finally, we obtain all of the field components as functions of the components labelled with ^{a}. To simplify the notations, we define the quantities E_0 and H_0 via the equations $E_0 = hE_0^a$, $H_0 = hH_0^a$, which represent the fields with respect to a unit vector $\mathbf{a} = \|\mathbf{e}_a\|^{-1}\mathbf{e}_a$. We can also define the scaled versions of the fields E_1 and H_1 in a similar manner. The compatibility conditions read

$$-i\left[\frac{\partial^2 H_1}{\partial \ell^2} + 2\left(\frac{\ell}{\rho} - \frac{\partial p}{\partial s}\right)H_1\right] - 2\tau E_0 + 2\frac{\partial H_0}{\partial s} + \frac{1}{h}\frac{\partial h}{\partial s}H_0 = 0, \tag{2.39}$$

$$-i\left[\frac{\partial^2 E_1}{\partial \ell^2} + 2\left(\frac{\ell}{\rho} - \frac{\partial p}{\partial s}\right)E_1\right] + 2\tau H_0 + 2\frac{\partial E_0}{\partial s} + \frac{1}{h}\frac{\partial h}{\partial s}E_0 = 0. \tag{2.40}$$

After changing to variable ν introduced by (2.20), we recognise the operator

$$L_0 = \frac{\partial^2}{\partial \nu^2} + (\nu - \xi),$$

appearing in (2.39) and (2.40). We then rewrite these equations as

$$i\left(\frac{2}{\rho}\right)^{2/3} L_0 H_1 = \frac{2}{\sqrt{h}}\frac{\partial}{\partial s}(\sqrt{h}H_0) - 2\tau E_0, \tag{2.41}$$

$$i\left(\frac{2}{\rho}\right)^{2/3} L_0 E_1 = \frac{2}{\sqrt{h}}\frac{\partial}{\partial s}(\sqrt{h}E_0) + 2\tau H_0. \tag{2.42}$$

We remark here that the convention we adopted for the torsion is

$$\frac{d\mathbf{n}}{ds} = -\frac{1}{\rho}\mathbf{s} + \tau\mathbf{a},$$

where **n, s, a** is the Frenet system of unit vectors. The opposite convention is sometimes adopted in the literature and affects the sign of the last terms in (2.41) and (2.42).

Resolution of the previous equations. For the magnetic creeping wave, $E_0 = 0$, and (2.41) becomes

$$i\left(\frac{2}{\rho}\right)^{2/3} L_0 H_1 = \frac{2}{\sqrt{h}}\frac{\partial}{\partial s}(\sqrt{h}H_0) \tag{2.43}$$

with the boundary conditions

$$\frac{\partial H_0}{\partial \nu} = 0, \qquad \frac{\partial H_1}{\partial \nu} = 0,$$

for the perfect conductor and

$$\frac{\partial H_0}{\partial \nu} + imZH_0 = 0, \qquad \frac{\partial H_1}{\partial \nu} + imZH_1 = 0$$

for the impedance case.

The field H_0 is given by (2.19), which is now rewritten as

$$H_0 = B_0(s,a)w_1(\xi_H - \nu).$$

The coefficient $B_0(s,a)$ can be determined by using Equation (2.43) by following the same procedure as used in Chapter 1 for scalar waves on the cylinder. Equation (2.43) with boundary and radiation conditions is a Sturm–Liouville problem, which has a solution only if its right-hand side is orthogonal to the solutions of the homogeneous problem. The orthogonality condition provides us with an equation for $B_0(s,a)$. In practice, we obtain this condition by multiplying (2.43) by the solution $w_1(\xi_H - \nu)$ of the homogeneous problem and integrating from 0 to $+\infty$ with respect to the variable ν. We obtain, after performing calculations similar to those presented before (1.87) in Chapter 1, the differential equation (prime here denotes derivative by s)

$$\frac{B_0'}{B_0} + \frac{1}{2}\frac{h'}{h} + \frac{1}{6}\frac{\rho'}{\rho} + \frac{1}{2}\frac{Q_H'}{Q_H} = 0, \tag{2.44}$$

which is valid along a geodesic, where a is constant. Let us recall that $h(s,a)$ represents the broadening (or the narrowing) of the geodesic pencil with respect to the axes of a. It may also be identified as the ratio $d\eta'/d\eta$ in the GTD, where $d\eta'$ is compiled at the observation point, while $d\eta$ is calculated at the attachment point. Also, $\rho(s,a)$ is the radius of curvature of the geodesics along s. Finally, Q_H is defined by the following equation (see (1.88) of Chapter 1)

$$Q_H = \xi_H(w_1(\xi_H))^2 - (w_1'(\xi_H))^2,$$

where ξ_H satisfies Equation (2.36). For a perfect conductor both ξ_H and Q are constants.

Solving (2.44) we find

$$B_0(s,a) = B_0(a)h^{-1/2}(s,a)\rho^{-1/6}(s,a)Q_H^{-1/2}(s,a). \tag{2.45}$$

Inserting (2.45) into (2.19), we obtain H_0, while (2.29) provides the phase $p_H(s,a)$. Inserting the above results in the asymptotic expansion (2.2), and neglecting terms of order $O(k^{-1/3})$, we obtain the following expression for the field of a magnetic creeping wave

$$\mathbf{H} \approx B_0(a)h^{-1/2}(s,a)\rho^{-1/6}(s,a)Q_H^{-1/2}(s,a)$$

$$\times \exp\left(iks + i\left(\frac{k}{2}\right)^{1/3}\int\limits_{s(a)}^{s}\xi_H(s',a)\frac{ds'}{\rho^{2/3}(s',a)}\right)w_1(\xi_H - \nu)\mathbf{e}_a. \quad (2.46)$$

Similarly, for the electric creeping wave, we can obtain

$$\mathbf{E} \approx A_0(a)h^{-1/2}(s,a)\rho^{-1/6}(s,a)Q_E^{-1/2}(s,a)$$

$$\times \exp\left(iks + i\left(\frac{k}{2}\right)^{1/3}\int\limits_{s(a)}^{s}\xi_E(s',a)\frac{ds'}{\rho^{2/3}(s',a)}\right)w_1(\xi_E - \nu)\mathbf{e}_a, \quad (2.47)$$

where

$$Q_E = \xi_E\,(w_1(\xi_E))^2 - (w_1'(\xi_E))^2,$$

and ξ_E satisfies Equation (2.35).

To summarise, we can say that the field calculation for each creeping mode, whether H or E-type, entails the implementation of the following steps. We have to determine:

- the root ξ of Equations (2.36) or (2.35)
- the function $B_0(a)$ or $A_0(a)$ that yields the amplitudes of the H or E fields on each geodesic
- the function $s(a)$, which is the lower bound of integration of the phase.

Each determination of ξ leads to a specific mode, usually labelled with its index p. For the case of the diffraction by an object, the functions $A_0(a)$, $B_0(a)$ and $s(a)$ are determined by matching the creeping wave solutions that we have just determined, to the solution in the neighbourhood of the light–shadow boundary, just as it was done for scalar waves in Chapter 1 [4]. The solution in the shadow zone is then completely specified.

Equations (2.46) and (2.47) provide the fields due to a creeping wave in the vicinity of the surface. One extracts the surface fields by setting $\nu = 0$ in these formulae. The case of the electric creeping wave for the perfect conductor represents a special case, because (2.47) reduces to $E_0 = 0$ on the surface.

2.1.7 Physical interpretation of the results

2.1.7.1 Impedance boundary condition

The expressions in (2.46) and (2.47) have a somewhat unexpected character because the only zeroth-order component in these equations is along **a**. One might wonder how the impedance condition, which imposes the ratio of the orthogonal components of $\mathbf{E}_t$ and $\mathbf{H}_t$, can be satisfied. The answer to the above puzzle is that if we consider the terms that are $O(k^{-1/3})$ the corresponding fields do indeed have components along the direction of propagation. As an example, for the magnetic creeping wave, there exists a component E_1^s along the direction of propagation, and, for the dominant order, the field E^s is equal to ZH_0. Similarly, for the electric creeping wave the field H^s, to the dominant order, is given by $Z^{-1}E_0$. These fields are $O(k^{-1/3})$ in the analysis, because Z is of order $k^{-1/3}$ for the magnetic creeping wave and is of order $k^{1/3}$ for the electric creeping wave.

2.1.7.2 Geometrical factors

The factor $h^{-1/2}(s,a)$ has a simple interpretation in the context of GTD. The width of a pencil of surface rays is proportional to $h(s,a)$. Let us now consider the thickness of the boundary layer, which is proportional to $k^{-2/3}\rho^{1/3}$. The creeping wave can be seen as a volumetric tube of rays with width proportional to h tangential to the surface, and height proportional to $k^{-2/3}\rho^{1/3}$ along the normal. The conservation of the power in this tube of rays leads to the law $h\rho^{1/3}A_0^2 = \text{const}$, $h\rho^{1/3}B_0^2 = \text{const}$, and enables us to determine the factor $h^{-1/2}\rho^{-1/6}$.

The factors $Q_H^{-1/2}$ and $Q_E^{-1/2}$ can be interpreted as indices of refraction. For ordinary waves, the amplitude is in fact proportional to $N^{-1/2}$, where N is the reflective index of the medium. We identify Q_H and Q_E with equivalent indices.

Finally, the lower bound $s(a)$ in the integrals appearing in (2.46) and (2.47) determines the point where the attenuation starts. Evidently, we expect this point to be at the intersection of the creeping ray and the light–shadow boundary.

Consequently, all of the factors appearing in expressions (2.46) and (2.47) may be interpreted by recognising the similarity of behaviour between the creeping and space waves.

2.1.7.3 Particular case of a perfect conductor

The electric creeping wave propagating on the surface of a perfect conductor represents a singular case, because the electric field, as given by (2.47), vanishes on the surface. The zeroth-order magnetic field, with respect to k, on the surface of the conductor, also goes to zero and the first non-vanishing components are H_1^s and H_1, which are of order $O(k^{-1/3})$. The component H_1^s is given by

$$H_1^s = i\frac{\partial E_0}{\partial \ell} = -iA_0(s,a)\left(\frac{2}{\rho}\right)^{1/3} w_1'(\xi_E - \nu),$$

where $E_0 = A_0(s,a)w_1(\xi_E - \nu)$, and $A_0(s,a)$ is defined in (2.47). Similarly, H_1 satisfies (2.41), which leads to

$$L_0 H_1 = 2i\tau\left(\frac{\rho}{2}\right)^{2/3} E_0,$$

and the boundary conditions (2.26)

$$\frac{\partial H_1}{\partial \ell} = 0.$$

The solution of the above equations satisfying the radiation condition is

$$H_1 = 2iA_0(s,a)\tau \left(\frac{\rho}{2}\right)^{2/3} w_1'(\xi_E - \nu).$$

Using the notation $m = (k\rho/2)^{1/3}$, we can write the $\mathbf{H}$ field as

$$\mathbf{H} = k^{-1/3} H_1^a \mathbf{a} + k^{-1/3} H_1^s \mathbf{s},$$

$$\mathbf{H} = \frac{i}{m} A_0 w_1'(\xi_E - \nu)\,(\tau\rho\mathbf{a} - \mathbf{s})\,.$$

The magnetic field on the surface is directed along the vector $\tau\rho\mathbf{a}-\mathbf{s}$. For the particular case of the electric creeping mode on a circular cylinder, the geodesics follow a helical path. The product $\tau\rho$ is equal to $\cot\beta$, where β is the angle of the helix with the axis of the cylinder. For this particular case, $\mathbf{s} - \tau\rho\mathbf{a} = \boldsymbol{\theta}/\sin\beta$ and, hence, we obtain the well-known result that the magnetic field is directed along the circumference. This follows from the fact that for TM polarisation, and for a metallic cylinder illuminated at oblique incidence, the surface field is TM, that is, directed along the circumference, and the surface currents flow along the generatrix of the cylinder.

2.1.8 Second-order term for the propagation constant

As seen in the previous sections, the main term of the phase function is given by

$$p(s,a) = 2^{-1/3} \int \xi\rho(s)^{-2/3}\, ds. \tag{2.48}$$

For the magnetic (i.e., with magnetic field $\mathbf{H}_0$ tangential to the surface) creeping wave, ξ is a solution of $w_1'(\xi) = imZw_1(\xi)$, where $m = (k\rho/2)^{1/3}$ is the usual Fock parameter and w_1 the Airy–Fock function.

Thus, at this order of approximation, the constant of propagation only depends on ρ. To get the influence of other geometrical parameters, such as torsion τ and transverse curvature $c = 1/\rho_t$, we have to compute $\mathbf{E}_1$ and $\mathbf{H}_1$, as in Reference 5, which we do by solving the equations at order $k^{-1/3}$.

It turns out that $\mathbf{E}_1$ and $\mathbf{H}_1$ can be written as the sum of a local amplitude term and a phase term. The latter term is the most interesting because it modifies the velocity and attenuation of creeping waves. It can be written as a correction $k^{-1/3}\psi$ to the phase. The corrected phase is therefore

$$iks + ik^{1/3}p(s,a) + ik^{-1/3}\psi(s,a), \tag{2.49}$$

$$\psi(s,a) = \int \left(\frac{2}{\rho}\right)^{1/3} S\, ds, \tag{2.50}$$

where S is the sum of five terms

$$S = S_0 + S_1 + S_2 + S_3 + S_4.$$

S_0 contains the second-order term effect of ρ, S_1 the effect of variation of radius of curvature ρ. The effect of transverse curvature is given by S_2. This term turns out to be of special interest, and a detailed analysis will be presented in Section 2.3 of this chapter. Contribution S_3 collects the effect of different torsion terms, and S_4 gives the effect of variation of impedance and related propagation constant ξ. We give below the different terms with some comments. We denote $q = (\xi + m^2 Z^{-2})^{-1}$. We consider the electric creeping wave (except for the effect of transverse curvature, where we deal with the magnetic creeping wave). The case of the magnetic creeping wave can be deduced by changing Z to $1/Z$ in the formulae ($q = (\xi + m^2 Z^2)^{-1}$ for the magnetic creeping wave). For a perfectly conducting body, $q = 0$ for the electric creeping wave, $q = \xi^{-1}$ for the magnetic creeping wave.

2.1.8.1 Effect of ρ and Z

The term S_0 mainly provides some more precision on the dependency over ρ and Z of the propagation constant

$$S_0 = \frac{1}{60\rho}\left(\xi^2 - 6q - 8i\frac{m}{Z}\xi q\right). \tag{2.51}$$

2.1.8.2 Effect of variation of ρ

We get

$$S_1 = \frac{2(\rho')^2 - 3\rho\rho''}{135\rho}\left(2\xi^2 - \frac{3}{4}q - i\frac{m}{Z}\xi q\right). \tag{2.52}$$

This term can either increase or decrease attenuation, depending on the sign of $2(\rho')^2 - 3\rho\rho''$. The correction term becomes the same order as the main term p when (ρ/L) is of order m, if L is the characteristic length of variation of ρ, that is, for rapid variation of ρ. Note that this quantity is assumed to be of order 1 in our analysis, which is verified in most practical applications.

2.1.8.3 Effect of transverse curvature

The contribution S_2 of transverse curvature turns out to be

$$S_2 = \frac{cq}{4}. \tag{2.53}$$

The effective propagation constant is therefore

$$\xi + \frac{c\rho}{4m^2}q = \xi + \frac{1}{2\kappa}q,$$

with $\kappa = 2m^2/\rho c$.

Let us deal with the magnetic creeping wave. If we only consider real impedances, the effect is maximal in the perfectly conducting case, $Z = 0$, nullifies for $Z = \infty$: transverse curvature has no effect at this order on the propagation constant of electric creeping waves. This result for a perfect conductor has been established by Hong [6], using boundary integral equations, and Voltmer [7], using the canonical problem method. The transition between these two cases is ruled by the multiplicative factor $q = (\xi + m^2Z^2)^{-1}$. For the least attenuated magnetic creeping ray, $\xi = a_0 e^{2i\pi/3}$, where $a_0 \approx 1.019$ is the negative of the first zero of the derivative of the Airy function. The attenuation is therefore

$$\frac{\sqrt{3}}{2}\left(a_0 - \frac{1}{2\kappa a_0}\right).$$

The attenuation decreases when c increases. The above formula gives negative attenuation for $\kappa < 1/2a_0^2$. Of course, this result is non-physical: our analysis postulates that $c\rho$ is of order 1, whereas the right-hand side of the above inequality is of order $k^{2/3}$. The case of strong transverse curvature, that is, $c\rho$ of order m and m^2, will be considered in Section 2.3 of this chapter.

2.1.8.4 Effect of torsion

We now analyse the contributions of different terms measuring the effect of torsion τ. The 'true' torsion term involves $\rho\tau^2$. The quantity

$$T = 2^{1/3}\frac{\rho^{-1/3}}{\xi}\frac{\partial p(s,a)}{\partial a},$$

is, in the case of circular cylinder, equal to torsion. So we analyse the effect of this quantity with the torsion term. Note, however, that T and τ are different for a general surface. Moreover the true torsion τ affects, as can be seen from the previous sections, the metric tensor, while T introduces an extra term in the boundary conditions (refer to Equation (2.37)).

For the perfect conductor, we get the following correction

$$S_3 = -\xi^2\rho\left(\left(\frac{2}{5} + \frac{7}{20\xi^3}\right)\tau^2 + \frac{2}{3}\frac{T\tau}{h} + \frac{1}{4}\frac{T^2}{h^2}\right), \tag{2.54}$$

for the magnetic creeping wave, and

$$S_3 = -\xi^2\rho\left(\frac{2}{5}\tau^2 + \frac{2}{3}\frac{T\tau}{h} + \frac{1}{4}\frac{T^2}{h^2}\right), \tag{2.55}$$

for the electric creeping wave.

If we suppose (which is the case on most convex bodies), that τ and T have the same sign, as $\arg(\xi) = \pi/3$, $\arg(iS_3) = \pi/6$, iS_3 has a positive real part and torsion decreases attenuation, for both types of waves. The effect is stronger for the electric creeping wave. It becomes significant when $\tau\rho \approx m$. From the physical point of view, this effect is less important than transverse curvature, because, for most objects $\tau\rho$ remains small, whereas $c\rho$ can become large, for example, for elongated bodies. The correction term improves precision. Note that a geodesic generally presents torsion, except for the sphere.

Note also that the boundary layer technique allows to separate the different contributions of τ^2, $T\tau$ and T^2. In the case of the circular cylinder, which is the canonical problem for creeping rays with torsion, all these terms are equal so that the different contributions get mixed.

For the general impedance case we get

$$S_3 = -\rho\tau^2\left(\frac{2\xi^2}{5} + \frac{7q}{20\xi^2} + i\frac{\xi^2 qZ + 4\xi Z}{5m} - i\frac{Z^3}{mq(Z^2-1)}\right)$$
$$-\rho\frac{T\tau}{h}\left(\frac{2}{3}\xi^2 - i\frac{\xi mq}{3Z} - i\frac{\xi Z}{m(Z^2-1)}\right) - \frac{1}{4}\frac{\rho T^2}{h^2}\left(1 - i\frac{q}{m(Z^2-1)}\right). \tag{2.56}$$

One can easily check that the terms where Z appears are null both for $Z = 0$ and for $Z = \infty$, and that one gets the previous results for a perfect conductor.

Note that the above correction is derived assuming $Z \neq 1$. A divergence appears in (2.56) for $Z = 1$. In this case, the torsion induces a coupling between electric and magnetic creeping waves, as shown in Section 2.2.3. Therefore, the effect of torsion appears in this case at order k^0, and not at order $k^{-1/3}$, as assumed in our analysis.

2.1.8.5 Effect of variation of ξ

The variations of Z and ρ induce variation of ξ, and a correction term

$$S_4 = \frac{1}{8}\rho(\xi')^2\left(3q^2 - 2 - 2i\frac{m}{Z}q^2\right) - \left(\frac{\xi''\rho}{6} + \frac{\xi'\rho'}{9}\right)\left(2\xi + \frac{3}{2}q^2 - i\frac{mq}{Z}\right)$$
$$+\xi'\frac{m}{Z}\left(\frac{m}{Z}\right)'\frac{\rho q}{2}.$$

2.1.9 *Conclusion*

By using asymptotic expansion we have been able to describe the creeping waves on a convex object characterised by an impedance condition. This technique enables us to deduce directly the expression of the creeping waves on a general surface, albeit at a cost of extensive and cumbersome calculations. The asymptotic technique does not rely as much on the solution of the canonical problems as does the conventional GTD approach. This is because the solution to the canonical problems is only used to guess an Ansatz of the solution (it is shown in Chapter 1 that the Ansatz can be somewhat derived).

The asymptotic solution is only uniform when the impedance $Z \to 0$ or $Z \to \infty$, in the sense that we regain the case of the perfect electric or magnetic conductor, respectively, for the above limiting values of Z. However, to obtain this result we implicitly assumed that the impedance is either of order $O(k^{1/3})$ or $O(k^{-1/3})$. By contrast, the assumption that $Z = O(1)$ does not lead to uniform results, which is illustrated in Section 2.2.2.

We have shown that all of the components of the fields $\mathbf{E}$ and $\mathbf{H}$ can be expressed in terms of components E^a and H^a, which can be considered as potentials. For the assumed surface impedances, we have shown that there exists two types of creeping waves, electric and magnetic. The principal tangential component of the electric and magnetic waves are the electric and magnetic fields, orthogonal to the geodesics followed by the creeping waves. These two creeping waves propagate with different velocities, they have different attenuation characteristics, and are uncoupled when Z is different from 1.

The advantage of the asymptotic expansion method is that we can recover, deductively, the results previously obtained in the context of GTD in a heuristic manner. Close to the surface, the dependence of the field along the normal is given by an Airy function, and the propagation constants ξ satisfies Equation (2.36) for the magnetic creeping wave, and Equation (2.35) for the electric one. The magnitude of the field is inversely proportional to the square root of the divergence factor of the geodesic pencil. We also draw attention to two effects that are difficult to obtain starting from the usual canonical problems, these being the $\rho^{-1/6}$ dependence on the radius of curvature ρ and the geometrical factor $Q^{-1/2}$, which depends upon the impedance. All of these effects can be interpreted physically by considering the creeping wave as a tube of rays of thickness $k^{-2/3}\rho^{1/3}$, with an index that depends on the impedance.

We have derived the correction terms of order $k^{-1/3}$ to the phase function for the creeping waves. The boundary-layer method allows us to deal with a general convex body satisfying an impedance boundary condition. Only partial results can be obtained by the canonical problem technique, because some terms are either zero, or get mixed with others, for canonical problems such as the sphere or the cylinder. The boundary-layer method is also more straightforward and more general than the approximate solution of an integral equation employed by Hong for perfectly conducting body and torsionless geodesic. We have checked that our results agree with all previously treated special cases. At this order, a number of parameters, either geometrical, or related to impedance, have an influence on the constant of propagation of creeping waves. We have tried as much as possible to provide some physical interpretation. The clearest physical effects come from the torsion and transverse curvature of geodesics. For a general perfectly conducting obstacle, torsion decreases the attenuation. Transverse curvature also decreases the attenuation. The effect of transverse curvature is maximal for the magnetic creeping wave on a perfectly conducting surface, and vanishes for the electric creeping wave. If we consider only real impedances, the parameter q, whose value is $1/\xi$ or 0 for the magnetic or electric creeping wave, respectively, provides a transition between these two cases. The method is valid for geometrical parameters such as $c\rho$, $\tau\rho$, ρ', $\rho\rho''$ of order 1 and smoothly varying impedance not too close to $Z = 1$. In Section 2.3, we will study what happens when transverse curvature c becomes large.

2.2 Special cases

2.2.1 Introduction

In the previous sections, we considered creeping waves on perfectly conducting or impedance surfaces in supposition that the impedance Z is either asymptotically large ($Z = O(k^{1/3})$), or asymptotically small ($Z = O(k^{-1/3})$). The idea of 'impedance stretching' allows the asymptotic expansions for creeping waves on a general surface to be written with some kind of uniformity, so that the case of a perfectly electrically conducting surface (as well as the case of a perfectly magnetic conductor) appears as the limiting case of $Z \to 0$ (or $1/Z \to 0$). However, the vicinity of $Z = 1$ is not properly handled by this approach. For example, in the case of the electric creeping wave given by formula (2.47) on an obstacle with non-vanishing torsion τ, we get the next order term of the magnetic field

$$k^{-1/3} H_1 = A_0(a)\tau\rho\left(\frac{i}{m} w_1'(\xi - \nu) + (Z - 1/Z)^{-1} w_1(\xi - \nu)\right).$$

The second term contains $(Z - 1/Z)^{-1}$, and is infinite at $Z = 1$. The same kind of singularity appears in the $k^{-1/3}$ term for the propagation constant (see Section 2.1.8). To address this problem introduced by impedance stretching, we use a twofold approach. First, we present in Section 2.2.2 the analysis for $Z = O(1)$. It is shown that the singularity at $Z = 1$ disappears, but the uniformity when $Z \to 0$ and $Z \to \infty$ is lost. For $Z = O(1)$ the creeping waves on the surface are coupled, which results in no explicit solution for the amplitudes. In a particular case when all the parameters are constants we demonstrate Rytov's law for polarisation rotation. Then, though the idea of 'impedance stretching' seems incompatible with the assumption of $Z = 1$, still we use in Section 2.2.3 this approach to get some partial results. The same effect of polarisation rotation is demonstrated in the case $Z = 1$.

Generalisation of the approach for the case of anisotropic impedance is undertaken in Section 2.2.4.

The effect of a caustic of geodesics on the field of creeping waves is analysed in Section 2.2.5. It is described by a similar asymptotic expression as in the case of a caustic of spatial rays derived in Chapter 1, Section 1.2.9.

2.2.2 Creeping waves on an impedance surface with $Z = O(1)$

2.2.2.1 Main formulae

If we suppose Z of order 1, the equations for the fields remain the same as in Section 2.1, namely, at order $k^{1/3}$ we have

$$L_0 H_0 = L_0 E_0 = 0,$$

which implies as previously

$$E_0 = A_0 w_1(\xi - \nu); \qquad H_0 = B_0 w_1(\xi - \nu).$$

The boundary condition at $\nu = 0$ becomes of Dirichlet type both for electric and magnetic fields, namely,

$$H_0 = E_0 = 0. \tag{2.57}$$

That is, the electric field in the principal order coincides with the electric field on a perfectly conducting surface, and the magnetic field is the same as on a perfect magnetic conductor. The boundary condition (2.57) implies

$$w_1(\xi) = 0.$$

Therefore, for any creeping wave, ξ is a zero of Fock–Airy function w_1, that is, impedance independent. Differently to the approach with 'impedance stretching' creeping waves cannot be separated into two different types (similar to TE and TM waves).

The effect of impedance now shows up only in the boundary condition for the next order

$$\frac{\partial H_0}{\partial \nu} + iZ\left(\frac{\rho}{2}\right)^{1/3} H_1 = \frac{\partial E_0}{\partial \nu} + \frac{i}{Z}\left(\frac{\rho}{2}\right)^{1/3} E_1 = 0 \qquad \text{for } \nu = 0. \tag{2.58}$$

The equations for the field components at this order coincide with the previous ones (2.41) and (2.42). Applying the usual procedure, we multiply these equations by $w_1(\xi - \nu)$ and integrate from 0 to ∞, which provides two coupled equations for A_0 and B_0. However, the boundary condition (2.58) for H_1 is now inhomogeneous. This introduces an extra term on the right-hand side of these equations, which now read

$$B_0' + \left(\frac{h'}{2h} + \frac{\rho'}{6\rho}\right) B_0 - \tau A_0 = \frac{B_0}{Z\rho} \tag{2.59}$$

and

$$A_0' + \left(\frac{h'}{2h} + \frac{\rho'}{6\rho}\right) A_0 + \tau B_0 = \frac{A_0 Z}{\rho}. \tag{2.60}$$

Note that, previously, there was a different situation. The boundary conditions for electric and for magnetic fields caused two different equations for the attenuation parameters and we introduced ξ_E and ξ_H as the solutions of these equations. As a result we had two possibilities, either to choose $\xi = \xi_E$ and $H_0 \equiv 0$, or to choose $\xi = \xi_H$ and $E_0 \equiv 0$. These defined TE and TM creeping waves correspondingly. For each of these waves we had only one operator taken on the spectrum and only one compatibility condition was derived for each wave (e.g., Equation (2.44) for TM creeping wave). Now both operators coincide and we have two compatibility conditions which result in coupled differential equations for the amplitudes A_0 and B_0 of each creeping wave. Therefore, in contrast to the previous section, we cannot obtain uncoupled equations such as (2.44).

Introducing

$$\mathcal{A}_0 = h^{1/2}\rho^{1/6}A_0 \qquad \text{and} \qquad \mathcal{B}_0 = h^{1/2}\rho^{1/6}B_0,$$

one can simplify the system (2.57), (2.60) to

$$\begin{pmatrix} \mathcal{B}'_0 \\ \mathcal{A}'_0 \end{pmatrix} = \begin{pmatrix} (Z\rho)^{-1} & \tau \\ -\tau & Z\rho^{-1} \end{pmatrix} \begin{pmatrix} \mathcal{B}_0 \\ \mathcal{A}_0 \end{pmatrix}. \tag{2.61}$$

In the general case, it is not possible to obtain a closed form solution of the above system. This is however possible in some special cases. If the surface has no torsion, the two equations split and one finds

$$B_0(s,a) = B_0(0,a)h^{-1/2}\rho^{-1/6}\exp\left(\int \frac{ds}{Z\rho}\right),$$

$$A_0(s,a) = A_0(0,a)h^{-1/2}\rho^{-1/6}\exp\left(\int \frac{Z}{\rho}ds\right).$$

The other case is the case when the matrix of the system is constant, that is, geometrical and impedance parameters of the surface are constant, namely $\rho' = \tau' = Z' = 0$. In this special case (circular cylinder with constant impedance), we introduce eigenvalues

$$\lambda_\pm = \frac{1}{2\rho}(Z + 1/Z \pm \sqrt{\Delta}), \tag{2.62}$$

and eigenvectors of the above system

$$\mathbf{V}_+ = \begin{pmatrix} 2\tau\rho \\ Z - 1/Z + \sqrt{\Delta} \end{pmatrix}, \qquad \mathbf{V}_- = \begin{pmatrix} Z - 1/Z + \sqrt{\Delta} \\ 2\tau\rho \end{pmatrix}.$$

Here

$$\Delta = \sqrt{(Z - 1/Z)^2 - 4\tau^2\rho^2}$$

and the square root is fixed such that $\sqrt{\Delta} = Z - 1/Z$ when $\tau\rho = 0$.

With the help of the above eigenvectors, we can represent any field as the sum of two modes

$$\begin{pmatrix} \mathcal{B}_0 \\ \mathcal{A}_0 \end{pmatrix} = C\mathbf{V}_+ + D\mathbf{V}_-.$$

If $\tau \neq 0$ both modes have non-zero components of both electric field E_0 and magnetic field H_0. If $\tau = 0$ these modes transform to TE and TM creeping waves, indeed $\mathbf{V}_\pm$ become

$$\begin{pmatrix} 0 \\ 2Z - 2/Z \end{pmatrix} \qquad \text{and} \qquad \begin{pmatrix} 2Z - 2/Z \\ 0 \end{pmatrix}.$$

In this case, we call the mode $C\mathbf{V}_+$ quasi-electric and the mode $D\mathbf{V}_-$ quasi-magnetic.

For the quasi-electric mode with components along $\mathbf{V}_+$, Equation (2.61) gives $dC/ds = \lambda_+ C$, that is,

$$C = C(0,a)\exp(\lambda_+ s). \tag{2.63}$$

For the quasi-magnetic mode with components along $\mathbf{V}_-$, we get

$$D = D(0,a)\exp(\lambda_- s). \tag{2.64}$$

The above formulae show that the two modes have different attenuation, as was the case for electric and magnetic creeping waves on a surface with small or large impedance. However, in contrast to this case, where effect of the impedance appears on the propagation constant, that is, at order $k^{1/3}$, impedance of order 1 modifies the amplitude, that is, the effect now appears at order k^0. Note that this effect can be either included in the propagation constant, as was done in Section 2.1.8, or in the amplitude, as in this section.

For the impedance close to 1 no singularities are present in the above formulae. Assuming $Z = 1+\varepsilon$ with $\varepsilon \ll 1$ allows the eigenvalues of the matrix to be simplified, namely

$$\lambda_\pm = \frac{1}{\rho} \pm i\tau + \frac{\varepsilon^2}{\rho} + \cdots .$$

The term $\pm i\tau$ in the above formula introduces rotation of polarisation in the modes described by Rytov's law.

We can also (partly) compute the second-order terms for the fields. Let us, for example, compute the second-order term for the magnetic field of the quasi-electric mode. In the principal order it has the components

$$E_0 = C(\rho\lambda_+ - 1/Z)\exp(\lambda_+ s)w_1(\xi_0 - \nu)$$

and

$$H_0 = C\tau\rho\exp(\lambda_+ s)w_1(\xi_0 - \nu).$$

Using Equation (2.41) for H_1 component of the field, which reads

$$i\left(\frac{2}{\rho}\right)^{2/3} L_0 H_1 = \frac{2}{\sqrt{h}}\frac{\partial\sqrt{h}H_0}{\partial s} - 2\tau E_0$$

and the inhomogeneous boundary condition (2.58), we get

$$k^{-1/3}H_1 = -iC(mZ)^{-1}\tau\rho\exp(\lambda_+ s)w_1'(\xi_0 - \nu) + Fm^{-1}w_1(\xi_0 - \nu), \tag{2.65}$$

where F cannot be determined at this order of analysis. It is defined by the compatibility conditions that appear in the smaller-order equations.

2.2.2.2 Comparison of the two approaches

Results obtained with and without 'stretching' of impedance are related, as explained below. We first compare the propagation constants.

Let Z be large, that is, of order $O(k^{1/3})$, or small (of order $O(k^{-1/3})$), in both cases the quantity $(Z - 1/Z)$ is of order $O(k^{1/3})$. Assuming $\tau\rho = O(1)$, we perform an expansion for large values of the parameter $(Z - 1/Z)$. In both cases of small and large impedance we find

$$\lambda_+ = \frac{Z}{\rho} + \frac{\tau^2\rho}{Z - 1/Z} + O(k^{-2/3}), \tag{2.66}$$

$$\lambda_- = \frac{1}{Z\rho} - \frac{\tau^2\rho}{Z - 1/Z} + O(k^{-2/3}). \tag{2.67}$$

We get the amplitudes of quasi-electric and quasi-magnetic creeping waves varying as

$$C(0)\exp\left(iks + im\xi_0\frac{s}{\rho} + \frac{Z}{\rho} + O(m^{-1})\right)$$

and

$$D(0)\exp\left(iks + im\xi_0\frac{s}{\rho} + \frac{1}{Z\rho} + O(m^{-1})\right).$$

This can be simulated by introducing equivalent attenuation parameters

$$\xi_+ = \xi_0 - \frac{iZ}{m} + O(m^{-2}) \qquad \text{and} \qquad \xi_- = \xi_0 - \frac{i}{mZ} + O(m^{-2}). \tag{2.68}$$

Now starting from 'stretched impedance' results, we compute the two-term expansion $\xi = \xi_0 + \xi_1/m$ of the magnetic creeping wave propagation constant. By using Equation (2.36), which is $w_1'(\xi) = imZw_1(\xi)$, we get

$$\xi = \xi_0 - \frac{i}{mZ} + O(m^{-2}),$$

which is identical to the expansion of ξ_- in (2.68), and, by the same token from Equation (2.35) we get

$$\xi = \xi_0 - \frac{iZ}{m} + O(m^{-2}),$$

which is identical to the expansion of ξ_+ in (2.68). In both equations, Z is assumed to be of order $O(1)$.

The $\tau^2\rho/(Z - 1/Z)$ term appearing in the asymptotic expansion (2.66) can be related in the same way to the last term in the first bracket of (2.56) for the correction S_3 for torsion in the second-order term of the propagation constant (see Section 2.1.8).

Let us now compare the next order term which is partly computed in (2.65) with the corresponding result of Section 1.1. Assume that Z is small, of order $O(k^{-1/3})$. Then, neglecting the factor $\exp(\lambda_+ s)$, and combining the two contributions for H, that becomes of the same order if $Z = O(k^{-1/3})$, we get

$$E_0 \approx -(C/Z)\, w_1(\xi - \nu),$$

$$H_0 + k^{-1/3} H_1 \approx \tau\rho \left(\frac{-i}{mZ} w_1'(\xi - \nu) + w_1(\xi - \nu) \right).$$

These expressions are consistent with the results obtained for the electric creeping wave on the surface with the impedance $Z = O(k^{-1/3})$, namely

$$E_0' = A_0 w_1(\xi - \nu) \approx -\frac{AZ}{C} E_0$$

and

$$H_0' = 0,$$

$$k^{-1/3} H_1' = A_0 \tau\rho \left(\frac{i}{m} w_1'(\xi - \nu) + (Z - 1/Z)^{-1} w_1(\xi - \nu) \right).$$

For small Z it gives

$$H_0' + k^{-1/3} H_1' \approx \tau\rho \left(\frac{i}{m} w_1'(\xi - \nu) - Z w_1(\xi - \nu) \right) A$$

$$\approx -\frac{AZ}{C}(H_0 + k^{-1/3} H_1)$$

as $\xi = \xi_0 + O(k^{-1/3})$. The fields in both approaches are the same, up to a multiplicative factor, which shows the consistency of the two approaches.

2.2.2.3 Conclusion

Thus, one can conclude that the two approaches, with and without 'impedance stretching', are consistent. The difference is that we describe the effect of impedance in the amplitude in the present approach (i.e., under the assumption of $Z = O(1)$), and in the propagation constant, in the 'impedance stretching' approach. This results in the following kinds of non-uniformity. As explained in the introduction the approach of impedance stretching gives incorrect results for Z close to 1. On the other hand, the approach of this section gives divergence $\lambda_- \to \infty$ for $Z \to 0$ for the quasi-magnetic mode which reflects the inability of the method to describe the magnetic creeping wave on a perfect conductor. The singularity now appears at $Z = 0$ instead of at $Z = 1$. A similar singularity ($\lambda_+ \to \infty$) appears at $Z \to \infty$ for the quasi-electric mode.

In our analysis we assumed the matrix of the system (2.5) to be constant. However one can accept it slowly varying and introduce an additional asymptotic parameter which measures this slowness of variation. Then the amplitudes $\mathcal{A}_0$ and $\mathcal{B}_0$ of the electric and magnetic vectors in quasi-electric and quasi-magnetic modes can be decomposed into asymptotic series by their new parameter. This approach is referred to as the adiabatic approximation [8]. One can also develop the approach based on canonical problems (for this see Reference 9).

2.2.3 Special case of the surface impedance $Z = 1$

When exploiting the idea of 'impedance stretching', the case of surface impedance $Z = 1$ is a special one, because the electric and magnetic creeping waves satisfy the same boundary condition

$$\frac{\partial H_0}{\partial \nu} + imH_0 = 0, \qquad \frac{\partial E_0}{\partial \nu} + imE_0 = 0.$$

Thus, we can no longer deduce that $H_0 = 0$ or $E_0 = 0$. For this particular case, it can be shown that the electric and magnetic creeping waves are coupled.

Starting from Equations (2.16) and (2.17) and setting

$$K_0 = H_0 + iE_0, \qquad J_0 = H_0 - iE_0,$$
$$K_1 = k^{-1/3}(H_1 + iE_1), \qquad J_1 = k^{-1/3}(H_1 - iE_1),$$

we obtain

$$L_0K_0 = L_0J_0 = 0. \tag{2.69}$$

Likewise, from (2.41) and (2.42) we have

$$L_0K_1 = \frac{\tau\rho}{m}K_0 - i\frac{\rho}{\sqrt{h}m}\frac{\partial}{\partial s}(\sqrt{h}K_0),$$

$$L_0J_1 = -\frac{\tau\rho}{m}K_0 - i\frac{\rho}{\sqrt{h}m}\frac{\partial}{\partial s}(\sqrt{h}J_0).$$

The boundary conditions are

$$\frac{\partial K_0}{\partial \nu} + imK_0 = 0, \tag{2.70}$$

$$\frac{\partial J_0}{\partial \nu} + imJ_0 = 0. \tag{2.71}$$

Using (2.37), (2.38) and the value of $\partial p/\partial s$ given by (2.27), we obtain the boundary conditions on K_1 and J_1, which read

$$\frac{\partial K_1}{\partial \nu} + imK_1 = \frac{\partial p}{\partial a}\frac{1}{hm}\left(\frac{\rho}{2}\right)^{2/3} K_0,$$
$$\frac{\partial J_1}{\partial \nu} + imJ_1 = \frac{\partial p}{\partial a}\frac{1}{hm}\left(\frac{\rho}{2}\right)^{2/3} J_0.$$

Equations (2.13)–(2.15) enable us to write

$$K_0 = K(s,a)w_1(\xi - \nu), \qquad J_0 = J(s,a)w_1(\xi - \nu),$$

where ξ is a solution of $w_1'(\xi) = imw_1(\xi)$.

In order to calculate amplitudes $K_0(s,a)$ and $J_0(s,a)$, we write the condition of orthogonality to the solution $w_1(\xi - \nu)$ of the homogeneous equation. This leads us to the following equation:

$$\frac{1}{\rho}\frac{\partial p}{\partial a}\frac{1}{mh}\left(\frac{\rho}{2}\right)^{2/3}\frac{w_1^2(\xi)}{(w_1'(\xi))^2-\xi w_1^2(\xi)}$$
$$=\frac{\rho\tau}{m}-i\frac{\rho}{m}\frac{\partial}{\partial s}\log\left(K_0(s,a)h^{1/2}\rho^{1/6}Q^{1/2}\right). \tag{2.72}$$

Similarly $J_0(s,a)$ is determined from the equation which differs from (2.72) by replacing τ with $-\tau$. Since $w_1'(\xi) = imw_1(\xi)$, the left-hand side of (2.72) is of order $O(m^{-3})$ and can be neglected compared with the right-hand side of order $O(m^{-1})$.

Finally, after solving (2.72) with a zero left-hand side, we obtain

$$K_0(s,a) = K_0(a)h^{-1/2}\rho^{-1/6}Q^{-1/2}\exp\left(-i\int \tau\, ds\right). \tag{2.73}$$

Similarly, for $J_0(s,a)$, we can write an expression different from (2.73) by changing τ into $-\tau$. These equations show, once again, that the factor $h^{-1/2}\rho^{-1/6}Q^{-1/2}$ is associated with the variation of the amplitude. The exponential factors correspond to rotation of the polarisation. The angle P of polarisation is given by $dP/ds = \tau$. For a positive torsion, the fields **E** and **H** move in the counterclockwise direction, and we recover the law for the ordinary waves.

To summarise, we present the following observations:

- For an impedance Z sufficiently different from 1, the electric and magnetic creeping waves are uncoupled.
- For an impedance Z equal to 1, the electric and magnetic modes satisfy the same boundary conditions and we extract one from the other by rotating **E** and **H** by $\pi/2$, the rotating axis being the tangent to the geodesic. The torsion of the geodesics thus induces a rotation of the fields **E** and **H** according to Rytov's law.

The above effect was also observed in Section 2.2.2. With the hypothesis $Z = O(1)$, we solved, to the leading order, the Dirichlet problem for both the electric and magnetic creeping waves, and we derived both the coupling and the rotation for them. However, when Z is sufficiently different from 1, one of the two modes attenuates much more rapidly than the other and the dominant effect is no longer the rotation. For instance, for the perfectly conducting case, the electric creeping wave attenuates much more than does the magnetic creeping wave, and the creeping waves can be treated separately. Thus, the approximation $Z = O(1)$ is well suited for describing the physics of the problem when the relative impedance of the surface is close to 1. However, it yields inaccurate results when the impedance deviates significantly from 1.

Finally, we should mention that one drawback of the method of impedance stretching is that the rotation effect in the creeping waves is only predicted for $Z = 1$ while, in reality, it also exists for Z in the neighbourhood of 1 and it would have certainly been desirable to be able to recover the torsion effect over a broader range of impedance than is possible with the impedance stretching method.

2.2.4 Anisotropic impedance case

The results of Section 2.1 can be extended to the case of anisotropic impedance. We assume that the impedance in the impedance boundary condition is now the matrix

$$\underline{Z} = \begin{bmatrix} Z_{ss} & Z_{sa} \\ Z_{sa} & Z_{aa} \end{bmatrix}.$$

Consistent with isotropic case, we take $mZ_{ss}, m/Z_{aa}$ of order 1. In order that the diagonal term influence appears at the same order, we suppose Z_{sa} of order 1.

Introducing the two components vector $\mathbf{V} = (H^a, E^a)^T$, the matrix impedance boundary condition can be rewritten as

$$\frac{\partial \mathbf{V}}{\partial \nu} + M\mathbf{V} = 0,$$

with

$$M = i\frac{m}{Z_{aa}} \begin{bmatrix} Z & -Z_{sa} \\ Z_{sa} & 1 \end{bmatrix},$$

and $Z = \det(Z) = Z_{ss}Z_{aa} - Z_{sa}^2$.

We now put the above formula in diagonal form

$$P^{-1}\frac{\partial \mathbf{V}}{\partial \nu} + (P^{-1}MP)P^{-1}\mathbf{V} = 0, \tag{2.74}$$

where matrix P, whose columns are the eigenvectors of M, diagonalise M, namely

$$P^{-1}MP = D = \begin{bmatrix} \lambda & 0 \\ 0 & \mu \end{bmatrix},$$

λ and μ are the eigenvalues of M:

$$\lambda = \frac{im}{2Z_{aa}}(Z + 1 + \sqrt{\Delta}), \qquad \mu = \frac{im}{2Z_{aa}}(Z + 1 - \sqrt{\Delta})$$

with $\Delta = (Z-1)^2 - 4Z_{sa}^2$.

Introducing $\mathbf{U} = P^{-1}\mathbf{V}$, Equation (2.74) becomes

$$\frac{\partial \mathbf{U}}{\partial \nu} + D\mathbf{U} = 0. \tag{2.75}$$

2.2.4.1 Determination of the propagation constant

The equations governing the field behaviour are the same as for isotropic impedance. At order $k^{1/3}$

$$L_0\mathbf{U}_0 = 0,$$

which implies, as in the isotropic case

$$\mathbf{U}_0 = \begin{pmatrix} \mathcal{A}_0(s) \\ \mathcal{B}_0(s) \end{pmatrix} w_1(\xi - \nu). \tag{2.76}$$

Inserting (2.76) in the impedance boundary condition (2.75), we get, at leading order,

$$\begin{aligned} &\mathcal{A}_0(s)(w_1'(\xi) - \lambda w_1(\xi)) = 0, \\ &\mathcal{B}_0(s)(w_1'(\xi) - \mu w_1(\xi)) = 0. \end{aligned} \tag{2.77}$$

As in the isotropic case, these equations can only be satisfied if $\mathcal{A}_0(s) = 0$ or $\mathcal{B}_0(s) = 0$.

We therefore obtain two kinds of creeping waves: one with $\mathcal{A}_0(s) \neq 0$, the other with $\mathcal{B}_0(s) \neq 0$. The first mode has a constant of propagation ξ defined by

$$w_1'(\xi) - \lambda w_1(\xi) = 0. \tag{2.78}$$

If Z_{sa} vanishes, (2.78) becomes $w_1'(\xi) - imZ_{ss}w_1(\xi) = 0$, and we get the magnetic creeping wave.

The second mode has a constant of propagation ξ defined by

$$w_1'(\xi) - \mu w_1(\xi) = 0. \tag{2.79}$$

If Z_{sa} vanishes, (2.79) becomes $w_1'(\xi) - imZ_{aa}^{-1}w_1(\xi) = 0$, and we get the electric creeping wave.

For non-vanishing Z_{sa}, both modes have non-vanishing E_0 and H_0

$$\begin{pmatrix} H_0 \\ E_0 \end{pmatrix} = P\mathbf{U}_0 = \begin{pmatrix} \mathcal{A}_0 \\ \alpha\mathcal{A}_0 \end{pmatrix} w_1(\xi - \nu)$$

for the first mode,

$$\begin{pmatrix} H_0 \\ E_0 \end{pmatrix} = P\mathbf{U}_0 = \begin{pmatrix} \alpha\mathcal{B}_0 \\ \mathcal{B}_0 \end{pmatrix} w_1(\xi - \nu)$$

for the second mode. Here

$$\alpha = \frac{Z - 1 - \sqrt{\Delta}}{2Z_{sa}}.$$

We now have to compute $\mathcal{A}_0$ and $\mathcal{B}_0$, which we do, as in the isotropic case, by using the next order, that is, k^0 equations (2.41) and (2.42).

2.2.4.2 Computation of the amplitude

Equations (2.41) and (2.42) in matrix form become

$$i\left(\frac{2}{\rho}\right)^{2/3} L_0\mathbf{V}_1 = 2\mathbf{V}_0' + \frac{h'}{h}\mathbf{V}_0 + 2\tau M\mathbf{V}_0, \tag{2.80}$$

with

$$M = \begin{bmatrix} 0 & -1 \\ 1 & 0 \end{bmatrix},$$

where prime denotes derivative with respect to s.

Multiplying by P^{-1} to put (2.80) in diagonal form

$$i\left(\frac{2}{\rho}\right)^{2/3} L_0\mathbf{U}_1 = 2\mathbf{U}_0' + \frac{h'}{h}\mathbf{U}_0 + 2P^{-1}P'\mathbf{U}_0 + \tau P^{-1}MP\mathbf{U}_0, \tag{2.81}$$

we define $\mathcal{H}_0 = \mathcal{A}_0(s)w_1(\xi - \nu)$ (respectively $\mathcal{H}_1$) as the first component of $\mathbf{U}_0$ (respectively $\mathbf{U}_1$), that is, $\mathbf{U}_0 = (\mathcal{H}_0, 0)^T$ implies

$$i\left(\frac{2}{\rho}\right)^{2/3} L_0\mathcal{H}_1 = 2\mathcal{H}_0' + \frac{h'}{h}\mathcal{H}_0 - 4\frac{\tau\alpha}{1-\alpha^2}\mathcal{H}_0 - 2\frac{\alpha\alpha'}{1-\alpha^2}\mathcal{H}_0 := L_1^{\mathrm{an}}\mathcal{H}_0. \tag{2.82}$$

Note that the first two terms on the right-hand side of (2.82), defining L_1^{an} operator for anisotropic case, are the usual L_1 operator. The last two account for anisotropy, and vanish for diagonal Z.

Equation (2.81) has a solution if its right-hand side is orthogonal to the solution of the homogeneous equation. This compatibility condition provides, as for the isotropic case, a differential equation for $\mathcal{A}_0(s)$. As in the isotropic case, we integrate from 0 to ∞. In the process, we need the boundary condition on $\mathcal{H}_1$, which turns out to be

$$\frac{\partial\mathcal{H}_1}{\partial\nu} + \lambda\mathcal{H}_1 = i\left(\frac{\rho}{2}\right)^{1/3}\frac{\partial p}{\partial a}\frac{1}{1-\alpha^2}\frac{2\alpha\mathcal{H}_0}{h}.$$

The computations are the same as in the isotropic case, and we finally get

$$\mathcal{A}_0' + \left(\frac{h'}{2h} + \frac{\rho'}{6\rho} + \frac{Q'}{2Q} - 2\frac{\tau\alpha}{1-\alpha^2} - \frac{\alpha\alpha'}{1-\alpha^2}\right)\mathcal{A}_0 = \frac{Th^{-1}\xi}{\xi-\lambda^2}\frac{\alpha}{1-\alpha^2}\mathcal{A}_0 \tag{2.83}$$

with, as previously, $Q = \xi w_1^2(\xi) - w_1'^2(\xi)$ and $T = 2^{1/3}\rho^{-1/3}\xi^{-1}\partial p/\partial a$.

We recognise the first four terms as those appearing in the equation for the isotropic case. The solution of (2.83) is

$$\mathcal{A}_0(s,a) = \mathcal{A}_0(0,a)h^{-1/2}\rho^{-1/6}Q^{-1/2}(1-\alpha^2)^{-1/2}$$
$$\times \exp\left(\int_{s_0}^{s}\frac{\alpha}{1-\alpha^2}\left(2\tau + \frac{\xi h^{-1}}{\xi-\lambda^2}T\right)ds\right). \tag{2.84}$$

Anisotropy introduces an extra amplitude factor $(1-\alpha^2)^{-1/2}$ and an exponential term.

Note that for diagonal Z, $\alpha = 0$, and we obtain the result for isotropic impedance Z.

Torsion τ and torsion-like T quantity appears in the integrand of the exponential term. Moreover, if we consider a matrix of the form

$$Z = \begin{bmatrix} 1 & Z_{sa} \\ Z_{sa} & 1 \end{bmatrix},$$

and let $Z_{sa} \to 0$, we get $\alpha = -i$, $\lambda = im$ and the exponential term becomes

$$\exp\left(\int_{s_0}^{s} \frac{\alpha}{1-\alpha^2}\left(2\tau + \frac{\xi h^{-1}}{\xi - \lambda^2}T\right) ds\right) = \exp\left(-i\int_{s_0}^{s}\left(\tau + \frac{1}{2h}\frac{\xi}{\xi + m^2}T\right) ds\right).$$

We recognise the Rytov rotation of polarisation effect induced by torsion, as described by (2.73). The additional term $\xi T/(2h(\xi + m^2))$, which is of order m^{-2}, namely $k^{-2/3}$ has to be neglected to be consistent with the precision of the above analysis.

The computations for the second mode are the same and the results are similar. We get for its amplitude $\mathcal{B}_0(s,a)$

$$\mathcal{B}_0(s,a) = \mathcal{B}_0(0,a)h^{-1/2}\rho^{-1/6}Q^{-1/2}(1-\alpha^2)^{-1/2}$$
$$\times \exp\left(-\int_{s_0}^{s} \frac{\alpha}{1-\alpha^2}\left(2\tau + \frac{\xi h^{-1}}{\xi - \lambda^2}T\right) ds\right). \qquad (2.85)$$

The special case $\alpha = 1$ requires separate analysis.

2.2.4.3 Conclusion

We have studied the creeping waves on a convex obstacle described by a matrix impedance boundary condition. If the impedance matrix is diagonal in (s,a) coordinates, we have, as in the isotropic case, a magnetic (respectively electric) creeping wave propagating with zero binormal electric (respectively magnetic) leading order field, and constant of propagation corresponding to impedance Z_{ss} (respectively Z_{aa}). When the extradiagonal term is present (which is, in general, if the impedance is generated by an anisotropic coating), both modes have non-zero binormal components at leading order. Moreover, this term introduces an extra amplitude factor and an exponential term involving torsion τ and torsion-like T quantity, which reduces to Rytov rotation for impedance equal to 1.

2.2.5 *Caustic of creeping rays*

The results of previous sections are valid if the geodesic system of coordinates is regular, that is, h parameter is not zero. For $h = 0$, the term $h^{-1/2}$ in the creeping wave amplitude diverges. This happens on the envelope, or caustic C of geodesics followed by creeping waves. For a general object, C is a line with one or several cusps. In this section we derive the asymptotic expansion of the field in the vicinity of a regular point of the caustic C, not too close to cusp points.

Similar to the case of a caustic of spatial rays (see Chapter 1, Section 1.2.9) the boundary layer is expected to have the width of order $k^{-2/3}$ in the direction across C. In that narrow strip around the curve C on the surface of the body we introduce the system of coordinates (s', t). Coordinate s' is the curvilinear abscissa along C, which is the s' axis of the coordinate system. Note that C is not the geodesics and therefore, s' and s are different quantities.

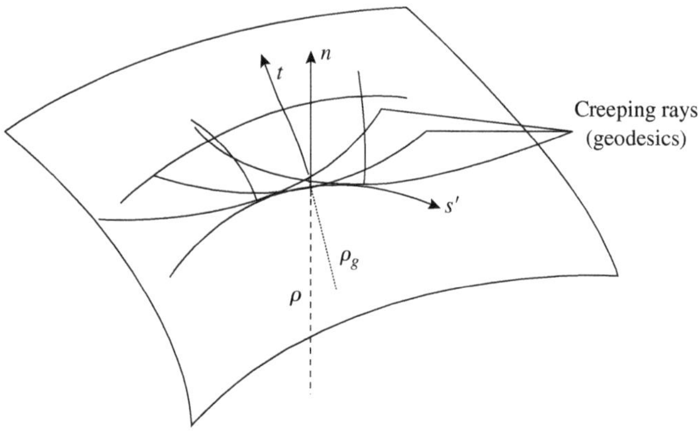

Figure 2.1 Caustic of creeping waves

To define the coordinate t we consider the family of geodesics orthogonal to C (see Figure 2.1), and measure t as the distance along the geodesic starting from C. Coordinates (s', t) is a geodesic system of coordinates on the surface. Their equivalent in the case of two-dimensional caustics are the curvilinear abscissa along the caustic and the normals to it.

Coordinate n is, as previously, the distance measured along the normal to the surface.

The metric matrix of the coordinate system (s', t, n) is

$$g_{ij} = \begin{pmatrix} 1 + \dfrac{2}{\rho_g}t + \dfrac{2}{\rho}n & 2\tau n & 0 \\ 2\tau n & 1 + \dfrac{2}{\rho_t}n & 0 \\ 0 & 0 & 1 \end{pmatrix} + O(n^2, t^2).$$

Here $\rho = \rho(s')$ and $\tau = \tau(s')$ are, respectively, the radius of curvature and the torsion of the geodesics tangential to caustic S at point $(s', 0)$, and ρ_t is the radius of transverse curvature of the surface at the point $(s', 0)$.

One can compare the matrix g_{ij} of the coordinate system (s', t, n) with formula (2.3). There is much in common between these two formulae. If one lets $h = 1$ in (2.3) and neglects smaller-order terms, the only difference will be in the g_{ss} element which now contains an additional term $2t/\rho_g$. This term, however, is natural in the coordinates (s', t, n) because the arc-length s' is measured along the curve C which is not a geodesic. Therefore, the derivative of tangent vector to C by s' has also the tangential transverse component which is given by the geodesic radius of curvature $\rho_g(s')$. This geodesic radius of curvature can be interpreted as the radius of curvature of the projection C' of the caustic C on the tangent plane to the surface in the point $(s', 0)$. Projected on the tangent planes, our problem looks therefore like the two-dimensional caustic (see Section 1.2.9 of Chapter 1).

The Ansatz has to match the creeping rays far enough from C, and to show some relationship to the two-dimensional caustic Ansatz. The form of the Ansatz, as well as the coordinate stretching, takes into account these requirements. We choose

$$E = \exp(iks' + k^{1/3} p(s')) \sum_{j=0}^{N} E_j k^{-j/3},$$

$$H = \exp(iks' + k^{1/3} p(s')) \sum_{j=0}^{N} H_j k^{-j/3},$$

where $E_j = E_j(\ell, \theta, s'), H_j = H_j(\ell, \theta, s')$ depend on the stretched coordinates $\ell = k^{2/3} n$, $\theta = k^{2/3} t$. In this section, we only determine the dependency on the stretched coordinates ℓ, θ. The dependency on s' is determined by matching to the creeping ray solution, by a procedure similar to matching to space rays for a spatial caustic (see Section 1.2.9 of Chapter 1).

Then, we follow the same steps as in the previous sections. Namely, we substitute the above Ansatz into Maxwell's equations in (s', t, n) coordinates, sort out terms in decreasing powers of $k^{1/3}$. From vectorial equation curl $\mathbf{E} = ik\mathbf{H}$, we get, as in Section 2.1.4, equations at different orders.

Order k

$$H_0^{s'} = 0,$$
$$E_0^n = -H_0^t,$$
$$H_0^n = E_0^t;$$

Order $k^{2/3}$

$$iH_1^{s'} = \frac{\partial E_0^n}{\partial \theta} - \frac{\partial E_0^t}{\partial \ell},$$
$$E_1^n = -H_1^t,$$
$$H_1^n = E_1^t;$$

Order $k^{1/3}$

$$iH_2^{s'} = \frac{\partial E_1^n}{\partial \theta} - \frac{\partial E_1^t}{\partial \ell}, \tag{2.86}$$

$$E_2^n + H_2^t = \left(\frac{\partial p}{\partial s'} - \ell \left(\frac{1}{\rho} + \frac{1}{\rho_t} \right) - \frac{\theta}{\rho_g} \right) H_0^t - \frac{\partial^2 H_0^t}{\partial \ell^2} + \frac{\partial^2 E_0^t}{\partial \ell \partial \theta}, \tag{2.87}$$

$$H_2^n - E_2^t = \left(\frac{\partial p}{\partial s'} + \ell \left(\frac{1}{\rho_t} - \frac{1}{\rho} \right) - \frac{\theta}{\rho_g} \right) E_0^t - \frac{\partial^2 E_0^t}{\partial \theta^2} + \frac{\partial^2 H_0^t}{\partial \ell \partial \theta}. \tag{2.88}$$

Equations of order k are the same as for the creeping ray, with $h = 1$. Extra terms involving derivatives with respect to θ appear because of stretching of transverse coordinate t in equations of order $k^{2/3}$ and $k^{1/3}$. Extra terms with θ/ρ_g come from the g_{ss} coefficient of the metric tensor. The second Maxwell equation, curl $\mathbf{H} = -ik\mathbf{E}$, provides another (dual) set of equations, which can be obtained from the above set by replacing E with H and H with $-E$ as in Section 2.1.4. The dual of Equation (2.88) reads

$$-E_2^n - H_2^t = \left(\frac{\partial p}{\partial s'} + \ell\left(\frac{1}{\rho_t} - \frac{1}{\rho}\right) - \frac{\theta}{\rho_g}\right) H_0^t - \frac{\partial^2 H_0^t}{\partial \theta^2} - \frac{\partial^2 E_0^t}{\partial \ell \partial \theta}. \tag{2.89}$$

Adding (2.87) and (2.89), we get

$$\left(2\frac{\partial p}{\partial s'} - \frac{2\ell}{\rho} - 2\frac{\theta}{\rho_g}\right) H_0^t - \frac{\partial^2 H_0^t}{\partial \theta^2} - \frac{\partial^2 H_0^t}{\partial \ell^2} = 0, \tag{2.90}$$

which is the sum of Equation (2.16) of Section 2.1.4, determining the dependence on coordinate n of creeping waves, and of equation $L_0 u_0 = 0$ of Section 1.2.9 of Chapter 1 about the caustic. Introducing variables ν and μ

$$\nu = \left(\frac{2}{\rho}\right)^{1/3} \ell, \qquad \mu = \left(\frac{2}{\rho_g}\right)^{1/3} \theta$$

and ξ as in (2.21), we get

$$\frac{\partial^2 H_0^t}{\partial \nu^2} + (\nu - \xi) H_0^t + \frac{\partial^2 H_0^t}{\partial \mu^2} + \mu H_0^t = 0. \tag{2.91}$$

Among the solutions of this equation, we choose the solution matching to magnetic creeping waves, namely

$$H_0^t = A_0(s) v(-\mu) w_1(\xi - \nu), \tag{2.92}$$

where ξ is, as for the creeping waves, the solution of $w_1'(\xi) = imZw_1(\xi)$ to satisfy the impedance boundary condition on the surface. We see that the parameter ξ is the same as for the locally tangent creeping ray. We thus obtain the same increment of phase function as for the creeping waves, and the phase function on C is given by

$$\frac{\partial p}{\partial s'} = 2^{-1/3} \rho(s')^{-2/3} \xi(s'). \tag{2.93}$$

Note that this equation defines p on the caustic C, that is, at points $(s', 0)$. Having in mind the necessity to match the above formulae to the asymptotics (2.46) and (2.47), we note that

$$p(s, a) = p(s') + O(k^{-2/3}).$$

Therefore, we can take $p(s') = (p_o + p_i)/2$, where p_o and p_i are, respectively, the attenuation factors for the outgoing (i.e., receding from C) and the incoming creeping waves.

We get similar result for electric creeping waves, replacing E with H, H with $-E$ and Z with $1/Z$ in the above formulas.

At this order, the problem of creeping waves description near the caustic of creeping rays decouples into the usual problem for creeping waves, which determines the normal dependence $w_1(\xi - \nu)$, and a two-dimensional caustic problem, which determines the binormal dependence $v(-\mu)$. The projection on the local tangent plane transforms (locally) the geodesics into straight lines, and the envelope of geodesics C into a curve with radius of curvature ρ_g. As for the two-dimensional caustic problem, the amplitude $A_0(s)$ is determined by matching to the ray solution (i.e. the creeping waves).

Though we have not studied this case, one can assume that the same kind of decoupling occurs for the cusp of creeping rays. This result may be obtained by an adaptation of the method of spectral reconstruction, namely with an integral Ansatz based on creeping waves, instead of plane waves.

2.3 Creeping waves on elongated objects

2.3.1 Introduction

Creeping waves play an important role in the process of high-frequency diffraction by a smooth convex body. Results presented in the preceding sections do not allow an explanation for some properties of creeping waves on strongly elongated bodies. Previously the asymptotic analysis was carried out in the supposition of single small parameter k^{-1}. If the body is strongly elongated another small parameter that measures the ratio of width to length may interfere with the asymptotic procedure.

In the ordinary asymptotics of creeping waves one finds radius of transverse curvature only in smaller order corrections. For a magnetic transversally polarised wave on an axisymmetric perfect conductor, for example, these corrections appear in the phase factor of order $O(k^{-1/3})$ (see (2.49) in the previous section). The current J is given by

$$J = h^{-1/2}\rho^{-1/6}\exp\left\{iks + i\left(\frac{k}{2}\right)^{1/3}\xi\int_0^s \rho^{-2/3}\,ds\right\}$$
$$\times \exp\left\{i\left(\frac{k}{2}\right)^{-1/3}\int_0^s \rho^{-4/3}\left[\xi^2\left(\frac{1}{60}-\frac{2}{45}\rho\ddot{\rho}+\frac{4}{135}\dot{\rho}^2\right)\right.\right.$$
$$\left.\left.-\frac{1}{\xi}\left(\frac{1}{10}-\frac{\rho}{4\rho_t}-\frac{\rho\ddot{\rho}}{60}+\frac{\dot{\rho}^2}{90}\right)\right]ds\right\}. \tag{2.94}$$

Here s, h and ρ stand, respectively, for the arc-length, divergence and the radius of curvature of the geodesics and ρ_t is the radius of transverse curvature. The ξ parameter is any zero of the equation

$$\dot{w}_1(\xi) = 0,$$

where w_1 is the Airy function.

If the transverse curvature is large and at the given frequency it is comparable to $k^{1/3}$, then the term with ρ_t in (2.94) appears to be of the same order as the leading term. This, generally speaking, signals that formula (2.94) loses its asymptotic character and may give wrong results for the field of creeping waves.

Below we present some asymptotic results obtained in the frame of the boundary-layer method for acoustic and electromagnetic creeping waves on ideal or impedance strongly elongated surface. Original papers are [14]–[17].

2.3.2 The Ansatz and types of elongated objects

We use the same geodesic coordinates (s, a, n), where s, a are geodesic coordinates on the surface of the body and n is external normal. The quadratic tensor of these coordinates is (see (2.3)).

$$g_{ij} = \begin{pmatrix} \left(1+\dfrac{n}{\rho}\right)^2 + \tau^2 n^2 & -h\tau\left(2n + n^2\left(\dfrac{1}{\rho}+\dfrac{1}{\rho_t}\right)\right) & 0 \\ -h\tau\left(2n + n^2\left(\dfrac{1}{\rho}+\dfrac{1}{\rho_t}\right)\right) & h^2\left(\left(1+\dfrac{n}{\rho_t}\right)^2 + \tau^2 n^2\right) & 0 \\ 0 & 0 & 1 \end{pmatrix}.$$

The Ansatz is taken in the form of usual asymptotics of creeping waves

$$u = \exp\{iks + ik^{1/3}p(s,a)\}\sum U_j(s,a,\ell)k^{-j/3}, \qquad \ell = k^{2/3}n. \tag{2.95}$$

Here ℓ is the stretched normal coordinate that determines the thickness of the boundary layer. The functions U_j are assumed not to depend rapidly on their arguments, that is, all derivatives of U_j are assumed to be of order $O(1)$.

Start with acoustic waves. The Helmholtz equation in coordinates s, a, n reads

$$\frac{1}{\sqrt{g}}\left\{\frac{\partial}{\partial s}\frac{\sqrt{g}}{(1+n/\rho)^2+\tau^2 n^2}\frac{\partial u}{\partial s} + \frac{\partial}{\partial a}\frac{\sqrt{g}h^{-1}}{(1+n/\rho_t)^2+\tau^2 n^2}\frac{\partial u}{\partial a} + \frac{\partial}{\partial n}\sqrt{g}\frac{\partial u}{\partial n}\right\}$$
$$+k^2 u = 0. \tag{2.96}$$

Here g is the determinant of the matrix g_{ij}.

The asymptotic procedure is standard. One substitutes the representation (2.95) into Equation (2.96) and equates terms of similar orders in k. The equations in the order k^2 and $k^{5/3}$ are satisfied automatically.

Depending on the order of transverse curvature $1/\rho_t$, three cases can be distinguished:

1. $1/\rho_t \ll k^{1/3}$. Then transverse curvature appears only in the equations of order $k^{2/3}$ (or less for acoustically soft surface) and yields smaller-order correction in the asymptotics of creeping waves.
2. $k^{1/3} \ll 1/\rho_t \ll k^{2/3}$. Transverse curvature appears in the order k^1 and modifies the amplitude factor in the asymptotics of creeping wave. We consider this case to be the case of a moderately elongated body.
3. $k^{2/3} \ll 1/\rho_t$. In the case of a strongly elongated body, transverse curvature appears already in the principal order $O(k^{4/3})$ and leads to total modification of the creeping waves' asymptotics.

When dealing with electromagnetic waves one can distinguish the same three cases.

2.3.3 *Moderately elongated body*

Consider first the case of a moderately elongated body, that is, assume that $\rho_t = \kappa_1 k^{-1/3}$ and $\kappa_1 = O(1)$. The recurrent system appears as follows:

$$\begin{aligned} &L_0 U_0 = 0, \qquad L_0 = \frac{\partial^2}{\partial \ell^2} + 2\left(\frac{\ell}{\rho} - \frac{\partial p}{\partial s}\right), \\ &L_0 U_1 = -L_1 U_0, \qquad L_1 = 2i\frac{\partial}{\partial s} + i\frac{1}{h}\frac{\partial h}{\partial s} + \frac{1}{\kappa_1}\frac{\partial}{\partial \ell}, \\ &L_0 U_2 = -L_1 U_1 - L_2 U_0, \ldots . \end{aligned} \tag{2.97}$$

In the leading order the operator L_0 is the same as in the usual case, therefore

$$U_0 = A w_1(\xi - \nu), \qquad \nu = \left(\frac{2}{\rho}\right)^{1/3} \ell.$$

The parameter ξ is defined from the boundary condition. For an impedance surface

$$\frac{\partial U}{\partial n} + ikZU = 0,$$

one finds

$$w_1'(\xi) = imZw_1(\xi), \qquad m = \left(\frac{k\rho}{2}\right)^{1/3}. \tag{2.98}$$

As usual the function U_0 is defined up to an amplitude multiplier A dependent on surface coordinates s and a that play the role of parameters in the principal Equation (2.97) and boundary condition (2.98). To define the amplitude factor, one needs to take into consideration the next order equations in the recurrent system. The boundary-value problem for determining function U_1 appears inhomogeneous and is solvable only for such right-hand sides that are orthogonal to the solution of the homogeneous problem. This orthogonality condition can be rewritten in the form of a differential equation for the amplitude A and allows its variation with s to be determined. Finally, in the principal order one finds

$$\begin{aligned} u(s,a,n) = {} & A(0,a)h^{-1/2}\rho^{-1/6}(\xi + m^2 Z^2)^{-1} \\ & \times \exp\left\{iks + i\left(\frac{k}{2}\right)^{1/3}\int_0^s \frac{\xi\, ds}{\rho^{2/3}}\right\} \\ & \times \exp\left\{-2^{-5/3} i \int_0^s \rho^{-1/3}\kappa_1^{-1}(\xi + m^2 Z^2)^{-1}\, ds\right\} w_1(\xi - \nu). \end{aligned} \tag{2.99}$$

One should note that the asymptotics (2.99) was derived in the supposition that all the parameters of the surface ρ, κ_1, τ are constant or slowly varying functions of coordinates s and a. For an ordinary body such supposition looks natural; however, for a moderately elongated body (and especially for a strongly elongated body) such supposition restricts considerations to revolution-like bodies. Let, for example, the surface of revolution be defined in cylindrical coordinates (r, φ, z) by the equation $r = r(z)$. We assume that

$$r, r'(z) \ll 1. \tag{2.100}$$

For such a body of revolution radii of curvature are expressed via the function $r(z)$ by the formulae

$$\frac{1}{\rho(s)} = \frac{r_{zz}}{(1 + r_z^2)^{3/2}}, \qquad \rho_t = r\frac{1 - r_z^2}{1 + r_z^2}.$$

Analysis of the asymptotics (2.99) shows that in the case of a moderately elongated body 'of revolution' the dependence of the creeping wave field on the normal n remains the same as in the case of an ordinary body and is given by the Airy function $w_1(\xi - \nu)$. The exponential factor expressing rapid oscillations of the field with arc-length s is also preserved. The difference in the asymptotics (2.99) from the standard creeping waves' asymptotics is only in the last exponential multiplier. If the radius ρ_t is of order $O(1)$, that is, the body is not elongated, this term tends to 1 and the asymptotics (2.99) reduces to the usual asymptotics of creeping waves.

2.3.4 *Waves on strongly elongated bodies*

2.3.4.1 General remarks

Similar to the failure of the usual asymptotics (2.94) in the case of a moderately elongated surface, asymptotics (2.99) is not valid for the case of a strongly elongated body. Therefore, the case of very large transverse curvature should be studied separately.

One can note that thickness of the boundary layer and the radius of transverse curvature on strongly elongated bodies are of the same order in k. This means that in the scaled coordinates the entire cross-section of the body appears in the boundary layer and the asymptotics loses its locality with respect to the a coordinate. That is, in order to construct the asymptotic expansion one needs to know precisely the shape of the body cross-section. We assume that we deal with the body of revolution and specify the factor

$$\exp(i\chi\varphi), \qquad \chi = 0, 1, 2, \ldots,$$

to present the field variation with the angle of revolution φ.

2.3.4.2 Acoustic creeping waves

Substitution of the form (2.95) in the Helmholtz equation rewritten in coordinates (s, φ, n) and equating similar order terms in k yields the system of recurrent equations consequently defining the functions U_j. The ratio n/ρ_t in the boundary layer is of order 1. As a result the transverse curvature appears in the first equation of the recurrent system

$$L_0 U_0 = 0, \qquad L_0 U_1 + L_1 U_0 = 0, \qquad \dots,$$

where the operators take the form

$$L_0 = \frac{\partial^2}{\partial \nu^2} + \frac{1}{\nu}\frac{\partial}{\partial \nu} + \left(\nu - \frac{\chi^2}{\nu^2} - \xi\right), \tag{2.101}$$

$$L_1 = \left(\frac{\rho}{2}\right)^{2/3}\left(2i\frac{\partial}{\partial s} + i\frac{\partial \rho_t/\partial s}{\rho_t}\frac{\kappa}{\nu}\right). \tag{2.102}$$

Here the shifted normal coordinate is introduced as

$$\nu = 2m^2\frac{\rho_t + n}{\rho}, \qquad \kappa = 2m^2/\rho c, \qquad c = \rho_t^{-1}$$

and the parameter ξ is related to the function $p(s)$ in (2.95) by the formula

$$p(s) = 2^{-1/3}\int_0^s \frac{\xi - \kappa}{\rho^{2/3}}\,ds.$$

The equation $L_0 U_0 = 0$ is considered on the semi-axis $\nu \geq \kappa$ with the boundary condition

$$\frac{\partial U_0(s,\kappa)}{\partial \nu} + imZU_0(s,\kappa) = 0 \tag{2.103}$$

and radiation condition at infinity. This equation defines the dependence of the creeping waves field on the normal coordinate

$$U_0(s,\nu) = A_0(s)V(\nu;\xi,\chi), \tag{2.104}$$

where function V is the solution of the Sturm–Liouville problem for the biconfluent Heun equation [10]

$$V'' + \frac{1}{\nu}V' + \left(\nu - \xi - \frac{\chi^2}{\nu^2}\right)V = 0. \tag{2.105}$$

Equation (2.105) has two singular points: one is regular in $\nu = 0$, the other is irregular at infinity. The asymptotics of an arbitrary solution for large ν contains two terms with the exponential multipliers $\exp(\pm(2i/3)\nu^{3/2})$. The radiation condition chooses the term with the exponential having the plus sign.

Condition (2.103) becomes the equation for the parameter ξ

$$V'(\kappa;\xi,\chi) + imZV(\kappa;\xi,\chi) = 0. \tag{2.106}$$

To achieve uniformity with respect to the impedance Z of the asymptotics that would allow getting solutions of the Dirichlet and Neumann problems by taking limits $Z \to 0$ and $Z \to \infty$ we consider the quantity mZ of order $O(1)$.

The equation in the next order in k allows the amplitude $A_0(s)$ in (2.104) to be defined. The Sturm–Liouville problem for the function U_1 is written at the eigenvalue and therefore is solvable only when the right-hand side is orthogonal to the solution of the homogeneous adjoint equation. First, let us rewrite the Heun equation in symmetric form containing no first derivative

$$L_0 V(\nu;\xi,\chi) = \nu^{-1/2}\left[\frac{d^2}{d\nu^2} + \left(\frac{1/4-\chi^2}{\nu^2} + \nu - \xi\right)\right](\nu^{1/2}V(\nu;\xi,\chi)) = 0. \tag{2.107}$$

Now the solvability condition can be written as

$$\int_{\kappa}^{+\infty} V(\nu;\xi,\chi)\left(2i\frac{\partial U_0}{\partial s} + i\frac{\partial\rho_t/\partial s}{\rho_t}\frac{\kappa}{\nu}U_0\right)\nu\, d\nu = 0.$$

By substituting the expression for U_0 and differentiating by s after a series of transformations the solvability condition can be rewritten as the differential equation for the amplitude $A_0(s)$. The solution of this equation is found explicitly

$$A_0(s) = A_0(0)\sqrt[3]{\frac{\rho(0)}{\rho(s)}}\sqrt{\frac{I_1(0)}{I_1(s)}}. \tag{2.108}$$

Here the integral

$$I_1 = \int_{\kappa}^{+\infty} V^2(\nu;\xi,\chi)\nu\, d\nu$$

is introduced.

The imaginary part of the parameter ξ characterising the attenuation of the creeping wave plays an important role in the asymptotics. Numerical results for Equation (2.106) are presented in Section 2.3.5. For a moderately elongated body in the assumption that $\kappa \gg 1$ the value ξ can be derived by considering the problem for creeping waves on the surface described by the impedance boundary condition with the effective impedance [11]

$$Z_{\text{eff}} = Z + \frac{i}{2\kappa m}.$$

2.3.4.3 Creeping waves for Maxwell's equations

The auxiliary vector functions $\mathbf{E} = \sqrt{\varepsilon}\mathcal{E}$ and $\mathbf{H} = \sqrt{\mu}\mathcal{H}$ are introduced. Then Maxwell's equations take the form

$$\text{curl}\,\mathbf{E} = ik\mathbf{H}, \qquad \text{curl}\,\mathbf{H} = -ik\mathbf{E},$$

where $k = \omega\sqrt{\varepsilon\mu}$. All the components of both vectors are searched in the form (2.95). It is natural to assume that the angular components E^{φ} and H^{φ} should be asymptotically large of order $O(k^{2/3})$ to give a tangential vector of order $O(1)$. Thus, it is convenient to introduce $E = k^{2/3}E^{\varphi}$ and $H = k^{2/3}H^{\varphi}$. Using the quadrature forms matrix, one rewrites Maxwell's equations for the components of the electric and magnetic vectors in the coordinates s, φ, n and, after substituting the representation (2.95), equates terms of similar orders in large parameter k. Part of these equations can be used to express the components E^n, E^s, H^n and H^s via E and H. We present here only the expressions for the two principal order terms

$$E_0^s = 0, \quad E_0^n = -(\kappa_1 + \nu_1)H_0, \quad E_1^s = -\chi E_0 - i(\kappa_1 + \nu_1)\frac{\partial H_0}{\partial \nu_1} - 2i H_0,$$

$$H_0^s = 0, \quad H_0^n = (\kappa_1 + \nu_1)E_0, \quad H_1^s = -\chi H_0 + i(\kappa_1 + \nu_1)\frac{\partial E_0}{\partial \nu_1} + 2i E_0.$$

Here $\kappa_1 = k^{2/3}\rho_t$. The other equations form a degenerated system whose compatibility conditions can be rewritten as ordinary differential equations for the components E and H. In the first order, one finds

$$\begin{aligned} &L_0 E_0 - \frac{2i\chi}{\nu^2} H_0 = 0, \\ &L_0 H_0 + \frac{2i\chi}{\nu^2} E_0 = 0, \end{aligned} \tag{2.109}$$

where

$$L_0 = \frac{d^2}{d\nu^2} + \frac{3}{\nu}\frac{d}{d\nu} + \left(\nu - \frac{\chi^2}{\nu^2} - \xi\right).$$

In the next order, the system becomes inhomogeneous

$$\begin{aligned} &L_0 E_1 - \frac{2i\chi}{\nu^2} H_1 = L_1 E_0, \\ &L_0 H_1 + \frac{2i\chi}{\nu^2} E_1 = L_1 H_0. \end{aligned} \tag{2.110}$$

Here

$$L_1 = i\left(\frac{\rho}{2}\right)^{2/3}\left\{2\frac{\partial}{\partial s} + 3\frac{\rho_t'}{\rho_t}\frac{\kappa}{\nu}\right\}.$$

The systems (2.109), (2.110) and all the following recurrent systems of equations are considered together with the radiation conditions that specify the exponential behaviour at infinity with the exponents $\exp(2i/3\nu^{3/2})$ and the boundary conditions. These boundary conditions follow from the conditions for the electric and magnetic fields on an impedance surface

$$\mathbf{E} - \mathbf{e}_n(\mathbf{E} \cdot \mathbf{e}_n) = Z\,\mathbf{e}_n \times \mathbf{H},$$

where $\mathbf{e}_n$ is the unit vector normal to the surface and Z is the impedance. Substituting the expressions for the components of the electric and magnetic vectors one can rewrite the boundary conditions in the form

$$\frac{dE_j}{d\nu} + \left(\frac{2}{\kappa} + i\frac{m}{Z}\right)E_j + \frac{i\chi}{\kappa}H_j = 0, \tag{2.111}$$

and

$$\frac{dH_j}{d\nu} + \left(\frac{2}{\kappa} + imZ\right) H_j - \frac{i\chi}{\kappa} E_j = 0, \tag{2.112}$$

where j takes the values of 0 and 1.

For $\chi = 0$ the system splits into two independent Sturm–Liouville problems for the components E and H. The principal terms of the asymptotics in this case can be constructed in the same manner as for scalar waves. One introduces the solution of the Heun equation

$$W''(\nu;\xi,0) + \frac{3}{\nu} W'(\nu;\xi,0) + (\nu - \xi) W(\nu;\xi,0) = 0, \tag{2.113}$$

satisfying the radiation condition. Then the attenuation parameter ξ is found from the boundary condition (2.111) or (2.112). In one case, the magnetic component H is zero in the principal order and the creeping wave is the transversally polarised electric creeping wave with the attenuation parameter defined from the equation

$$W'(\kappa;\xi,0) + \left(\frac{2}{\kappa} + i\frac{m}{Z}\right) W(\kappa;\xi,0) = 0. \tag{2.114}$$

For the other wave the electric component E is zero in the principal order and the wave is a transversally polarised magnetic creeping wave. The attenuation parameter is defined from a similar equation with Z^{-1} replaced by Z.

The amplitudes of these creeping waves are determined by the compatibility conditions for Equations (2.110). Similar to the scalar case, the compatibility condition is rewritten as a differential equation for the amplitudes and its solution is expressed explicitly. We present the formula for the magnetic creeping wave only

$$H_0 = A_0(0)\rho^{-2/3} I_3^{-1/2} W(\nu;\xi,0), \tag{2.115}$$

where

$$I_3 = \int_{\kappa}^{+\infty} W^2(\nu;\xi,0)\nu^3 \, d\nu.$$

The system of differential equations and the boundary conditions do not split into two independent Sturm–Liouville problems if $\chi \neq 0$. This is natural as in this case the waves run along geodesics that have a torsion of order $k^{2/3}$ and the terms that prohibit splitting of the system are proportional to that torsion. Nevertheless, by introducing functions $\Psi_\pm$ using the formulae

$$\begin{pmatrix} H_0 \\ E_0 \end{pmatrix} = A_0(s) \begin{pmatrix} i \\ 1 \end{pmatrix} \frac{\Psi_+(\nu;\xi,\chi)}{\Psi_+(\kappa;\xi,\chi)} + B_0(s) \begin{pmatrix} -i \\ 1 \end{pmatrix} \frac{\Psi_-(\nu;\xi,\chi)}{\Psi_-(\kappa;\xi,\chi)},$$

one can split the system (2.109) into two independent equations for the functions $\Psi_\pm$

$$\left\{ L_0 \pm \frac{2\chi}{\nu^2} \right\} \Psi_\pm(\nu;\xi,\chi) = 0. \tag{2.116}$$

The following system for the amplitudes $A_0(s)$ and $B_0(s)$ results from the boundary conditions

$$\begin{aligned}&\left(\frac{\Psi'_+}{\Psi_+}+\frac{2-\chi}{\kappa}+iQ\right)A_0+iPB_0=0,\\ &iPA_0+\left(\frac{\Psi'_-}{\Psi_-}+\frac{2+\chi}{\kappa}+iQ\right)B_0=0,\end{aligned} \tag{2.117}$$

where

$$Q=\frac{m}{2}\left(1/Z+Z\right),\qquad P=\frac{m}{2}\left(1/Z-Z\right).$$

The non-zero solution of system (2.117) exists only if the determinant of its matrix is equal to zero. That gives the equation for the parameter ξ involved in the operator L_0

$$\left(\frac{\Psi'_+(\kappa;\xi,\chi)}{\Psi_+(\kappa;\xi,\chi)}+\frac{2-\chi}{\kappa}+iQ\right)\left(\frac{\Psi'_-(\kappa;\xi,\chi)}{\Psi_-(\kappa;\xi,\chi)}+\frac{2+\chi}{\kappa}+iQ\right)+P^2=0. \tag{2.118}$$

Here the functions $\Psi_\pm$ are supposed to satisfy Equations (2.116) and radiation conditions. For the case of a perfectly conducting body analogous derivations bring to the system

$$\begin{aligned}&A_0+B_0=0,\\ &\left(\frac{\Psi'_+(\kappa;\xi,\chi)}{\Psi_+(\kappa;\xi,\chi)}+\frac{2}{\kappa}\right)A_0-\left(\frac{\Psi'_-(\kappa;\xi,\chi)}{\Psi_-(\kappa;\xi,\chi)}+\frac{2}{\kappa}\right)B_0=0,\end{aligned} \tag{2.119}$$

and equation

$$\frac{\Psi'_+(\kappa;\xi,\chi)}{\Psi_+(\kappa;\xi,\chi)}+\frac{\Psi'_-(\kappa;\xi,\chi)}{\Psi_-(\kappa;\xi,\chi)}+\frac{4}{\kappa}=0, \tag{2.120}$$

for the parameter ξ. Below, only that case will be considered.

In the next order, the vector Sturm–Liouville problem becomes inhomogeneous and the dependence of the amplitudes A_0 and $B_0=-A_0$ on s can be found from the compatibility conditions. To write these compatibility conditions and the transport equation for the amplitude $A_0(s)$ one needs to rewrite the operator of the problem in symmetric form

$$\nu^{-3/2}\left\{\frac{d^2}{d\nu^2}+\nu-\frac{\chi^2+3/4\mp 2\chi}{\nu^2}-\xi\right\}\left(\nu^{3/2}\Psi_\pm(\nu;\xi,\chi)\right)=0. \tag{2.121}$$

The compatibility condition can be written as a differential equation for the amplitude $A_0(s)$. After cumbersome derivations employing integration by parts and the properties of the function $\xi(s)$ as the solution of (2.120), this equation can be rewritten in the following simple form

$$A_0'(s)+\left(\frac{2}{3}\frac{\rho'(s)}{\rho(s)}+\frac{1}{2}\frac{J_3'(s)}{J_3(s)}\right)A_0(s)=0,$$

where

$$J_3(s) = \int_{\kappa}^{+\infty} \nu^3 \left(\frac{\Psi_+^2(\nu;\xi,\chi)}{\Psi_+^2(\kappa;\xi,\chi)} + \frac{\Psi_-^2(\nu;\xi,\chi)}{\Psi_-^2(\kappa;\xi,\chi)} \right) d\nu.$$

Finally, in the principal order, the asymptotics of the electromagnetic field takes the form

$$\begin{pmatrix} H \\ E \end{pmatrix} = A_0(0) \exp \left\{ iks + i\chi\varphi + ik^{1/3} \int_0^s \frac{\xi(s) - \kappa(s)}{2^{1/3}\rho^{2/3}(s)} ds \right\}$$
$$\times \rho^{-2/3}(s) J_3(s)^{-1/2} \left\{ \begin{pmatrix} i \\ 1 \end{pmatrix} \frac{\Psi_+(\nu;\xi,\chi)}{\Psi_+(\kappa;\xi,\chi)} + \begin{pmatrix} i \\ -1 \end{pmatrix} \frac{\Psi_-(\nu;\xi,\chi)}{\Psi_-(\kappa;\xi,\chi)} \right\}. \tag{2.122}$$

The amplitude $A_0(0)$ remains an arbitrary constant in (2.122). It is defined by the incident field or by the source.

2.3.5 Numerical analysis

Let us examine the dispersion equations for the attenuation parameter ξ. For the case of scalar waves, this is Equation (2.106). For non-rotating ($\chi = 0$) electromagnetic creeping waves on a perfect conductor, the attenuation parameter is the solution of Equation (2.114) with $Z^{-1} = 0$ for an electric wave and with $Z = 0$ for a magnetic wave. The parameter ξ for rotating waves satisfies Equation (2.120). All these equations contain Heun functions that have no explicit integral representations.

The computations were performed for scalar waves with $\chi = 0$ and electromagnetic waves with $\chi = 0$ and $\chi = 1$. The dependences of the imaginary part of the parameter ξ on the normalised radius of transverse curvature κ are presented in Figure 2.2. Curves (*b*) and (*d*) correspond to scalar waves on acoustically soft and rigid surfaces, correspondingly. Curves (*c*) and (*e*) are for electric and magnetic creeping waves with $\chi = 0$ and curves (*a*) and (*f*) are for electromagnetic creeping waves with $\chi = 1$. One can note that for large values of κ the attenuation parameter $\xi - \kappa$ tends to zeros of the Airy function w_1 or of its derivative. (b') and (e') curves represent the asymptotics of ξ simulated by the effective impedance boundary condition (compare with (*b*) and (*e*)).

The formulae of the previous section were written in the supposition that functions $\Psi_\pm \neq 0$ on the surface $\nu = \kappa$. However, for the limiting case of large κ one can find that one part of zeros of Equation (2.120) tends to zeros of the Airy function $w_1(\xi)$ (curve (*a*)), while the other part approaches the zeros of its derivative (curve (*f*)). Corresponding waves become electric and magnetic transversally polarised waves when $\kappa \to \infty$. For small values of normalised transverse curvature the magnetic wave with $\chi = 0$ and quasi-magnetic wave with $\chi = 1$ become almost non-attenuated.

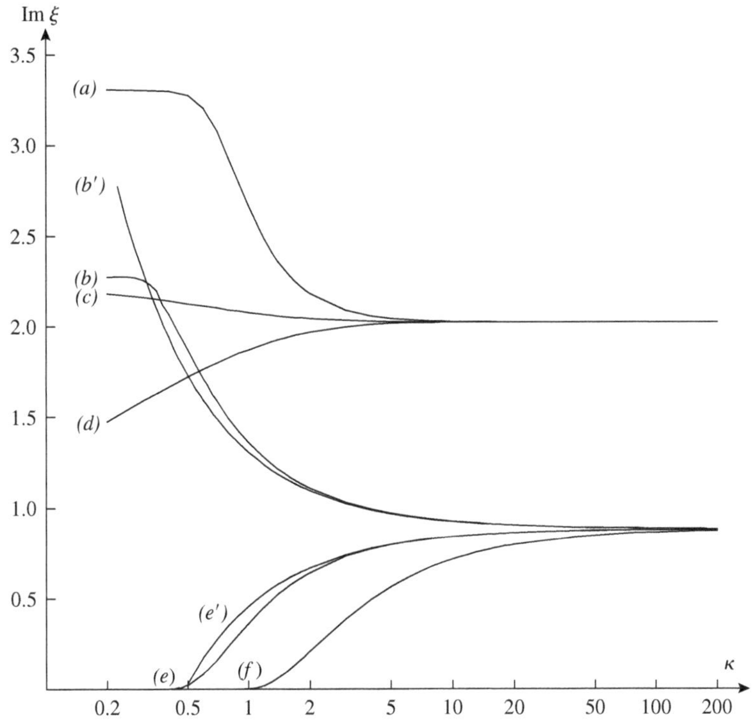

Figure 2.2 Im ξ as a function of normalised radius of transverse curvature κ

2.4 Creeping and whispering gallery waves at interfaces

2.4.1 Introduction

Curvilinear boundaries of acoustic or electromagnetic objects carry specific waves. On convex surfaces these are creeping waves and on concave surfaces these are whispering gallery waves. Penetrable surfaces in some cases can be described with the Leontovich impedance boundary condition. Here we examine creeping and whispering gallery waves running along curvilinear interfaces of acoustic and electromagnetic media. Depending on the parameters of that media and depending on the type of considered waves one can or cannot use the Leontovich impedance boundary condition.

We discover that if the contrast of parameters in the two media is sufficiently large, then waves of creeping and of whispering gallery types are possible. Both can be simulated by an impedance boundary condition. If the contrast is small, then waves appear of mixed type. Such waves do not satisfy an impedance boundary condition, their attenuation parameters are subject to a more complicated dispersion equation.

We examine the excitation of interface waves both for large and small contrast and found that waves of creeping type are well excited while excitation coefficients for whispering gallery waves are exponentially small. Original papers are [18,19].

2.4.2 Scalar waves

2.4.2.1 The problem formulation and the Ansatz

Let Ω denote the transparent body. The creeping waves on its surface satisfy the Helmholtz equation

$$\Delta u + k^2 N^2 u = 0. \tag{2.123}$$

The index is $N = 1$ outside Ω. Inside Ω it is a function of coordinates. We consider N depending only on the external normal n to the surface and represent it in the form

$$N^2(n) = N_1^2 - \beta n + \cdots, \qquad n < 0. \tag{2.124}$$

The boundary conditions on $\partial\Omega$ express the continuity of displacements and velocities

$$\lim_{n\to+0} u = \lim_{n\to-0} u, \qquad \lim_{n\to+0} \frac{\partial u}{\partial n} = \kappa \lim_{n\to-0} \frac{\partial u}{\partial n}, \tag{2.125}$$

where κ is the density ratio of acoustic media inside and outside Ω. The radiation conditions for $n \to \pm\infty$ conclude the formulation of the problem.

The creeping waves in high-frequency diffraction have the form

$$u = \exp(iks + ik^{1/3} p(s,a)) \sum_{j=0}^{\infty} U_j(s,\ell,a) k^{-j/3}. \tag{2.126}$$

Here we introduce the geodesic coordinates (s, a, n), where s is the arc-length of the geodesics followed by the wave. The normal coordinate n is stretched

$$\ell = \begin{cases} k^{2/3} n & \text{for } n > 0, \\ k^q n & \text{for } n < 0. \end{cases} \tag{2.127}$$

The factor q, as it will be shown later, depends on the order of constants in the decomposition (2.124). The functions $p(s,a)$ and $U_j(s,\ell,a)$ can be found when the representation (2.126) is substituted into Equation (2.123) and conditions (2.125). We shall only be interested in the leading term of the asymptotics (2.126), that is, in finding functions p and U_0 only.

2.4.2.2 The case of large contrast

Let us start with the Helmholtz equation in the exterior region. Standard derivations yield the following system of recurrent equations

$$\mathcal{L}_0 U_0 = 0, \qquad \mathcal{L}_0 U_1 + \mathcal{L}_1 U_0 = 0, \qquad \ldots, \tag{2.128}$$

where

$$\mathcal{L}_0 = \frac{\partial^2}{\partial \ell^2} + 2\left(\frac{\ell}{\rho} - \varphi'\right),$$

$$\mathcal{L}_1 = 2i\frac{\partial}{\partial s} + \frac{h'}{h}.$$

The radius of curvature of the geodesic ρ and the divergence of the geodesics h depend on the surface coordinates s and a, prime here and below denotes derivative by s. The boundary conditions for $\ell = 0$ will be considered later, when the field in the interior of Ω will be examined.

The solution of the first equation in (2.128) is given by the formula

$$U_0 = A_0(s,a)w_1(\xi - \nu), \qquad \nu = \left(\frac{2}{\rho}\right)^{1/3} \ell. \tag{2.129}$$

It satisfies the leading order equation in (2.128) and the radiation condition for $\ell \to +\infty$ for any $\xi(s,a)$ and $A_0(s,a)$. The variable parameter ξ is used to satisfy the conditions on $\partial\Omega$ and the function $p(s,a)$ is then calculated as an integral along the geodesic ray

$$p(s,a) = 2^{-1/3} \int_0^s \frac{\xi(s,a)}{\rho^{2/3}(s,a)}\, ds.$$

For the interior of Ω the order of difference of the wave numbers $(N_1 - 1)$ and order of the parameter β play an important role. They define the stretching parameter q in (2.127). If the index N of the body on its surface differs sufficiently from 1, that is, $N_1^2 - 1 = O(k^{-2/3+\varepsilon})$, $\varepsilon > 0$, then the principal terms of the equation for U_0 inside Ω are

$$\frac{\partial^2 U_0}{\partial n^2} \qquad \text{and} \qquad k^2(N_1^2 - 1)U_0.$$

Thus q should be taken equal to 1. That brings to the following recurrent equations:

$$\mathcal{M}_0 U_0 = 0, \qquad \mathcal{M}_0 U_1 = 0, \qquad \mathcal{M}_0 U_2 - 2p' U_0 = 0, \qquad \ldots, \tag{2.130}$$

where

$$\mathcal{M}_0 = (N_1^2 - 1) + \frac{\partial^2}{\partial \ell^2}.$$

The solutions U_0 and U_1 of Equations (2.130) are the exponentials

$$U_j = B_j \exp(-iD\ell), \quad D = \sqrt{N_1^2 - 1}, \qquad j = 0, 1. \tag{2.131}$$

The sign '$-$' is chosen in (2.131) to satisfy the radiation condition for $\ell \to -\infty$.

Substituting from (2.129) and (2.131) into the first condition (2.125) one finds that

$$B_0 = A_0 w_1(\xi). \tag{2.132}$$

The second condition (2.125) shows that the parameter ξ should be chosen from the equation

$$w_1(\xi) = 0.$$

Thus, $\xi = \text{const}$ and from (2.132) follows $B_0 = 0$. In the next order of k the condition for U_1 becomes inhomogeneous

$$\lim_{\ell \to +0} U_1 = -i \left(\frac{2}{\rho}\right)^{1/3} \frac{1}{D\kappa} A_0 w_1'(\xi).$$

The solution U_1 consists of the function

$$U_1^{(1)} = -i \left(\frac{2}{\rho}\right)^{1/3} \frac{A_0}{D\kappa} w_1'(\xi - \nu)$$

and general solution $U_1^{(0)}$ of the inhomogeneous Airy equation

$$\mathcal{L}_0 U_1^{(0)} = -\mathcal{L}_0 U_1^{(1)} - \mathcal{L}_1 U_0 \tag{2.133}$$

with the boundary condition $U_1^{(0)}(s, \alpha, 0) = 0$ and radiation condition for $\nu \to +\infty$. As this Sturm–Liouville problem is unsolvable for the arbitrary right-hand side of (2.133) the function A_0 should satisfy some compatibility condition. To find this compatibility condition, following Reference 12, one can multiply (2.133) by the solution $w_1(\xi - \nu)$ of the adjoint homogeneous problem $\mathcal{L}_0 f = 0$ and integrate from $\nu = 0$ to $+\infty$. The integrals containing the Airy function and its derivative can be calculated explicitly and that gives the differential equation

$$A_0' + \left(\frac{1}{6}\frac{\rho'}{\rho} + \frac{1}{2}\frac{h'}{h} - \frac{1}{\rho D\kappa}\right) A_0 = 0.$$

Finally, the creeping wave field outside the body Ω is given by the expression

$$u = A_0(0, a) \left(\frac{\rho(0, a)}{\rho(s, a)}\right)^{1/6} \left(\frac{h(0, a)}{h(s, a)}\right)^{1/2} \times \exp\left\{iks + i\left(\frac{k}{2}\right)^{1/3} \int_0^s \frac{\xi \, ds}{\rho^{2/3}} + \frac{1}{D\kappa} \int_0^s \frac{ds}{\rho}\right\} w_1(\xi - \nu). \tag{2.134}$$

The result (2.134) can also be obtained using the Leontovitch impedance boundary condition with the impedance $Z = D\kappa$. This impedance is the same as that corresponding to a plane wave incident on a planar interface between two materials at grazing incidence from the media characterised by wave number k_1 (see Figure 2.3).

Analogously, one can search for a solution of the problem in the form of a whispering gallery wave running along the boundary $\partial\Omega$ on the concave side. All the derivations are performed in a similar way, but now n is stretched with a factor $k^{2/3}$ inside Ω and with a factor k^1 outside Ω. We present only the final formula for the field

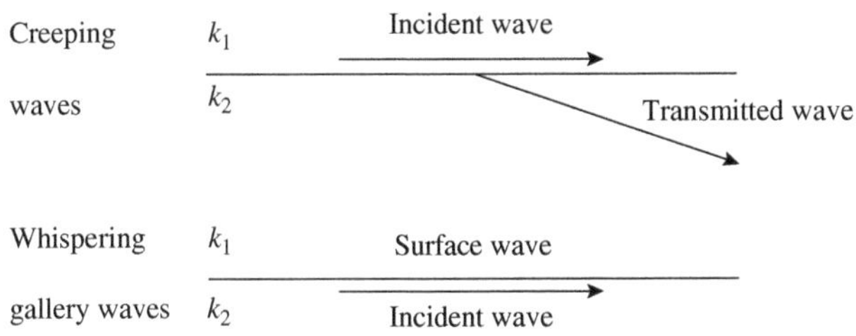

Figure 2.3 Equivalent planar problems

outside the body

$$u = A_0(0,a)\left(\frac{\tilde{\rho}(0,a)}{\tilde{\rho}(s,a)}\frac{h(0,a)}{h(s,a)}\right)^{1/2}$$
$$\times \exp\left\{ikN_1 s - kDn + i\left(\frac{kN_1}{2}\right)^{1/3}\int_0^s \frac{\eta\,ds}{\tilde{\rho}^{2/3}} + i\frac{\kappa N_1}{D}\int_0^s \frac{ds}{\tilde{\rho}}\right\}. \quad (2.135)$$

Here $1/\tilde{\rho} = 1/\rho - \beta/2$ is the effective curvature of the boundary when considered from inside Ω. The parameter η is defined from the equation

$$v(\eta) = 0,$$

and is real valued. For $N_1 > 1$ whispering gallery waves (2.135) propagate along the boundary $\partial\Omega$ without attenuation. If $N_1 < 1$ the dependence on n becomes oscillatory which corresponds to waves going from the boundary. Simultaneously the decay of waves appears which is described by the factor

$$\exp\left(-\frac{\kappa N_1}{\sqrt{1-N_1^2}}\int_0^s \frac{ds}{\tilde{\rho}}\right).$$

Formula (2.135) can be obtained using the Leontovitch impedance boundary condition when the exterior media is replaced by the impedance condition with the impedance $Z = -iD/(N_1\kappa)$. This impedance is the same as for a spatial wave incident at grazing incidence on the interface from the media characterised by wave number k_2 (see Figure 2.3).

2.4.2.3 The case of small contrast

Formulae (2.134) and (2.135) are obtained in the supposition that $D = O(1)$. They are also valid for large values of D. For small D, the formulae remain valid until $D > O(k^{-1/3})$. As for $D = O(k^{-1/3})$ the last two terms in the exponential factors in (2.134) and in (2.135) become of similar order, and all the following terms which are not present in the decompositions become of the same order in k, that is, formulae (2.134) and (2.135) lose their asymptotic character.

To find the asymptotic expansion for the case of small difference of wave numbers one should consider D to be small when performing stretching of the normal coordinate in the interior of Ω. For further derivations it is convenient to introduce the notation $D^2 = k^{-2/3}\delta$ and consider $\delta = O(1)$. Stretching the normal coordinate in the interior of Ω in the same way as in its exterior ($q = \frac{2}{3}$) one can find by equating terms of similar orders in k the following recurrent system

$$\mathcal{N}_0 U_0 = 0, \qquad \mathcal{N}_0 U_1 + \mathcal{L}_1 U_0 = 0, \qquad \ldots,$$

$$\mathcal{N}_0 = \frac{\partial^2}{\partial \ell^2} + \left(\frac{2}{\rho} - \beta\right)\ell + \delta - 2p'. \tag{2.136}$$

The solution in the principal term is given by the Airy function v (this choice is due to the radiation condition)

$$U_0 = B_0 v(\tilde{\xi} - \tilde{\nu}), \tag{2.137}$$

where $\tilde{\xi}$ and $\tilde{\nu}$ are defined by the formulae

$$\tilde{\xi} = (2p' - \delta)\left(\frac{\tilde{\rho}}{2}\right)^{2/3} = (\xi - d)\left(\frac{\tilde{\rho}}{\rho}\right)^{2/3}, \qquad d = \delta\left(\frac{\rho}{2}\right)^{2/3},$$

$$\tilde{\nu} = \left(\frac{2}{\tilde{\rho}}\right)^{1/3}\ell.$$

Using the boundary condition for u we substitute

$$B_0 = A_0 \frac{w_1(\xi)}{v(\tilde{\xi})},$$

and find the equation

$$\frac{w_1'(\xi)}{w_1(\xi)} = \left(\frac{\rho}{\tilde{\rho}}\right)^{1/3} \kappa \frac{v'(\tilde{\xi})}{v(\tilde{\xi})}, \tag{2.138}$$

which defines parameter ξ.

To find the amplitude A_0 one should consider the equations in the next order of small parameter. For this it is convenient to consider the equations in the external and internal regions simultaneously. Substituting from (2.129) and (2.137) into the equation for U_1 and integrating it with the solution

$$W(\nu) = \begin{cases} w_1(\tilde{\xi} - \tilde{\nu}), & \nu > 0 \\ \kappa \dfrac{w_1(\xi)}{v(\xi)} v(\tilde{\xi} - \tilde{\nu}), & \nu < 0 \end{cases}$$

of the adjoint homogeneous problem one can find the compatibility condition

$$A_0' + \left(\frac{1}{6}\frac{\rho'}{\rho} + \frac{1}{2}\frac{h'}{h} + \frac{1}{2}\frac{Q'}{Q}\right) A_0 = 0, \tag{2.139}$$

where

$$Q = \left(\left(1 - \kappa\frac{\tilde{\rho}}{\rho}\right)\xi + \kappa d\left(\frac{\tilde{\rho}}{\rho}\right)^{1/3}\right) w_1^2(\xi) + \left(\frac{\tilde{\rho}}{\rho}\frac{1}{\kappa} - 1\right)(w_1'(\xi))^2. \quad (2.140)$$

Finally, it gives

$$u = A_0(0,a)\exp\left\{iks + i\left(\frac{k}{2}\right)^{1/3}\int_0^s \frac{\xi}{\rho^{2/3}}\,ds\right\}\left(\frac{h(0,a)}{h(s,a)}\right)^{1/2}$$

$$\times\left(\frac{\rho(0,a)}{\rho(s,a)}\right)^{1/6}\left(\frac{Q(0,a)}{Q(s,a)}\right)^{1/2}\begin{cases} w_1(\xi-\nu), & n>0,\\ \dfrac{w_1(\xi)}{v(\xi)}v(\tilde{\xi}-\tilde{\nu}), & n<0.\end{cases}$$

2.4.3 *Electromagnetic waves*

2.4.3.1 The problem formulation

It is convenient to introduce the vectors $\mathbf{E} = \sqrt{\varepsilon}\mathcal{E}$ and $\mathbf{H} = \sqrt{\mu}\mathcal{H}$, where ε and μ are the electric and magnetic permeabilities and $\mathcal{E}$ and $\mathcal{H}$ are the usual electric and magnetic fields. Maxwell's equations for these vectors have the form

$$\begin{cases}\operatorname{curl}\mathbf{E} = ik\mathbf{H}\\ \operatorname{curl}\mathbf{H} = -ik\mathbf{E}\end{cases} \qquad k = \omega\sqrt{\varepsilon\mu}.$$

The parameters ε and μ are supposed to be equal to one outside Ω and depend only on the normal n inside Ω. It is convenient to introduce the following quantities:

$$\kappa_E = \sqrt{1/\varepsilon}, \qquad \kappa_H = \sqrt{1/\mu}.$$

Taking into account that $\mathbf{E}$ and $\mathbf{H}$ differ by multiplier from the real electric and magnetic fields one can find that at the boundary $\partial\Omega$ the following conditions should be satisfied

$$E^a|_{n=+0} = \kappa_E E^a|_{n=-0}, \qquad H^a|_{n=+0} = \kappa_H H^a|_{n=-0},$$

$$E^s|_{n=+0} = \kappa_E E^s|_{n=-0}, \qquad H^s|_{n=+0} = \kappa_H H^s|_{n=-0},$$

$$E^n|_{n=+0} = \frac{1}{\kappa_E}E^n|_{n=-0}, \qquad H^n|_{n=+0} = \frac{1}{\kappa_H}H^n|_{n=-0}.$$

2.4.3.2 Ansatz and recurrent equations outside Ω

Both vectors $\mathbf{E}$ and $\mathbf{H}$ are searched in the form (2.126)

$$\begin{pmatrix}\mathbf{E}\\ \mathbf{H}\end{pmatrix} = \exp(iks + ik^{1/3}p(s,a))\sum_{j=0}^{\infty}\begin{pmatrix}\mathbf{E}_j(s,\ell,a)\\ \mathbf{H}_j(s,\ell,a)\end{pmatrix}k^{-j/3}. \quad (2.141)$$

Substituting from (2.141) to Maxwell's equations and equating terms of similar orders in k one can obtain for the outer region, the standard formulae [4,13]

$$
\begin{aligned}
&H_0^s = E_0^s = 0, \quad H_0^n = hE_0^a, \quad E_0^n = -hH_0^a,\\
&H_1^s = ih\frac{\partial E_0^a}{\partial \ell}, \quad E_1^s = -ih\frac{\partial H_0^a}{\partial \ell}, \quad H_1^n = hE_1^a, \quad E_1^n = -hH_1^a,\\
&H_2^s = ih\frac{\partial E_1^a}{\partial \ell} - p'H_0^a, \quad E_2^s = -ih\frac{\partial H_1^a}{\partial \ell} - p'E_0^a,
\end{aligned}
$$

$$
E_2^a - \frac{1}{h}H_2^n = \left(p' - \frac{\ell}{\rho} - \frac{\ell}{\rho_t}\right)E_0^a - \frac{\partial^2 E_0^a}{\partial \ell^2} = -\left(p' - \frac{\ell}{\rho} - \frac{\ell}{\rho_t}\right)E_0^a, \tag{2.142}
$$

$$
H_2^a + \frac{1}{h}E_2^n = \left(p' - \frac{\ell}{\rho} - \frac{\ell}{\rho_t}\right)H_0^a - \frac{\partial^2 H_0^a}{\partial \ell^2} = -\left(p' - \frac{\ell}{\rho} - \frac{\ell}{\rho_t}\right)H_0^a, \tag{2.143}
$$

$$
\begin{aligned}
E_3^a - \frac{1}{h}H_3^n &= \left(p' - \frac{\ell}{\rho} - \frac{\ell}{\rho_t}\right)E_1^a - \frac{\partial^2 E_1^a}{\partial \ell^2} - 2i\tau H_0^a - 2i\tau\ell\frac{\partial H_0^a}{\partial \ell}\\
&= -\left(p' - \frac{\ell}{\rho} - \frac{\ell}{\rho_t}\right)E_1^a - p'\frac{\partial H_0^a}{\partial \ell},
\end{aligned} \tag{2.144}
$$

$$
\begin{aligned}
H_3^a - \frac{1}{h}E_3^n &= \left(p' - \frac{\ell}{\rho} - \frac{\ell}{\rho_t}\right)H_1^a - \frac{\partial^2 H_1^a}{\partial \ell^2} + 2i\tau E_0^a + 2i\tau\ell\frac{\partial E_0^a}{\partial \ell}\\
&= -\left(p' - \frac{\ell}{\rho} - \frac{\ell}{\rho_t}\right)H_1^a + p'\frac{\partial E_0^a}{\partial \ell}.
\end{aligned} \tag{2.145}
$$

Here τ and ρ_t are the torsion and the radius of binormal curvature.

The compatibility of expressions (2.142) and (2.143) brings to the Airy equations $\mathcal{L}_0 H_0^a = 0$ and $\mathcal{L}_0 E_0^a = 0$. So

$$
E_0^a = A_0 w_1(\xi - \nu), \qquad H_0^a = B_0 w_1(\xi - \nu). \tag{2.146}
$$

The parameter ξ can be found from the boundary conditions on $\partial\Omega$. The form of the equation for ξ depends on the order of the parameters and these equations for different cases will be obtained later.

For the next terms of E^a and H^a decomposition one can find from (2.144) and (2.145) the following system of equations:

$$
-i\mathcal{L}_0 H_1^a - 2\tau E_0^a + 2\frac{\partial H_0^a}{\partial s} + 3\frac{h'}{h}H_0^a = 0, \tag{2.147}
$$

$$
-i\mathcal{L}_0 E_1^a + 2\tau H_0^a + 2\frac{\partial E_0^a}{\partial s} + 3\frac{h'}{h}E_0^a = 0. \tag{2.148}
$$

2.4.3.3 Creeping waves in the case of large contrast

Let us consider the case for $D \equiv \sqrt{(k_1/k_2)^2 - 1} = \sqrt{(\kappa_E \kappa_H)^2 - 1}$ of order 1. For that case analogously to the previously examined creeping waves in acoustics the stretching parameter q should be taken equal to 1. Substituting from (2.141) into Maxwell's equations one can find in the three leading orders of k the following equalities:

$$H_j^s = ih\kappa_E\kappa_H \frac{\partial E_j^a}{\partial \ell}, \qquad E_j^s = -ih\kappa_E\kappa_H \frac{\partial H_j^a}{\partial \ell},$$

$$H_0^n = h\kappa_E\kappa_H E_0^a, \qquad E_0^n = -h\kappa_E\kappa_H H_0^a,$$

$$\frac{\partial E_j^s}{\partial \ell} - iE_j^n - \frac{h}{\kappa_E\kappa_H} H_j^a = 0, \qquad \frac{\partial H_j^s}{\partial \ell} - iH_j^n + \frac{h}{\kappa_E\kappa_H} E_j^a = 0.$$

The index j in these formulae is 0, 1 and 2. The formulae mean that transverse components of electric and magnetic vectors satisfy the equation

$$\left(\frac{\partial^2}{\partial \ell^2} + D^2\right)\begin{pmatrix} E_j^a \\ H_j^a \end{pmatrix} = 0. \tag{2.149}$$

The radiation condition for $\ell \to -\infty$ selects the following solution of (2.150):

$$E_j^a = \frac{C_j}{\kappa_E} \exp(-iD\ell), \qquad H_j^a = \frac{D_j}{\kappa_H} \exp(-iD\ell). \tag{2.150}$$

To calculate the coefficients C_j and D_j of the solutions and the propagation constant ξ, one should use the conditions on the surface $\partial\Omega$. The boundary conditions for the s components of electric and magnetic vectors for (2.146) and (2.150) give the following values of the coefficients C_0, C_1 and D_0, D_1 of the solutions

$$C_0 = 0, \qquad C_1 = -\frac{i}{\kappa_E^2 D}\left(\frac{2}{\rho}\right)^{1/3} B_0 w_1'(\xi),$$

$$D_0 = 0, \qquad D_1 = -\frac{i}{\kappa_H^2 D}\left(\frac{2}{\rho}\right)^{1/3} A_0 w_1'(\xi).$$

The conditions for the normal and transverse components both give

$$A_j w_1(\xi) = C_j, \qquad B_j w_1(\xi) = D_j.$$

Thus, in the leading term transverse components of the electric and magnetic vectors satisfy the Dirichlet boundary condition, and to have non-zero field one should take ξ as a root of the equation

$$w_1(\xi) = 0. \tag{2.151}$$

In the next order, the conditions for E_1^α and H_1^α are inhomogeneous. Calculating the compatibility conditions for (2.147) and (2.148) one can find the system of equations for the amplitudes A_0 and B_0

$$2A_0' + \left(\frac{1}{3}\frac{\rho'}{\rho} + 3\frac{h'}{h} - \frac{2}{\rho\kappa_H^2 D}\right) A_0 + 2\tau B_0 = 0,$$
$$2B_0' + \left(\frac{1}{3}\frac{\rho'}{\rho} + 3\frac{h'}{h} - \frac{2}{\rho\kappa_E^2 D}\right) B_0 - 2\tau A_0 = 0. \tag{2.152}$$

Formulae (2.152) can also be obtained if the penetrable body Ω is replaced by a surface with the following matrix impedance boundary condition:

$$\begin{pmatrix} E^s \\ E^a \end{pmatrix} = \begin{pmatrix} 0 & D\kappa_E^2 \\ D^{-1}\kappa_H^{-2} & 0 \end{pmatrix} \begin{pmatrix} H^s \\ H^a \end{pmatrix}.$$

As in the acoustic case this matrix impedance coincides with that obtained for the incidence of spatial waves at grazing incidence on the interface. The impedance is $D^{-1}\kappa_H^{-2}$ if the polarisation of the incident wave is TE and it is equal to $D\kappa_E^2$ if the incident wave is TM polarised.

For an arbitrary surface it is impossible to get an explicit asymptotic formula, and Equations (2.152) can be solved only numerically.

For the surface without torsion the system (2.152) splits into two independent differential equations. In this case one can find

$$A_0(s,a) = A_0(0,a)\left(\frac{h(0,a)}{h(s,a)}\right)^{3/2}\left(\frac{\rho(0,a)}{\rho(s,a)}\right)^{1/6} \exp\left(\frac{1}{\kappa_H^2 D}\int_0^s \frac{ds}{\rho}\right),$$

$$B_0(s,a) = B_0(0,a)\left(\frac{h(0,a)}{h(s,a)}\right)^{3/2}\left(\frac{\rho(0,a)}{\rho(s,a)}\right)^{1/6} \exp\left(\frac{1}{\kappa_E^2 D}\int_0^s \frac{ds}{\rho}\right).$$

When substituting formulae (2.146) and (2.150) into the boundary conditions on $\partial\Omega$, we supposed that the parameters κ_E and κ_H are of order k^0, so the previous formulae are not valid for the case of small differences of electric or magnetic permeabilities of the materials inside and outside Ω. Repeating the previous derivations for $\kappa_H = O(k^{-1/6})$ one can find that in this case the boundary condition for E_0^α becomes

$$\frac{\partial E_0^a}{\partial n} + \frac{ikE_0^a}{Z} = 0$$

with the impedance $Z = -1/(\kappa_H^2 D)$. That is, two types of creeping waves propagate along the surface of such a transparent body. One is the tangential electric wave with

$E_0^a \neq 0, H_0^a = 0$ and propagating factor ξ_E given by the equation

$$w_1'(\xi_E) - iDm\kappa_H^2 w_1(\xi_E) = 0, \qquad m = \left(\frac{k\rho}{2}\right)^{1/3}.$$

The other is the tangential magnetic wave with $E_0^a = 0, H_0^a \neq 0$ and propagating factor ξ given by Equation (2.151). The amplitudes of these waves are the following

$$A_0 = A_0(0,a)\left(\frac{h(0,a)}{h(s,a)}\right)^{3/2}\left(\frac{\rho(0,a)}{\rho(s,a)}\right)^{1/6}\frac{1}{\sqrt{(w_1'(\xi_E))^2 - \xi_E w_1^2(\xi_E)}}, \qquad B_0 = 0,$$

and

$$A_0 = 0, \qquad B_0 = B_0(0,a)\left(\frac{h(0,a)}{h(s,a)}\right)^{3/2}\left(\frac{\rho(0,a)}{\rho(s,a)}\right)^{1/6}.$$

For the case $\kappa_E = O(k^{-1/6})$ the magnetic component satisfies the impedance condition

$$\frac{\partial H_0^a}{\partial n} + ikZH_0^a = 0$$

with the impedance $Z = -\kappa_E^2 D$. The asymptotics of such waves are

$$A_0 = A_0(0,a)\left(\frac{h(0,a)}{h(s,a)}\right)^{3/2}\left(\frac{\rho(0,a)}{\rho(s,a)}\right)^{1/6}, \qquad B_0 = 0,$$

and

$$A_0 = 0,$$

$$B_0 = B_0(0,a)\left(\frac{h(0,a)}{h(s,a)}\right)^{3/2}\left(\frac{\rho(0,a)}{\rho(s,a)}\right)^{1/6}\frac{1}{\sqrt{(w_1'(\xi_H))^2 - \xi_H w_1^2(\xi_H)}}.$$

For the second wave ξ_H is a root of the equation

$$w_1'(\xi) - iDm\kappa_E^2 w_1(\xi) = 0.$$

2.4.3.4 Creeping waves in the case of $D = O(k^{-1/3})$

For the creeping waves on the surface of a transparent body with the wave number approximately equal to that in the exterior when $D = O(k^{-1/3})$, the stretching parameter should be chosen equal to $\frac{2}{3}$. Analogous to the acoustics case, that results in replacing the operator $\mathcal{L}_0$ by $\mathcal{N}_0$ (see formula (2.136)). The formulae are the following:

$$E_0^s = H_0^s = 0, \qquad E_1^s = -ih\kappa_E\kappa_H \frac{\partial H_0^a}{\partial \ell}, \qquad H_1^s = ih\kappa_E\kappa_H \frac{\partial E_0^a}{\partial \ell},$$

$$E_2^s = -ih\kappa_E\kappa_H \frac{\partial H_1^a}{\partial \ell} - p' E_0^a, \qquad H_2^s = ih\kappa_E\kappa_H \frac{\partial E_1^a}{\partial \ell} - p' H_0^a,$$

$$E_0^n = -\frac{h}{\kappa_E\kappa_H} H_0^a, \qquad H_0^n = \frac{h}{\kappa_E\kappa_H} E_0^a, \qquad E_1^n = -\frac{h}{\kappa_E\kappa_H} H_1^a,$$

$$H_1^n = \frac{h}{\kappa_E\kappa_H} E_1^a,$$

$$(\mathcal{N}_0 + \delta) E_0^a = 0, \qquad (\mathcal{N}_0 + \delta) H_0^a = 0,$$

$$(\mathcal{N}_0 + \delta) E_1^a + 2i\tau\kappa_E\kappa_H H_0^a + 2i \frac{\partial E_0^a}{\partial s} + 3i \frac{h'}{h} E_0^a = 0,$$

$$(\mathcal{N}_0 + \delta) H_1^a - 2i\tau\kappa_E\kappa_H E_0^a + 2i \frac{\partial H_0^a}{\partial s} + 3i \frac{h'}{h} H_0^a = 0.$$

In the first order, it gives the solution

$$E_0^a = \frac{C_0}{\kappa_E} v(\tilde{\xi} - \tilde{\nu}), \qquad H_0^a = \frac{D_0}{\kappa_H} v(\tilde{\xi} - \tilde{\nu}).$$

When substituting this solution into the boundary conditions it is necessary to take into account that $(\kappa_E\kappa_H)^2 - 1 = O(k^{-2/3})$. One can find formulae connecting C_0, D_0 with A_0, B_0

$$A_0 w_1(\xi) = \kappa_E^2 C_0 v(\tilde{\xi}), \qquad B_0 w_1(\xi) = \kappa_H^2 D_0 v(\tilde{\xi}),$$
$$A_0 w_1'(\xi) = C_0 \left(\frac{\rho}{\tilde{\rho}}\right)^{1/3} v'(\tilde{\xi}), \qquad B_0 w_1'(\xi) = D_0 \left(\frac{\rho}{\tilde{\rho}}\right)^{1/3} v'(\tilde{\xi}). \tag{2.153}$$

Formulae (2.153) mean that in the case under examination the two types of creeping waves exist. One is the tangentially polarised electric wave with $E_0^a \neq 0$ and $H_0^a = 0$, the other is the tangentially polarised magnetic wave with $H_0^a \neq 0$ and $E_0^a = 0$. These waves propagate with different parameters ξ_E and ξ_H given by the equation

$$w_1'(\xi) v(\tilde{\xi}) = \kappa \left(\frac{\rho}{\tilde{\rho}}\right)^{1/3} w_1(\xi) v'(\tilde{\xi}) \tag{2.154}$$

with $\kappa = \kappa_E^2$ for ξ_E and with $\kappa = \kappa_H^2$ for ξ_H.

Taking into account that $H_0^a = 0$ for the electric wave and $E_0^a = 0$ for the magnetic wave, one can note that relations (2.153) hold in the next order too. That is, the continuity conditions in the first and in the second order are similar. Comparing the formulae for electric and magnetic creeping waves with those for creeping waves in acoustics one can find that they differ in replacing κ in (2.138) by κ_E^{-2} or κ_H^{-2}, respectively, and term $h'/2h$ in (2.139) by $3h'/2h$ (note that the norm of vector **a** is equal to h). The final asymptotics, thus, can be obtained immediately

$$\begin{pmatrix} E_0^a \\ H_0^a \end{pmatrix} \simeq \begin{pmatrix} A_0 \\ B_0 \end{pmatrix} \exp\left(iks + i\left(\frac{k}{2}\right)^{1/3} \int_0^s \frac{\xi}{\rho^{2/3}}\, ds \right) \left(\frac{h(0,a)}{h(s,a)}\right)^{3/2} \left(\frac{\rho(0,a)}{\rho(s,a)}\right)^{1/6}$$

$$\times \left(\frac{Q(0,a)}{Q(s,a)}\right)^{1/2} \begin{cases} w_1(\xi - \nu), & n > 0, \\ \dfrac{w_1(\xi)}{v(\tilde{\xi})} v(\tilde{\xi} - \tilde{\nu}), & n < 0. \end{cases}$$

The function Q is defined by (2.140) where the above mentioned replacements should be made. Note that $A_0 \neq 0$, $B_0 = 0$ (rep $A_0 = 0$, $B_0 \neq 0$) for the electric (rep magnetic) creeping wave.

2.4.4 *Excitation of waves at interfaces*

Let us examine the small vicinity of the light–shadow boundary on the surface of the body. The symmetry of the formulae for surface waves in deep shadow allows only electric transversally polarised waves to be considered. The formulae for the magnetic waves can be obtained by this symmetry principle.

Let us introduce the coordinates (s, a, n) in such a way that the direction of s coincides with the direction of the incident wave in the point C of the light–shadow boundary on the surface $\partial\Omega$. The Fock region in the vicinity of C has small dimensions for the k parameter. The asymptotics of the solution is searched in the form of Ansatz

$$u = \exp(iks) \sum_j U_j(\varsigma, \nu) k^{-j/3}, \tag{2.155}$$

where ς and ν are stretched coordinates. It is convenient to introduce

$$m_0 = \left(\frac{k\rho_0}{2}\right)^{1/3},$$

where ρ_0 is the radius of curvature of the surface $\partial\Omega$ in the point C. Then the stretched coordinates for the Fock region can be written as

$$\begin{aligned} \varsigma &= m_0 \frac{s}{\rho_0}, \\ \nu &= 2m_0^2 \frac{n}{\rho_0} \qquad \text{for } n > 0, \\ \nu &= k^q n \qquad \text{for } n < 0. \end{aligned} \tag{2.156}$$

The parameter q depends on the order of the difference $D = \sqrt{N_1^2 - 1}$ similar to previous sections.

The procedure of the asymptotics derivation consists of substituting the Ansatz (2.155) into Maxwell's equations and equating terms of similar orders of k. In the highest order, one gets the parabolic equation

$$i\frac{\partial U_0}{\partial \varsigma} + \frac{\partial^2 U_0}{\partial \nu^2} + \nu U_0 = 0.$$

Applying the Fourier transform one finds the representation

$$\widehat{U_0} = \widehat{A_0}(\xi) w_1(\xi - \nu)$$

outside the body, for $\nu > 0$. The incident field has the Fourier transform

$$\frac{1}{\sqrt{\pi}} v(\xi - \nu)$$

and is assumed zero inside Ω.

The representation for $\widehat{U_0}$ inside Ω depends on the order of D. Let us examine the case of large contrast when $D = O(1)$. One finds $q = 1$ and

$$\widehat{U_0} = \widehat{C_0}(\xi) e^{-iD\nu}, \qquad \nu < 0.$$

The amplitudes $\widehat{A_0}$ and $\widehat{C_0}$ can be found from the boundary conditions. In the principal order the electromagnetic field does not penetrate inside the body, that is, $\widehat{C_0} = 0$. For $\widehat{A_0}$ the formula becomes the same as in the case of electric waves on a perfectly conducting body

$$\widehat{A_0}(\xi) = -\frac{1}{\sqrt{\pi}} \frac{v(\xi)}{w_1(\xi)}.$$

This gives for the field the asymptotics expressed by the Fock function

$$u^s \approx -\frac{1}{\sqrt{\pi}} e^{iks} \int_{-\infty}^{+\infty} e^{i\varsigma\xi} \frac{v(\xi)}{w_1(\xi)} w_1(\xi - \nu)\, d\xi. \tag{2.157}$$

For large ς, the integral in (2.157) can be computed approximately as the residue in the pole with the minimal imaginary part. That gives the excitation coefficients for the surface waves

$$c_p = -2\sqrt{\pi} i \frac{v(\xi_p)}{w_1'(\xi_p)}, \tag{2.158}$$

where ξ_p is the pth zero of $w_1(\xi)$. Thus for the case of large contrast the excitation of an electric creeping wave appears the same as for the perfectly conducting body.

The formula for the magnetic creeping wave is exactly the same. That is, the magnetic creeping wave is excited as if the surface is a perfect magnetic conductor.

For the case of small contrast when $D = O(k^{-1/3})$ the parameter q should be taken equal to $\frac{2}{3}$ and it is convenient to extend formula (2.156) for negative n. One finds

$$\widehat{A_0}(\xi) = -\frac{1}{\sqrt{\pi}} \frac{\kappa \rho_0^{1/3} v'(\tilde{\xi}) v(\xi) - \tilde{\rho}_0{}^{1/3} v(\tilde{\xi}) v'(\xi)}{\kappa \rho_0^{1/3} w_1(\xi) v'(\tilde{\xi}) - \tilde{\rho}_0{}^{1/3} w_1'(\xi) v(\tilde{\xi})}. \tag{2.159}$$

Calculating the residue in the zero of the denominator (2.159), which coincides with the dispersion equation for surface waves (2.164), one finds the excitation coefficients

$$c_p = \frac{2\sqrt{\pi}i}{Q_p} \tag{2.160}$$

with ξ_p hidden in Q_p, satisfying Equation (2.154) where $\rho = \rho_0$.

For the transverse electric creeping wave one lets $\xi = \xi_E$ and $\kappa = \kappa_E^2$ and for the magnetic creeping wave $\xi = \xi_H$, $\kappa = \kappa_H^2$.

For the case of constant index inside Ω and $\kappa \approx 1$, formula (2.160) simplifies to

$$c_p = \frac{2\sqrt{\pi}i}{w_1^2(\xi_p)d}, \tag{2.161}$$

where the attenuation parameter ξ_p satisfies the equation

$$w_1'(\xi_p)v(\xi_p - d) = w_1(\xi_p)v'(\xi_p - d). \tag{2.162}$$

The solutions of this dispersion equation and the values of the excitation coefficients are presented in the next section.

2.4.5 Numerical results

The dispersion equation for the case of small contrast between the refraction indices in the outer space and on the surface of the transparent body permits no explicit solution. The numerical procedure based on Newton's iteration method is used. The attenuation parameters Im ξ_p defined from Equation (2.162) as functions of the contrast d are presented in Figure 2.4.

Here the solutions given by curves -2 and -1 correspond to waves of creeping type. When d tends to infinity and the contrast becomes large, these solutions tend to the zeros of the Airy function $w_1(\xi)$. The solutions marked as 1, 2, 3 are the

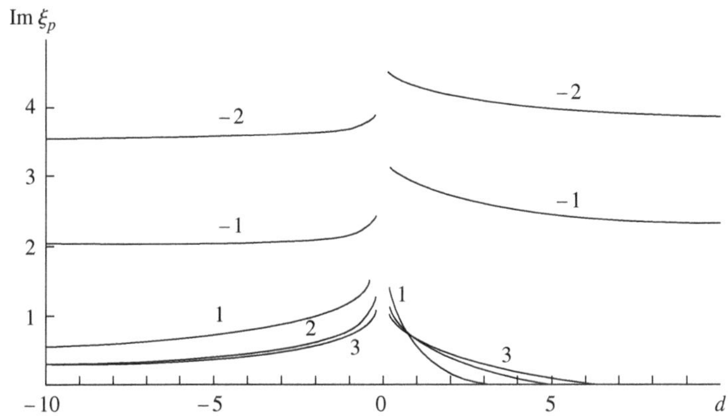

Figure 2.4 Attenuation parameters Im ξ_p

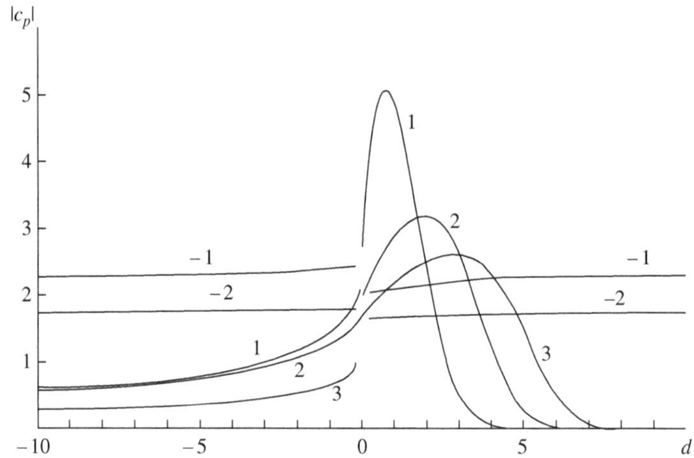

Figure 2.5 Excitation coefficients c_p

attenuation parameters for the surface waves of whispering gallery type. For large d these solutions approach the zeros of the Airy function $v(\xi - d)$. When d approaches zero, solutions of the dispersion equation tend to infinity, which is natural as for $d = 0$ the interface disappears.

The excitation coefficients for the above mentioned waves are presented in Figure 2.5. One notes that the waves of whispering gallery type are weakly excited if d is large.

Let us examine the expression (2.161) for the excitation coefficient in the case of asymptotically large positive d. First, it is necessary to analyse the dispersion equation which gives values of ξ_p. For asymptotically large values of d the solutions of the dispersion equation separate into two series. One with $\xi_p = O(1)$ corresponds to waves of creeping type. For other solutions $\xi_p - d = O(1)$ and such waves are of whispering gallery type.

Consider first the zeros from the first series. Using the asymptotics of the Airy function $v(\xi_p - d)$ and its derivative permits us to rewrite the dispersion equation in the form

$$w_1(\xi_p) \approx \frac{-i}{\sqrt{d}} w_1'(\xi_p).$$

Its solution has the asymptotics

$$\xi_p = \xi_p^0 - \frac{i}{\sqrt{d}} + \cdots ,$$

where ξ_p^0 is the zero of the Airy function w_1. Finally, the excitation coefficient can be found coincident with (2.158).

Consider now the zeros of the second series. Replacing the Airy function $w_1(\xi_p)$ and its derivative by the leading terms of their asymptotics yields

$$v(\eta_p) \approx \frac{1}{\sqrt{d}} v'(\eta_p).$$

The asymptotics of the solution η_p is

$$\eta_p = \eta_p^0 + \frac{1}{\sqrt{d}} + \cdots,$$

where η_p^0 is the zero of the Airy function v.

Substituting this asymptotics into the formula for the excitation coefficients yields exponentially small value

$$c_p \sim 2\sqrt{\pi} i d^{-1/2} \exp\left(-\tfrac{4}{3} d^{3/2} - 2d^{1/2}\eta_p^0 - 2\right).$$

This result is not reproduced by the ordinary asymptotic considerations when power series with large parameters are constructed only because the excitation coefficients are exponentially small and are therefore equal to zero at all orders. However, the surface waves of the whispering gallery type run along the inner boundary of the body almost without attenuation and at large distances from the light–shadow boundary can give a field comparable with the field of decreasing surface waves of creeping type.

2.5 References

1 FOCK, V. A.: 'Electromagnetic wave propagation and diffraction problems' (Pergamon, Oxford, 1960)

2 GOUYON, R.: 'Tensorial calculus' (Librairie Vuibert 63, Boulevard Saint-Germain, Paris, 1963) (in French)

3 ANDRONOV, I. V., and BOUCHE, D.: 'Computation of the second order term for the propagation parameter of creeping waves by boundary layer method', *Annales des Télécommunications*, 1994, **49** (3–4), pp. 199–204

4 BOUCHE, D., MOLINET, F., and MITTRA, R.: 'Asymptotic methods in electromagnetics' (Springer-Verlag, Berlin, 1997)

5 ANDRONOV, I., and BOUCHE, D.: 'Theoretical analysis of creeping waves', *Annales des Télécommunications*, 1994, **49** (3–4), pp. 193–210

6 HONG, S.: 'Asymptotic theory of electromagnetic and acoustic diffraction by smooth convex surfaces of variable curvature', *Journal of Mathematical Physics*, 1967, **8** (6), p. 1223

7 VOLTMER, D. R.: 'Diffraction by doubly curved convex surfaces' (PhD Thesis, Ohio State University, Ohio, 1970)

8 FEDORYUK, M. V. 'Asymptotic methods for linear ordinary differential equations' (Springer-Verlag, Berlin, 1993)

9 HUSSAR, P. E., and ROWLAND, E. S.: 'An asymptotic solution for boundary-layer fields', *Journal of Electromagnetic Waves and Applications*, 2002, **16** (2), pp. 185–200

10 SLAVYANOV, S. YU., LAY, W., and SEEGER, A.: 'The classification of Heun's equation and its special and confluent cases', in RONVEAUX, A. (Ed.): 'Heun equation' (Oxford University Press, Oxford, 1995)
11 ANDRONOV, I. V., and BOUCHE, D.: 'Creeping waves on strongly prolate bodies'. Proceedings of the URSI international symposium on *Electromagnetic theory*, St. Petersburg, Russia, 1995, pp. 404–6
12 BABICH, V. M., and KIRPICHNIKOVA, N. YA.: 'The boundary-layer method in diffraction problems' (Springer-Verlag, Berlin, 1979)
13 BOUCHE, D.: 'Study of creeping rays on an impedance convex body using asymptotic expansion method', *Annales des Télécommunications*, 1992, **47**, pp. 400–12
14 ANDRONOV, I., and BOUCHE, D.: 'Asymptotics of creeping waves on a strongly prolate body', *Annales des Télécommunications*, 1994, **49** (3–4), pp. 205–10
15 ANDRONOV, I. V.: 'Recent achievements in high frequency diffraction'. Proceedings of 10-e *Journées Internationales de Nice sur les Antennes*, Nice, France, 1998, pp. 25–32
16 ANDRONOV, I. V.: 'Some recent results on high frequency diffraction by 3D bodies'. Annales des 'Journées Maxwell'97', Bordeaux, France, 1997 (CEA/DAM/CESTA, 1998), pp. 311–14
17 ANDRONOV, I. V.: 'Waves on the surface of strongly stretched bodies', *Journal of Communications Technology & Electronics*, 1998, **43** (1)
18 ANDRONOV, I. V., and BOUCHE, D.: 'The creeping and whispering gallery waves on the surface of a transparent body', *Journal of Electromagnetic Waves and Applications*, 1995, **9** (4), pp. 503–20
19 BULDYREV, V. S.: 'Short wave interference in the problem of diffraction on inhomogeneous arbitrary-section cylinder', *Radiofizika USSR*, 1967, **10** (5), pp. 699–711

Chapter 3

Hybrid diffraction coefficients

3.1 Introduction

A hybrid diffraction coefficient is a combination of the diffraction coefficients corresponding to two types of diffraction phenomena.

We consider here the combination of edge diffraction with the launching of creeping waves or the reciprocal situation of a creeping wave diffracted by an edge.

The edge is either a discontinuity of the first derivatives of the surface coordinates (edge of a wedge) or a discontinuity of the second derivatives of the surface coordinates (line of discontinuity of the curvature). In both cases, we suppose that the faces involving surface diffraction are convex and that the source and the observation point are located far from the edge.

The field diffracted by the edge of a curved wedge undergoes several transition regions (which are also boundary layers), as shown in Figure 3.1.

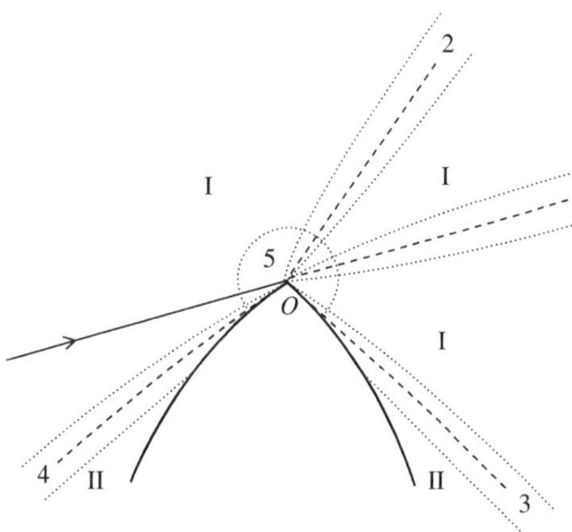

Figure 3.1 Transition regions for a curved wedge

For an incident ray, there are in general four transition regions in the far zone diffracted field. Since the shadow boundaries of the direct and reflected rays (regions 1 and 2) are also present in the canonical or modal problem of the straight wedge, the field in these regions has been deduced by Kouyoumjian and Pathak [1] from this model problem and is given by the uniform theory of diffraction (UTD).

For the shadow boundaries of the edge diffracted rays (regions 3 and 4) and the deep shadow region (region II), there is no model problem giving directly the analytical form of the solution in the far zone. However, in the near zone close to the edge (region 5), the straight wedge can still be used as a model problem for the curved wedge, since in this region, the tangent plane approximation gives the correct behaviour of the dominant term of the asymptotic expansion of the solution for a curved wedge. This behaviour constitutes the basis used by several authors for constructing hybrid diffraction coefficients for a curved wedge. Starting with an approximate solution of the straight wedge problem at the periphery of region 5, which they put in the form of the field radiated by an equivalent line source, Pathak and Kouyoumjian [2] derived a hybrid diffraction coefficient by reducing the problem to the radiation of a line source on a smooth curved surface. Their solution recovers some hybrid diffraction coefficients established earlier by Albertsen [3] for the diffraction of a creeping wave in the deep shadow (region II) and by Molinet [4] for the diffraction of an incident magnetic field parallel to the edge, propagating in region I and diffracted in the transition regions 2 or 3. Hybrid diffraction coefficients for first- and second-order discontinuities of two-dimensional (2D) scatterers, valid in the deep shadow region have also been proposed by Albertsen and Christiansen [5].

The equivalent line source technique has also been applied by Deschamps and his coworkers [6] to construct a hybrid diffraction coefficient for a line of discontinuity of the curvature. This solution is valid in the transition regions of edge diffracted rays (regions 2 and 3 of Figure 3.2).

In all the approaches mentioned so far, the surface diffraction follows the edge one. We will present a solution due to Michaeli [7] in which the edge diffraction is considered following the surface diffraction. The advantage of this method is that it applies to both problems – the curved wedge and the curvature discontinuity. The surface field close to the edge is decomposed into a spectrum of local inhomogeneous

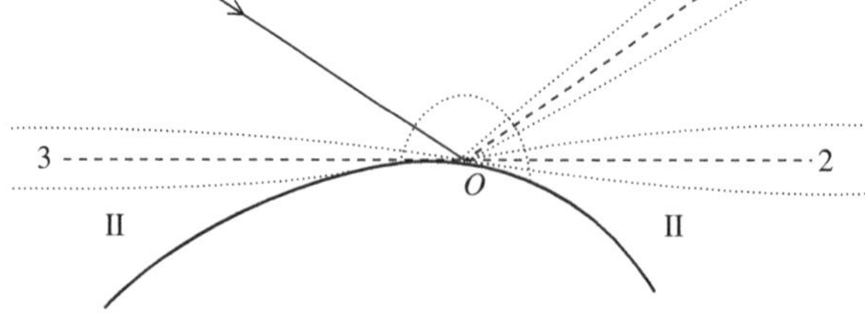

Figure 3.2 Transition regions for a curvature discontinuity

plane waves and the diffracted field is obtained by applying to each component of this spectrum the formulae established for the diffraction of a plane wave by the edge of a wedge or by a discontinuity in curvature. This procedure is also known as the spectral theory of diffraction (STD) [8].

A critical situation appears at grazing incidence when regions 1 and 3 in the wedge problem (Figure 3.1) or regions 1 and 2 in the curvature discontinuity problem (Figure 3.2), overlap. In this case, the special functions describing the field are Fock type integrals containing an incomplete Airy function in the integrand. This behaviour can be deduced from the solution of the canonical problem of a curved sheet established by Idemen and Felsen [9]. Following Michaeli [7], we will show that his method can also be adapted to this situation.

We consider first the curved wedge and treat successively the (two-dimensional) 2D problem with perfectly conducting surfaces and then extend the solutions to the three-dimensional (3D) case and finally to coated 2D surfaces. The curvature discontinuity is treated in the same way. For both problems we present also solutions valid for overlapping penumbra regions conditions.

3.2 Spectral representation of the Fock field on a smooth surface

The cylindrical surface is supposed to be perfectly conducting and convex. Let P' be the location of a line source and P an observation point close to the surface at a distance n from its orthogonal projection P_N on this surface (see Figure 3.3).

We denote by s the arc length from the launching point Q' of the creeping wave to P_N and suppose that P is in the vicinity of a fixed point O on the ray which will be the origin in our spectral representation of the field.

The lengths l and l'_0 of the curvilinear arcs along the creeping ray are positive on the right of the origin O and negative on the left so that $s = l'_0 - l$ (Figure 3.3).

Using the results valid in the Fock domain, presented in Chapter 1, Section 1.2.4 (Equation 1.81), the zeroth-order term of the asymptotic expansion of the total field for the soft (s) and hard (h) boundary conditions is given by:

$$u_s(\sigma, \nu) = u^i(Q')\frac{e^{iks}}{\sqrt{\pi}}\int_{-\infty}^{+\infty}\left[v(\xi-\nu) - \frac{v(\xi)}{w_1(\xi)}w_1(\xi-\nu)\right]e^{i\sigma\xi}\,d\xi, \tag{3.1}$$

$$u_h(\sigma, \nu) = u^i(Q')\frac{e^{iks}}{\sqrt{\pi}}\int_{-\infty}^{+\infty}\left[v(\xi-\nu) - \frac{v'(\xi)}{w'_1(\xi)}w_1(\xi-\nu)\right]e^{i\sigma\xi}\,d\xi, \tag{3.2}$$

where

$$\nu = \frac{kn}{m}, \qquad m = \left(\frac{k\rho}{2}\right)^{1/3}, \qquad \sigma = \frac{ms}{a},$$

and where a designates the radius of curvature ρ of the surface at the origin O.

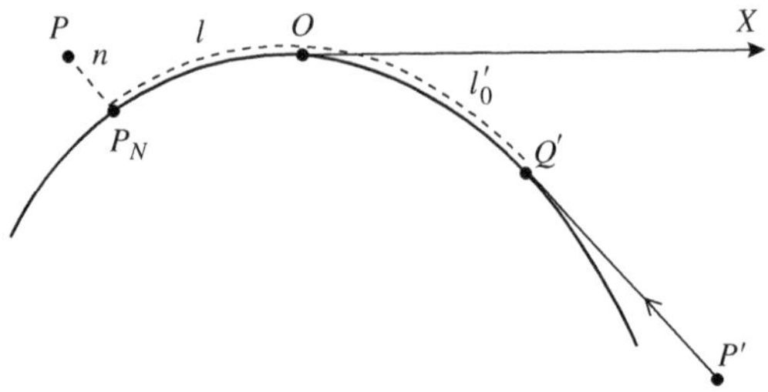

Figure 3.3 Geometry for the representation of the Fock field on a smooth surface

Formulae (3.1) and (3.2) are valid in the penumbra region where the surface around O can be replaced by its circular cylinder approximation.

By expanding the Fock functions $v(\xi - \nu)$ and $w_1(\xi - \nu)$ in Taylor series in the parameter ν which is supposed to be small, and taking into account that $w_1''(\xi) = \xi w_1(\xi)$ and $v''(\xi) = \xi v(\xi)$, we obtain:

$$u_s = u^i(Q') \int_{-\infty}^{+\infty} e^{i(l_0'-l)(k+\xi(m/\rho))} \frac{\nu + \nu^3(\xi/6) + O(\xi^4)}{\sqrt{\pi}\, w_1(\xi)}\, d\xi, \tag{3.3}$$

$$u_h = u^i(Q') \int_{-\infty}^{+\infty} e^{i(l_0'-l)(k+\xi(m/\rho))} \frac{1 + \nu^2(\xi/2) + O(\xi^3)}{\sqrt{\pi}\, w_1'(\xi)}\, d\xi. \tag{3.4}$$

These formulae were first established by Pathak [10].

The functions $w_1(\xi)$ and $w_1'(\xi)$ in the denominators of (3.3) and (3.4) increase as $\exp(\frac{2}{3}|\xi|^{3/2})$ on the contour $(\infty\, e^{2i\pi/3}, +\infty)$ into which the path of integration can be deformed. Hence the dominant contribution to the integrals comes from $|\xi| \le 1$, and hence $|\nu\sqrt{\xi}| \ll 1$. We can therefore write

$$\begin{aligned} \nu + \nu^3 \frac{\xi}{6} &\cong \frac{1}{i\sqrt{\xi}} \sin(i\nu\sqrt{\xi}), \\ 1 + \nu^2 \frac{\xi}{2} &\cong \cos(i\nu\sqrt{\xi}), \end{aligned} \tag{3.5}$$

and replace the sine and cosine functions by exponentials. This yields:

$$u_{s,h}(l,n) = \int_{-\infty}^{+\infty} A_{s,h}(\xi)[C_\xi^i(l,n) \pm C_\xi^r(l,n)]\, d\xi, \tag{3.6}$$

where

$$C_\xi^{i,r}(l,n) = \exp\left[-ikl\left(1 + \frac{\xi}{2m^2}\right) \mp ikni\sqrt{\frac{\xi}{m}}\right], \tag{3.7}$$

$$\begin{aligned} A_s(\xi) &= u^i(Q')\frac{\exp[il_0'(k+\xi(m/a))]}{2\sqrt{\pi\xi}\; w_1(\xi)}, \\ A_h(\xi) &= u^i(Q')\frac{\exp[il_0'(k+\xi(m/a))]}{2\sqrt{\pi}\; w_1'(\xi)}. \end{aligned} \tag{3.8}$$

Michaeli [7] was the first who recognised that the first terms of the Taylor series are also the series expansions of the sine and cosine functions as shown by (3.5).

We see that Equation (3.7) represents quasi-plane waves since if we state $\theta = i\sqrt{\xi}\, m^{-1}$ then, for $|\xi| \leq 1$ we can write

$$\frac{i\sqrt{\xi}}{m} \cong \sin\theta, \qquad 1 + \frac{\xi\, m^{-2}}{2} \cong \cos\theta \tag{3.9}$$

and

$$C_\xi^{i,r}(l,n) = \exp[-ik(l\cos\theta \pm n\, \sin\theta)] \tag{3.10}$$

which means that C_ξ^i is an incident wave from the complex direction θ and in the tangent plane approximation $l \cong x, n \cong y$, it is an inhomogeneous plane wave and $C_\xi^r(l,n)$ is the reflected plane wave associated to $C_\xi^i(l,n)$.

Hence (3.6) is a spectral representation of the Fock fields and $A_{s,h}$ given by (3.8) are the spectral amplitudes. The latter are rapidly decreasing with $|\xi|$ along the integration path. We will now apply these results to the construction of hybrid diffraction coefficients for a creeping wave diffracted either by the edge of a curved wedge or by a line of discontinuity of the curvature.

3.3 Hybrid diffraction coefficients for a curved wedge

3.3.1 *Two-dimensional perfectly conducting wedge*

We use the same notations as those defined in Figure 3.3, but now the origin O is taken at the edge of the wedge and the observation point P is supposed far away from O in the direction ϕ (see Figure 3.4).

We suppose that P' is in the penumbra region corresponding to the reciprocal problem. The angular extension of this region is of order m^{-1}.

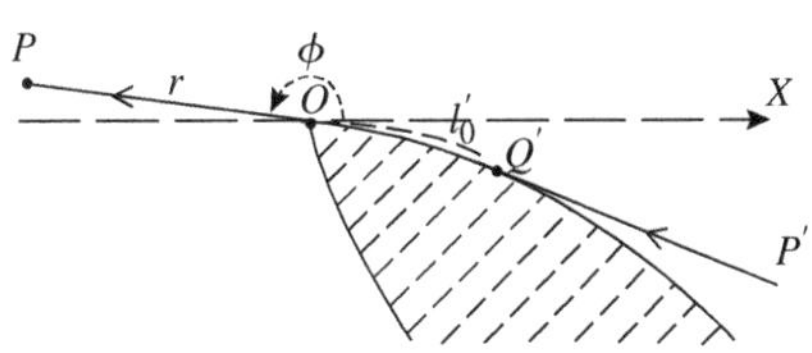

Figure 3.4 Creeping wave diffracted by the edge into a space wave

Applying the STD, we obtain directly from (3.6), by spectral synthesis the edge diffracted field at P:

$$U_h^s(P) = \Pi(r) \int_{-\infty}^{+\infty} A_{s,h}(\xi) D_{s,h}(\phi, \theta)\, d\xi, \tag{3.11}$$

where

$$\Pi(r) = \exp\left(ikr + \frac{i\pi}{4}\right)(8\pi kr)^{-1/2} \tag{3.12}$$

and where $D_{s,h}$ are the analytical continuation of Keller's edge-diffraction coefficients $d_{s,h}$ for a wedge, divided by the source factor S for an isotropic line source

$$D_{s,h} = \frac{d_{s,h}}{S}, \qquad S = (8\pi k)^{-1/2} \exp\left(i\frac{\pi}{4}\right). \tag{3.13}$$

For a wedge with exterior angle $N\pi$, Keller's diffraction coefficients are given by

$$\begin{aligned} d_{s,h}(\phi, \theta) = \frac{e^{i\pi/4} \sin(\pi/N)}{N\sqrt{2\pi k}} &\left[\frac{1}{\cos(\pi/N) - \cos((\phi - \theta)/N)}\right. \\ &\left.\mp \frac{1}{\cos(\pi/N) - \cos((\phi + \theta)/N)}\right]. \end{aligned} \tag{3.14}$$

Since these coefficients depend only on trigonometric functions, they vary slowly with $\theta(\xi)$ as compared to $A_{s,h}(\xi)$ for ϕ not close to π, which means away from the situation where the transition regions of the direct and diffracted rays, in the reciprocal problem, overlap.

Outside this region, which can be defined by $\pi - \phi \geq O(m^{-1})$, we can therefore replace Keller's diffraction coefficients by the first term of their Taylor series expansion with respect to θ:

$$\begin{aligned} D_s(\phi, \theta) &\cong \theta \left[\frac{\partial}{\partial \theta} D_s(\phi, \theta)\right]_{\theta=0} + O(m^{-3}), \\ D_h(\phi, \theta) &\cong D_h(\phi, 0) + O(m^{-2}), \end{aligned} \tag{3.15}$$

where $\theta = i\sqrt{\xi}/m$.

Inserting (3.15) into (3.11) yields

$$\begin{aligned} U_s(P) &= \frac{1}{2} u^i(Q') \frac{i}{m} f(\sigma_0')\, e^{ikl_0'} \left[\frac{\partial}{\partial \theta} D_s(\phi, \theta)\right]_{\theta=0} \Pi(r), \\ U_h(P) &= \tfrac{1}{2} u^i(Q') g(\sigma_0')\, e^{ikl_0'}\, D_h(\phi, 0) \Pi(r), \end{aligned} \tag{3.16}$$

where $\sigma_0' = ml_0'/a$, $l_0' = \text{arc } Q'O$, a is the radius of curvature at O, and where $f(x)$ and $g(x)$ are the Fock functions defined by

$$f(x) = \frac{1}{\sqrt{\pi}} \int_\Gamma \frac{e^{ixt}}{w_1(t)}\, dt, \qquad g(x) = \frac{1}{\sqrt{\pi}} \int_\Gamma \frac{e^{ixt}}{w_1'(t)}\, dt, \tag{3.17}$$

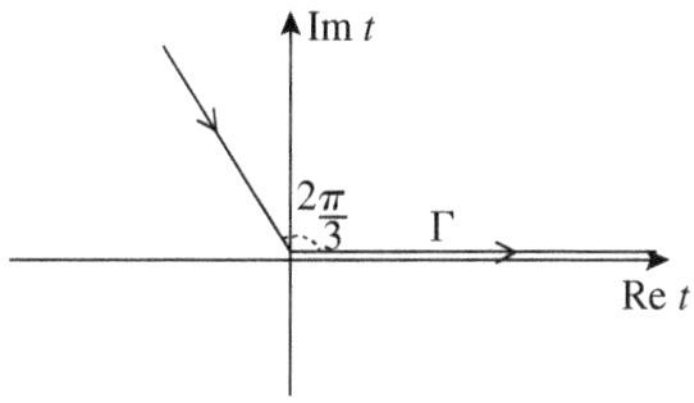

Figure 3.5 Contour of integration

where Γ is the contour $(-\infty + i\varepsilon, \infty + i\varepsilon)$ with ε arbitrarily small. Since the zeros of $w_1(t)$ and $w'(t)$ lie in the first quadrant, the initial contour can also be deformed into the contour $(\infty e^{2i\pi/3}, +\infty + i\varepsilon)$. In the following sections, we denote by Γ this deformed contour (see Figure 3.5). Formulae (3.16) are in agreement with the result of Pathak and Kouyoumjian [2].

By reciprocity, Equation (3.16) also describes the diffracted field at P' for a line source at P. Interchanging the position of P and P', and accordingly, the primed parameters with the unprimed ones and using the symmetry properties of $D_{s,h}(\phi, \theta)$, one obtains

$$U_s(P) = \frac{1}{2} u^i(O) \frac{i}{m} f(\sigma_0)\, e^{ikl_0} \left[\frac{\partial}{\partial \theta'} D_s(\theta', \phi) \right]_{\theta'=0} \Pi(s),$$
$$U_h(P) = \tfrac{1}{2} u^i(O) g(\sigma_0)\, e^{ikl_0} D_h(0, \phi) \Pi(s), \tag{3.18}$$

where $s = QP$ (counterpart of $Q'P'$).

In order to extend the STD treatment to the case where the source and the observation point are either both located in region 3 or in region 4 of Figure 3.1 or separately located in one and in the other of these regions, we extend analytically (3.18) to complex angles of incidence. By substituting to ϕ the complex angle $\theta(\xi) = i\sqrt{\xi}/m$ defining each spectral component of the incident field at O due to a source at P' (Figure 3.6) and performing the spectral synthesis, we yield then the result for the diffraction of a creeping wave into another creeping wave propagating on the same face or on the other face of the wedge.

If Q' and Q are on the same face, the solution is given by

$$U_s(P) = \frac{1}{4} u^i(Q') \left[\frac{\partial^2}{\partial \theta\, \partial \theta'} D_s(\theta', \theta) \right]_{\theta=\theta'=0} \frac{i}{m} f(\sigma_0') e^{ikl_0'} \frac{i}{m} f(\sigma_0) e^{ikl_0} \Pi(s),$$
$$U_h(P) = \tfrac{1}{4} u^i(Q') D_h(0, 0) g(\sigma_0') e^{ikl_0'} g(\sigma_0)\, e^{ikl_0} \Pi(s), \tag{3.19}$$

where $s = |\overrightarrow{QP}|$, $\sigma_0 = ml_0/a$ and $\sigma_0' = ml_0'/a$.

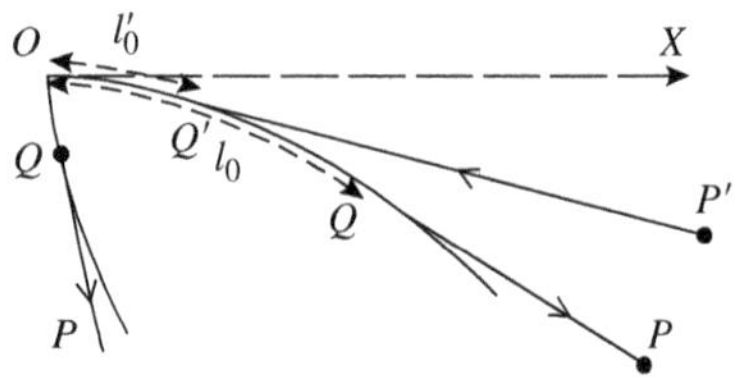

Figure 3.6 Creeping wave diffracted by the edge into another creeping wave propagating on the same face of the wedge or on the other face

If Q' and Q are not on the same face of the wedge, the solution is obtained from (3.19) by replacing

$$D_h(0,0) \to D_h(0, N\pi),$$

$$\left[\frac{\partial^2}{\partial\theta\partial\theta'} D_s(\theta',\theta)\right]_{\theta=\theta'=0} \to \left[\frac{\partial^2}{\partial\theta\partial\theta'} D_s(\theta',\theta)\right]_{\theta=0,\theta'=N\pi}.$$

If in Figure 3.4, P' lies in the deep shadow zone of the reciprocal ray, then $\sigma_0' \gg 1$ and $f(\sigma_0), g(\sigma_0)$ can be replaced by their residue series corresponding to creeping modes. However, the radius of curvature of the geodesic followed by the creeping ray is no longer constant so that in formulae (3.16), σ_0' must be replaced by

$$\sigma_0' = \int_{Q'}^{O} \frac{m(s)\,ds}{\rho(s)}, \qquad m(s) = \left[\frac{k\rho(s)}{2}\right]^{1/3}.$$

In addition, the diffracted field has to be multiplied by the factor $(\rho(s)/a)^{1/6}$.

These factors which permit us to extend the asymptotic solution giving the behaviour of the creeping waves on a circular cylinder at normal incidence to a 2D convex surface with variable curvature, at cylindrical incidence normal to the generatrix, have been rigorously justified in Chapter 1, Section 1.2.5.

A similar modification has to be introduced for σ_0 if both P and P' lie in the deep shadow. If P is in the illuminated part of the penumbra region, then the ray coordinate system for creeping waves can be extended by introducing the pseudo-ray path $P'Q'O$ defined in Figure 3.7.

The illuminated face of the wedge is extended continuously behind the edge by a fictitious surface having the same curvature as the real face at O. The incident ray is tangent at Q' to this surface. The Fock field in the vicinity of O is still given by the spectral decomposition (3.6)–(3.8), but now $l_0' < 0$ and consequently $\sigma_0' < 0$. With this modification, formulae (3.14) remain valid and the edge-diffracted field is continuous when P' crosses the tangent OX to the illuminated face. In order that this solution recovers the GTD result, when P' goes in the illuminated region, one has to choose the phase reference at O on the edge. Since for P' at the periphery of the penumbra region, the radius of curvature at Q' is approximately equal to that at O,

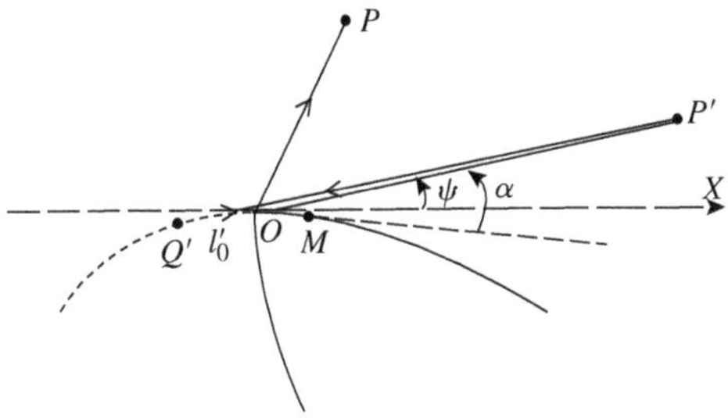

Figure 3.7 Pseudo-ray path

using simple trigonometric relations and the condition $|\ell_0'|/a \ll 1$ we get

$$Q'P' = OP' + a \sin \frac{|l_0'|}{a} \approx OP' + |l_0'| - \frac{|l_0'|^3}{6a^2}.$$

Hence, for $l_0' < 0$, we have

$$u^i(Q')\, e^{ikl_0'} = u^i(O)\, e^{i\sigma_0'^3/3}. \tag{3.20}$$

At a point M close to O on the illuminated face, the phase difference with respect to O is given by $\phi(M) - \phi(O) \cong -kl \cos\alpha, l > 0$, where α is the angle between OP' and the tangent at M to the surface. With a phase reference at O, the arc length l in (3.3) and (3.4) has therefore to be replaced by $l \cos\alpha$. Moreover, if M is at a distance n from the surface, the phase difference $\phi(M) - \phi(O) \cong -k(l \cos\alpha + n \sin\alpha)$. Hence (3.10) has to be modified in the following way:

$$C_\xi^{i,r}(l, n) = \exp[-ik(l \cos(\alpha + \theta) + n \sin(\alpha + \theta))],$$

where θ is defined by (3.9). For $\alpha = 0$, this equation is identical to (3.10), hence the continuity of the field through the shadow boundary of the diffracted rays is still satisfied.

With these modifications, and the fact that α tends to the angle ψ between OP' and OX when M tends to O, the solution valid in the illuminated part of the transition region is given by

$$\begin{aligned} U_s(P) &= \frac{1}{2} u_i(O) \frac{i}{m} f(\sigma_0')\, e^{i\sigma_0'^3/3} \left[\frac{\partial}{\partial\theta} D_s(\phi, \psi + \theta)\right]_{\theta=0} \Pi(r), \\ U_h(P) &= \tfrac{1}{2} u_i(O) g(\sigma_0')\, e^{i\sigma_0'^3/3}\, D_h(\phi, \psi)\, \Pi(r). \end{aligned} \tag{3.21}$$

Moreover, taking into account that

$$\left[\frac{\partial}{\partial\theta} D_s(\phi, \psi + \theta)\right]_{\theta=0} = \frac{\partial}{\partial\psi} D_s(\phi, \psi) \tag{3.22}$$

and noting that for P' located in the transition region, ψ is small ($\psi \cong m^{-1}$), we have

$$\left[\frac{\partial}{\partial\theta} D_s(\phi, \psi + \theta)\right]_{\theta=0} \cong \frac{D_s(\phi, \psi)}{\psi}. \tag{3.23}$$

When the source P' moves into the illuminated region, we have

$$\begin{aligned} &-\sigma_0' \gg 1, \\ &f(x) \cong 2ix \exp\left(-\frac{ix^3}{3}\right), \qquad g(x) \cong 2\exp\left(-\frac{ix^3}{3}\right). \end{aligned} \tag{3.24}$$

Inserting (3.24) into (3.21) and taking into account (3.23) we obtain the GTD result if we replace $\sigma_0' = ml_0'/a$ by

$$\sigma_0' = -m\psi. \tag{3.25}$$

Equation (3.25) is verified in the transition region. If we extend it outside this region we obtain from (3.21) modified by (3.23) the following uniform solution:

$$\begin{aligned} U_s(P) &= \frac{1}{2} u_i(O) \frac{i}{m} f(\sigma_0')\, e^{i\sigma_0'^3/3} \frac{D_s(\phi, \psi)}{\psi} \Pi(r), \\ U_h(P) &= \tfrac{1}{2} u_i(O) g(\sigma_0')\, e^{i\sigma_0'^3/3}\, D_h(\phi, \psi)\Pi(r), \end{aligned} \tag{3.26}$$

in which σ_0' is given by (3.25) inside and outside the transition region.

This solution is also continuous through the shadow boundary of the edge diffracted rays.

3.3.2 *Three-dimensional wedge*

The extension of the asymptotic solution established in Section 3.3.1 to the 3D case involves several aspects. First, the incident wave direction, which in Section 3.3.1 is restricted to the normal to the edge, can now be oblique. As a result, a 3D solution must be explicitly vectorial. Second, the incident wave front can now be of a general astigmatic type. Third, the edge itself may now be curved.

In a first step we consider the case of oblique incidence of a plane wave on a 2D wedge (see Figure 3.8).

We start with the reciprocal situation to that considered for the normal incidence at the beginning of Section 3.3.1. For normal incidence, the asymptotic solution of this problem has been deduced from (3.16) by applying the reciprocity theorem. The result is given by (3.18).

For oblique incidence, the incident electric field $\vec{E}^i$ may be decomposed into a component parallel to the plane of incidence and a component perpendicular to the plane of incidence. Using the ray fixed coordinates introduced by Pathak and Kouyoumjian in the straight wedge problem [11], we write

$$\vec{E}^i = E^i_{\parallel}\hat{\beta}' + E^i_{\perp}\hat{\phi}', \qquad \vec{H}^i = H^i_{\perp}\hat{\beta}' + H^i_{\parallel}\hat{\phi}'. \tag{3.27}$$

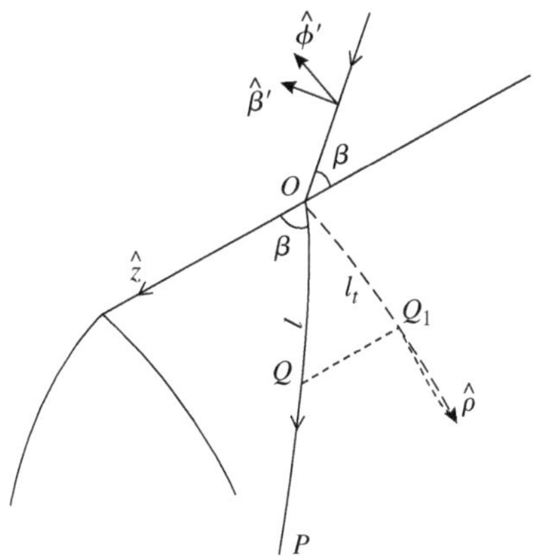

Figure 3.8 Oblique incidence of a plane wave on a 2D wedge

The $\hat{z}$ components of $E^i_{\|}$ and $E^i_{\perp}$ are

$$E^i_{\|z} = (\hat{\beta}' \cdot \vec{E}^i) \sin\beta, \qquad E^i_{\perp z} = 0. \tag{3.28}$$

Similarly

$$H^i_{\|z} = 0, \qquad H^i_{\perp z} = -\sqrt{\frac{\varepsilon}{\mu}} (\hat{\phi}' \cdot \vec{E}^i) \sin\beta. \tag{3.29}$$

The total electric field $\vec{E}$ and the total magnetic field $\vec{H}$ in the region surrounding the 2D wedge may be decomposed into total transverse to $\hat{z}$ fields $(\vec{E}_t, \vec{H}_t)$ and total axial fields (E_z, H_z), respectively. Thus,

$$\vec{E} = \vec{E}_t + \hat{z}\, E_z, \qquad \vec{H} = \vec{H}_t + \hat{z}\, H_z. \tag{3.30}$$

A travelling wave dependence in $\hat{z}$ allows one to deduce E_z and H_z from (3.18) by multiplying both expressions by $e^{ik_z z}$ and by replacing k by $k_t = k \sin\beta$. The result is

$$\begin{aligned} E_z(P) &= E^i_z(O) \frac{1}{2} \frac{i}{m_t} f(\sigma_0)\, e^{ik_t l_t} \left[\frac{\partial}{\partial \theta'_t} D_s(\phi, \theta'_t) \right]_{\theta'_t = 0} \Pi(k_t \rho)\, e^{ik_z z} \\ H_z(P) &= H^i_z(O) \tfrac{1}{2} g(\sigma_0)\, e^{ik_t l_t} D_h(\phi, 0) \Pi(k_t \rho)\, e^{ik_z z} \end{aligned} \tag{3.31}$$

in which

$$\begin{aligned} m_t &= \left(\frac{k_t a}{2} \right)^{1/3}, \qquad \sigma_0 = \frac{m_t l_t}{a}, \qquad \rho = QP \sin\beta, \\ l_t &= l \sin\beta, \qquad \theta'_t = i\sqrt{\xi} m_t^{-1}, \end{aligned} \tag{3.32}$$

where a is the radius of curvature at O of the face on which the creeping wave is launched and l is the geodesic path OQ followed by the creeping ray.

For a 2D structure, if one knows the solutions for the axial fields E_z and H_z then one can evaluate $\vec{E}_t$ and $\vec{H}_t$ in terms of the axial fields by applying the following relations which are a direct consequence of Maxwell's equations for a 2D problem:

$$\begin{aligned}\vec{E}_t &= \frac{1}{k_t^2}[ik_z\nabla_t E_z - i\omega\mu\hat{z}\times\nabla_t H_z],\\ \vec{H}_t &= \frac{1}{k_t^2}[ik_z\nabla_t H_z + i\omega\mu\hat{z}\times\nabla_t E_z].\end{aligned} \tag{3.33}$$

Returning to (3.31) and noting that the ray coordinates ℓ_t and $\ell_t + \rho$ are orthogonal, we have

$$\nabla_t E_z = \hat{\rho}\frac{\partial E_z}{\partial\rho} + \hat{\phi}\frac{a_1}{\rho}\frac{\partial E_z}{\partial l_t}, \qquad \nabla_t H_z = \hat{\rho}\frac{\partial H_z}{\partial\rho} + \hat{\phi}\frac{a_1}{\rho}\frac{\partial H_z}{\partial l_t}, \tag{3.34}$$

where a_1 is the radius of curvature of the projected trajectory at Q_1 and where $\hat{\rho}$ and $\hat{\phi}$ are, respectively, the unit vectors tangential to the projected trajectory and normal to the surface at Q_1 (Figure 3.8). For an observation point situated at large distance from the edge, a_1/ρ is small, hence the second term on the right-hand side of (3.14) can be neglected. We obtain therefore

$$\nabla_t E_z = \hat{\rho}\frac{\partial}{\partial\rho}E_z, \qquad \nabla_t H_z = \hat{\rho}\frac{\partial}{\partial\rho}H_z. \tag{3.35}$$

Since only $\Pi(k_t\rho)$ depends on ρ the dominant term in the derivative stems from the factor $\exp(ik_t\rho)$. Hence

$$\nabla_t E_z = ik_t E_z\hat{\rho}, \qquad \nabla_t H_z = ik_t H_z\hat{\rho}. \tag{3.36}$$

Inserting (3.36) into (3.33), and then (3.33) into (3.30), we get

$$\begin{aligned}\vec{E} &= \frac{1}{\sin\beta}(-\cos\beta E_z\hat{\rho} + Z_c H_z\hat{n}) + \hat{z}E_z,\\ \vec{H} &= \frac{1}{\sin\beta}(-\cos\beta H_z\hat{\rho} - Y_c E_z\hat{n}) + \hat{z}H_z,\end{aligned} \tag{3.37}$$

where we have used $\hat{z}\times\hat{\rho} = \hat{n}$ and where Z_c and Y_c are, respectively, the characteristic impedance and admittance of the free space ($Z_c = Y_c^{-1} = \sqrt{\mu_0/\varepsilon_0}$).

Replacing $\hat{\rho}$ and $\hat{z}$ in (3.37) by

$$\begin{aligned}\hat{\rho} &= \hat{b}\cos\beta + \hat{s}\sin\beta,\\ \hat{z} &= -\hat{b}\sin\beta + \hat{s}\cos\beta,\end{aligned} \tag{3.38}$$

where $\hat{b}$, $\hat{n}$ and $\hat{s}$ are, respectively, the unit vectors of the binormal, the principal normal and the tangent to the geodesic at Q, with $\hat{b} \times \hat{n} = \hat{s}$, we get

$$\vec{E} = \frac{1}{\sin\beta}(-\hat{b}E_z + Z_c\hat{n}H_z),$$
$$\vec{H} = \frac{1}{\sin\beta}(-\hat{b}H_z + Y_c\hat{n}E_z). \tag{3.39}$$

Now, (3.31) may be written in the form:

$$E_z(P) = E_z^i(O)F, \qquad H_z(P) = H_z^i(O)G, \tag{3.40}$$

where

$$F = \frac{1}{2}\frac{i}{m_t} f(\sigma_0)\, e^{ik_t l_t} \left[\frac{\partial}{\partial\theta_t'} D_s(\phi, \theta_t')\right]_{\theta_t'=0} \Pi(k_t\,\rho)\, e^{ik_z z},$$
$$G(P) = \tfrac{1}{2} g(\sigma_0)\, e^{ik_t l_t} D_h(\phi, 0)\Pi(k_t\,\rho)\, e^{ik_z z}. \tag{3.41}$$

and using (3.28) and (3.29) we get

$$E_z(P) = (\hat{\beta}' \cdot \vec{E}^i)\sin\beta F, \qquad H_z(P) = -Y_c(\hat{\phi}' \cdot \vec{E}^i)\sin\beta G. \tag{3.42}$$

Since in the shadow region the total field is identical to the diffracted field, we get therefore from (3.37):

$$\vec{E}(P) = \vec{E}^i(O) \cdot (-\hat{\beta}'\hat{b}F - \hat{\phi}'\hat{n}G),$$
$$\vec{H}(P) = Y_c\vec{E}^i(O) \cdot (-\hat{\phi}'\hat{b}G - \hat{\beta}'\hat{n}F), \tag{3.43}$$

Since

$$\theta_t' = \frac{\theta'}{\sin\beta}, \qquad \frac{\partial}{\partial\theta_t'} = \sin\beta\frac{\partial}{\partial\theta'} \qquad \text{with } \theta' = i\sqrt{\xi}m^{-1}$$

and

$$l_t = l\sin\beta, \qquad k_z = k\cos\beta, \qquad k_t = k\sin\beta, \qquad \rho = s\sin\beta,$$
$$z = (l+s)\cos\beta, \qquad \rho_g = \frac{a}{\sin^2\beta},$$

where ρ_g is the radius of curvature of the geodesic, (3.41) may also be written

$$F = \frac{1}{2}\frac{i}{m} f(\sigma_0)\, e^{ikl} \left[\frac{\partial}{\partial\theta'}\frac{1}{\sin\beta} D_s(\phi, \theta')\right]_{\theta'=0} \Pi(ks),$$
$$G = \frac{1}{2} g(\sigma_0)\, e^{ikl} \frac{1}{\sin\beta} D_h(\phi, 0)\Pi(ks), \tag{3.44}$$

where

$$m = \left(\frac{k\rho_g(O)}{2}\right)^{1/3}, \qquad \sigma_0 = \frac{ml}{\rho_g(O)} \tag{3.45}$$

When the observation point P is in the deep shadow zone, we have to adapt (3.44) in order to take into account the variation of the curvature of the surface, by introducing the factor

$$\left[\frac{\rho_g(O)}{\rho_g(Q)}\right]^{1/6}$$

and by replacing σ_0 by

$$\sigma_0 = \int_0^l \frac{m(\ell)\, d\ell}{\rho_g(\ell)}.$$

For a general astigmatic incident ray field, we first construct the diffracted rays using Fermat's principle and then apply the results of Chapter 1 in order to calculate the divergence factors on the surface and away from it. The general expressions for the divergence factor of the creeping rays and the space rays emitted at Q in the shadow zone remain the same as those presented in Chapter 1, Section 1.1.8 and are, respectively, given by

$$\sqrt{\frac{\rho_g(Q_E)}{\rho_g(Q)}}, \qquad \sqrt{\frac{\rho^d}{s(\rho^d + s)}},$$

where Q_E is the point of diffraction on the edge and ρ^d is the distance to the caustic of the space rays emitted in the shadow region, in the plane tangent to the surface at Q.

With these modifications, the final result for the 3D case may be written for an observation point in the shadow region

$$\vec{E}^d(P) = \vec{E}^i(Q_E) \cdot \left\{ -\hat{\beta}'\hat{b}\frac{1}{2}\left[\frac{\partial}{\partial\theta} d_s(\phi,\theta,\beta)\right]_{\theta=0} S(\sigma_0) - \frac{1}{2}\hat{\phi}\hat{n} d_h(\phi,0,\beta) \right.$$
$$\left. \times H(\sigma_0) \right\} e^{ikl} \sqrt{\frac{d\eta(Q_E)}{d\eta(Q)}} \left[\frac{\rho_g(Q_E)}{\rho_g(Q)}\right]^{1/6} \sqrt{\frac{\rho^d}{s(\rho^d + s)}}\, e^{iks} \tag{3.46}$$

with

$$S(\sigma_0) = \frac{i}{m} f(\sigma_0), \qquad H(\sigma_0) = g(\sigma_0), \qquad \sigma_0 = \int_0^l \frac{m\, d\ell}{\rho_g}. \tag{3.47}$$

Similarly, when the observation point is in the illuminated region, we obtain from (3.26) by using the same procedure

$$\vec{E}^d(P) = \vec{E}^i(Q_E) \cdot \left\{ -\hat{\beta}'\hat{\beta}\frac{d_s(\phi,\phi',\beta)}{2(\hat{n}\cdot\hat{s})} S^\ell(\sigma_0) - \frac{1}{2}\hat{\phi}'\hat{\phi}\, d_h(\phi,\phi',\beta)\, H^\ell(\sigma_0) \right\}$$
$$\times \sqrt{\frac{\rho^e}{s(\rho^e + s)}}\, e^{iks}, \tag{3.48}$$

where

$$S^{\ell}(\sigma_0) = \frac{i}{m} f(\sigma_0)\, e^{i\sigma_0^3/3}, \qquad H^{\ell}(\sigma_0) = g(\sigma_0)\, e^{i\sigma_0^3/3},$$

$$\sigma_0 = -m(Q_E)\cos\theta^i \tag{3.49}$$

with $\cos\theta^i = \hat{n}\cdot\hat{s}$, where $\hat{n}$ and $\hat{s}$ are, respectively, the principal normal to the geodesic followed by creeping rays and the unit vector of the edge diffracted ray, at Q_E. ρ^e is the radius of curvature of the edge diffracted wave in the plane of diffraction. In (3.46) and (3.48)

$$d_{s,h}(\phi, \theta, \beta) = \frac{1}{\sin\beta} d_{s,h}(\phi, \theta),$$

where $d_{s,h}(\phi, \theta)$ is Keller's diffraction coefficient for a wedge at normal incidence, given by (3.14) and ϕ, θ or ϕ' are the angles of the projections of the diffracted and incident ray on the plane orthogonal to the edge at Q_E, with the illuminated face.

These 3D solutions are uniform through the transition region of the edge diffracted rays. They have been presented for the first time by Hill and Pathak [12] in 1987.

3.4 Hybrid diffraction coefficients for a curvature discontinuity

The asymptotic theory of diffraction by a discontinuity in curvature on a smooth convex perfectly conducting surface was initially restricted to situations where both the incident and the diffracted rays were bound to make sufficiently wide angles with the tangent plane at the point of diffraction on the discontinuity [13–15]. More recently, the theory has been extended by Michaeli [16] to cover the case where one of the two rays is nearly tangent to the surface or represents a creeping ray. A further augmentation of the theory has been established by Michaeli and Molinet [17] who extended the solution to the case where both rays are now permitted to be nearly tangent to the surface or represent a creeping ray, provided both of them are situated on the same side of the discontinuity in question. This situation excludes forward scattering, but includes back scattering.

We will follow exactly the same analysis as in Section 3.3, and consider first the 2D geometry depicted in Figure 3.9 with origin O at the point of discontinuity in curvature.

A grazing incident ray at Q' excites a creeping ray field which propagates to O and undergoes diffraction there. To find the resulting diffracted field at P in the lit region of the rays diffracted at O, we can apply the spectral theory of diffraction technique of Sections 3.2 and 3.3, the only modification needed is to replace the diffraction coefficient for an ordinary edge by the one for a second-order edge given by

$$D_{h,s}(\phi, \phi') = \frac{2}{ik}\left(\frac{1}{a_2} - \frac{1}{a_1}\right)\frac{1 + \cos(\phi + \phi') \pm 1 \pm \cos(\phi' - \phi)}{(\cos\phi + \cos\phi')^3}, \tag{3.50}$$

where ϕ' and ϕ are the polar angles for the incident (in the case of illuminated point O) and diffracted rays, respectively, and where a_1 and a_2 are the radii of curvature of the

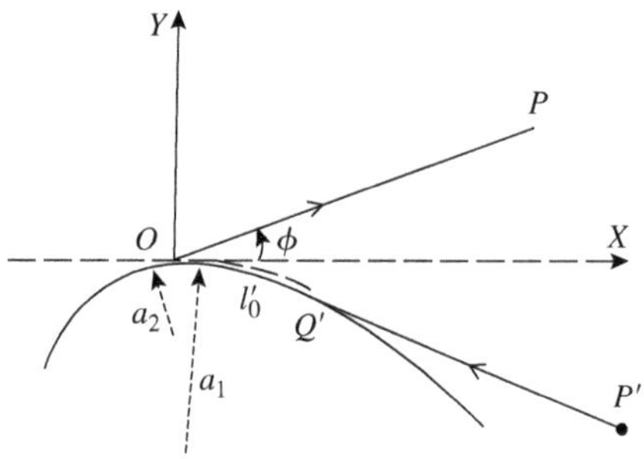

Figure 3.9 Geometry associated with the diffraction by a curvature discontinuity

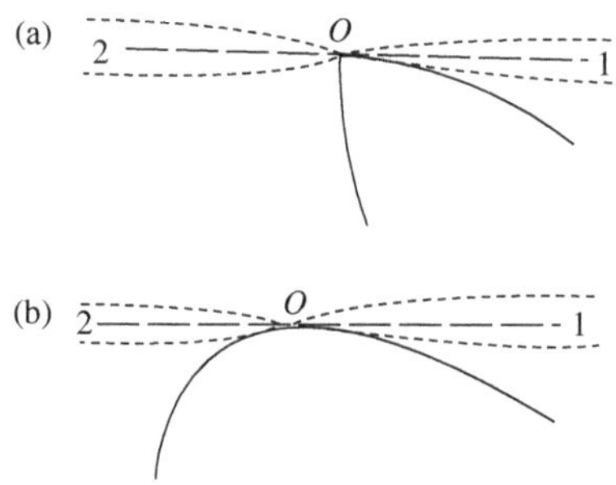

Figure 3.10 Grazing incidence and observation on: (a) a wedge and (b) a surface having a curvature discontinuity

surface on both sides of O. We obtain therefore the same results as in Section 3.3, namely formula (3.16) when the source is located in the shadow region and the observation point is in the lit region and formula (3.19) when both the source and the observation point are located in the shadow region. The only modification consists in substituting into these formulae the expressions (3.50) for the diffraction coefficients.

The extension of these formulae to a 3D geometry follows also the same steps as in Section 3.3.2 and leads to the same results.

3.5 Solution valid at grazing incidence and grazing observation

We now consider the case of forward scattering when the transition regions corresponding to the shadow boundaries of the edge diffracted rays overlap with the transition regions corresponding to the reflected and direct rays. This situation happens when the source is located in region 1 and the observation point is located in region 2 of Figures 3.10(a) and (b). For this case, the diffraction coefficients become very large since they have a singularity when $\phi' = 0$ and $\phi = \pi$. Hence, the spectral diffraction coefficients in the integrals (3.11) are no longer slowly varying as functions of ξ.

In addition, the validity of the tangent plane approximation used in the spectral method is not evident since the reflected field of each spectral component has a caustic close to the edge.

One way to treat the problem is to find the solution of a canonical or model problem in order to start with a general analytical form of the solution (also called an Ansatz) justified by the analytical form of the asymptotic expansion of the model problem solution valid in the domain of the overlapping transition regions.

3.5.1 Two-dimensional perfectly conducting wedge

In 1981, Idemen and Felsen [9] solved a new canonical problem: the diffraction of a whispering gallery mode by a truncated cylindrical shell. Their approach, based on the solution of a Hilbert (or Wiener–Hopf) problem, applies also to a line source illuminating the same structure and exciting creeping waves on the convex face of the cylindrical shell. Several years later, Michaeli [18] obtained an asymptotic expansion of their solution in which the diffracted field at grazing incidence is described by an expression which involves Fock functions and a pair of new transition functions different from the usual Fock functions. These new universal functions are obtained by replacing the Airy function in the integrand of the Fock functions, by an incomplete Airy function.

The analytical form of the asymptotic solution of the model problem being rather complicated, it was not possible to extract directly from this solution an Ansatz with unknown parameters to be determined by imposing the continuity conditions with the solutions established in this chapter, when the source or the observation point leave regions 1 or 2. An intermediate step was necessary. This step consists in recovering the asymptotic form of the solution of the model problem by decomposing the diffraction problem in elementary diffraction processes for which a solution is available. Then, knowing the correspondence of these diffraction processes with the terms of the asymptotic expansion of the solution of the model problem, the same strategy is applied to the original problem.

This procedure has been followed by Michaeli [19] who decomposed the truncated cylindrical shell problem into two parts: (1) the radiation due to the currents on the surface as predicted by physical optics (PO) and an Ufimtsev-type correction [20] (see also References 21 and 22) describing the radiation of the fringe current due to the presence of the edge. Here, the PO currents are the currents which would exist on the same surface if the edge were absent. The radiation of the fringe currents can be calculated by the spectral synthesis technique, since the corresponding spectral diffraction coefficients $D^u_{s,h}(\phi,\theta)$ given by

$$D^u_{s,h}(\phi,\theta) = D_{s,h}(\phi,\theta) - D^{PO}_{s,h}(\phi,\theta), \tag{3.51}$$

where $D_{s,h}(\phi,\theta)$ is related to Keller's diffraction coefficient by (3.13) and where

$$D^{PO}_{s,h}(\phi,\theta) = \pm\frac{2}{\cos\theta+\cos\phi}\begin{Bmatrix}\sin\theta\\ \sin\phi\end{Bmatrix} \tag{3.52}$$

are not singular when $\theta = 0, \phi = \pi$.

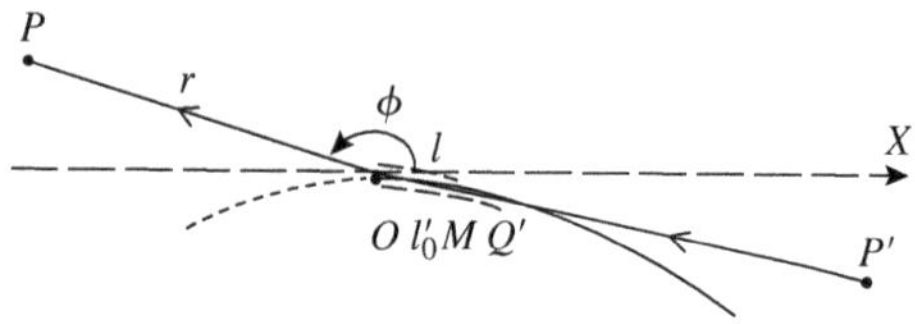

Figure 3.11 *Geometry of the scattering by a truncated sheet*

The radiation of the PO currents is given by

$$u_{s,h}(P) = \mp \frac{i}{4} \int_{0}^{\infty} \Omega_{2,1} H_0^{(1)}(k|\vec{r} - \vec{\rho}|) \Omega_{1,2} u_{s,h}^{\infty}(l, 0)\, dl, \tag{3.53}$$

where $\Omega_1 = \partial/\partial n$, $\Omega_2 = 1$ and $u_{s,h}^{\infty}(M)$ is the field that would exist at a point M of the truncated curved sheet if the edge were absent, and where we have set (Figure 3.11)

$$\overrightarrow{OP} = \vec{r}, \quad l = \text{arc}\, OM \quad \text{and} \quad l_0' = \text{arc}\, OQ'.$$

The Debye approximation of the Hankel function cannot be used to calculate the integral in (3.53) via the stationary phase method. Indeed, there is a caustic effect along the direction of observation due to the fact that, at grazing incidence, the surface currents near the edge radiate in phase towards the point P. Consequently, there is no stationary point in this case as the phase behaviour is essentially constant.

A procedure to circumvent this difficulty consists in rewriting the integral in (3.53) in the form:

$$\int_{0}^{\infty} (\cdot)\, dl = \int_{0}^{-\infty} (\cdot)\, dl + \int_{-\infty}^{+\infty} (\cdot)\, dl, \tag{3.54}$$

where the integration is carried out on the sheet extended beyond the edge through continuity.

The second integral on the right-hand side of (3.54) corresponds to the diffraction by a regular surface for which a uniform asymptotic solution is known (see Pathak [10]). Since we consider the radiation of the currents, one must subtract the incident field from the asymptotic solution.

Since the field on the regular surface obtained by extending the sheet beyond the edge, decreases exponentially, the dominant contribution of the first integral on the right-hand side of (3.54) arises from the neighbourhood of $l = 0$. This integral corresponds to the diffraction effect associated with the truncated surface and thus describes the diffraction by the edge which we denote by u^e.

By using the Debye approximation of the Hankel function

$$\frac{i}{4} H_0^{(1)}(k|\vec{r} - \vec{\rho}|) \approx \Pi(r)\, e^{-ik\hat{r}\cdot\vec{\rho}} \tag{3.55}$$

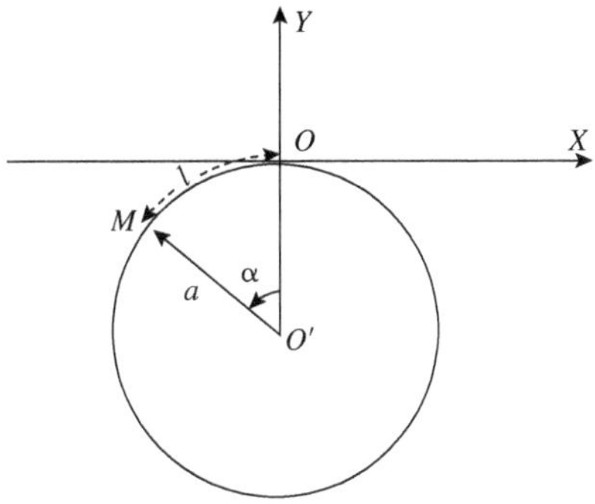

Figure 3.12 Local cylindrical approximation of the surface

and a cylindrical rather than a tangent plane approximation of the surface (see Figure 3.12)

$$\alpha = \frac{l}{a}, \qquad l < 0,$$
$$\vec{\rho} = \overrightarrow{OM} = \hat{x} a \sin\frac{l}{a} - \hat{y} a \left(1 - \cos\frac{l}{a}\right), \tag{3.56}$$
$$\hat{n} = \hat{x}\sin\frac{l}{a} - \hat{y}\cos\frac{l}{a},$$

the first integral in (3.54) can be expressed (without saddle point integration) in the following form:

$$u^e_{s,h}(P) = \Pi(r) \int_0^{-\infty} \begin{Bmatrix} -ik(\hat{r}\cdot\hat{n}) \\ 1 \end{Bmatrix} e^{ik\hat{r}\cdot\vec{\rho}}\, \Omega_{1,2} u^\infty_{s,h}(l,0)\, dl \tag{3.57}$$

with

$$\hat{r}\cdot\hat{n} = \sin\left(\psi - \frac{l}{a}\right), \qquad \hat{r}\cdot\vec{\rho} = a\left[\sin\left(\psi - \frac{l}{a}\right) - \sin\psi\right]. \tag{3.58}$$

Next we set $\eta = \psi - l/a$ and replace $u^\infty_{s,h}$ by its spectral decomposition in the vicinity of O given by (3.6)–(3.8), evaluated at $n = 0$ after having performed the differentiation with respect to the normal coordinate n for the TE polarisation case (case s).

Interchanging the order of integration and using the approximations

$\sin\psi \cong \psi$ in the amplitude terms

$$\left.\begin{aligned} \sin\eta &\cong \eta - \frac{\eta^3}{6} \\ \sin\psi &\cong \psi - \frac{\psi^3}{6} \end{aligned}\right\} \quad \text{in the phase terms}$$

which are valid when ψ is small, we obtain

$$u_h^e(P) = \Pi(r)u^i(Q')e^{ikl_0'}e^{-\sigma_0^3/3}\frac{1}{\sqrt{\pi}}\int_\Gamma \frac{e^{ik(\sigma_0'-\sigma_0)\xi}}{w_1'(\xi)}\,d\xi$$

$$\int_\psi^\infty ik\,a\eta e^{i((m\eta)^3/3)+im\eta\xi}\,d\eta \tag{3.59}$$

with $\sigma_0 = m\psi$ and $\sigma_0' = ml_0'/a$ and a similar expression for $u_s^e(P)$.

Introducing the change of variable $m\eta = s$, the internal integral may be written as

$$2im\int_{\sigma_0}^\infty se^{is^3/3+is\xi}\,ds = 2m\frac{\partial}{\partial\xi}I_1(\xi,\sigma_0), \tag{3.60}$$

where I_1 is an incomplete Airy function given by

$$I_1(\xi,Y) = \int_Y^\infty e^{is^3/3+is\xi}\,ds. \tag{3.61}$$

Finally, following the notations of Michaeli, the field diffracted by the edge can be written as

$$u_{s,h}^e(P) = u^i(Q')\,e^{ikl_0'}S_{s,h}(\sigma_0',\sigma_0)\Pi(r), \tag{3.62}$$

where

$$S_{s,h}(\sigma_0',\,\sigma_0) = 4m\sqrt{\pi}\,e^{-i\sigma_0^3/3}M_{s,h}(\sigma_0'-\sigma_0,\,\sigma_0) \tag{3.63}$$

and the new transition functions are defined by

$$M_h(X,Y) = \frac{1}{2\pi}\int_\Gamma \frac{e^{iX\xi}}{w_1'(\xi)}\frac{\partial}{\partial\xi}I_1(\xi,Y)\,d\xi, \tag{3.64}$$

$$M_s(X,Y) = \frac{1}{2\pi}\int_\Gamma \frac{e^{iX\xi}}{w_1(\xi)}I_1(\xi,Y)\,d\xi. \tag{3.65}$$

The complete asymptotic solution obtained by this procedure is in agreement with the asymptotic solution of the model problem. Its extension to a 2D wedge with curved faces is straightforward and leads to the same formulae, since the wedge angle appears only in (3.51) through Keller's diffraction coefficient for the wedge.

Another approach which has some similarities with the preceding one had been published in the form of a PhD thesis one year earlier by Liang [23], Pathak being his supervisor. This approach relies upon an Ansatz derived directly from the uniform asymptotic solution of the curved screen which was originally developed by Chuang and Liang [24].

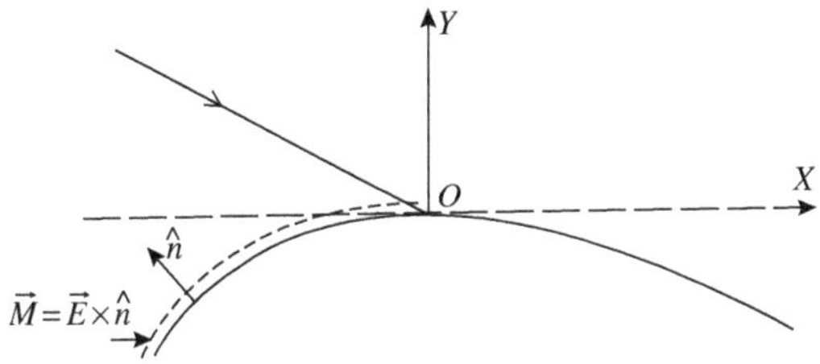

Figure 3.13 *Equivalence principle applied to the curved screen*

In Reference 24, the authors treat the reciprocal situation of a plane wave originating from the region $X < O$ (Figure 3.11). In this region, the total field close to the edge can be approximated by the Sommerfeld solution corresponding to the diffraction by a half-plane. Hence, if we extend the screen beyond the edge through continuity, the total field in this fictitious surface is known. Applying the equivalence principle, the fictitious surface can then be replaced by a perfectly conducting surface which supports an equivalent magnetic current $\vec{M} = \vec{E} \times \hat{n}$ where $\hat{n}$ is the normal directed outward from the surface (Figure 3.13) and $\vec{E}$ is the total electric field ($\vec{E} = \vec{E}^i + \vec{E}^d$), which is given by the Sommerfeld solution and is identical to the UTD solution for the half-plane. It is important to note that the expression of the field remains exact even when the observation point comes close to the edge of the half-plane.

The problem can now be divided into two parts: (i) the diffraction of a plane wave by a regular, perfectly conducting surface, and (ii) the radiation by a sheet of magnetic current, also on a regular perfectly conducting surface. The solution to both these problems is known.

The solution to problem (ii) involves Fock functions and phase terms which depend upon the position of the dipole and consequently they appear in the integrals over the sheet of magnetic current. The authors in Reference 24 have decomposed these integrals into different terms that involve new transition functions. To derive the final result, it is necessary to extract the contribution of the direct incident field from the magnetic current $\vec{M}$ by writing $\vec{M} = \vec{M}^i + \vec{M}^d$ where $\vec{M}^i = \vec{E}^i \times \hat{n}$ and $\vec{M}^d = \vec{E}^d \times \hat{n}$. Indeed, the current $\vec{M}^d$ decreases rapidly when the observation point recedes from the origin O. The asymptotic expansion of the corresponding integral then reduces to a contribution arising from the end-point of the path of integration. On the other hand, the current $\vec{M}^i$ does not decrease with an increase of the observation distance, and in general, the integral does not reduce to a contribution from the end-point. Its contribution is grouped with certain terms of the solution of problem (i).

More precisely, when the point of reflection or the launching point of the creeping ray is on the real surface, the integral over the current sheet $\vec{M}^i$ has only an end-point contribution and its evaluation for both TM and TE polarisations yields a Fresnel integral and two new transition functions which reduce to Michaeli's transition functions when $m|\phi| \leq 1$ and $m|\pi - \phi'| \leq 1$. The contribution of the current sheet $\vec{M}^d$

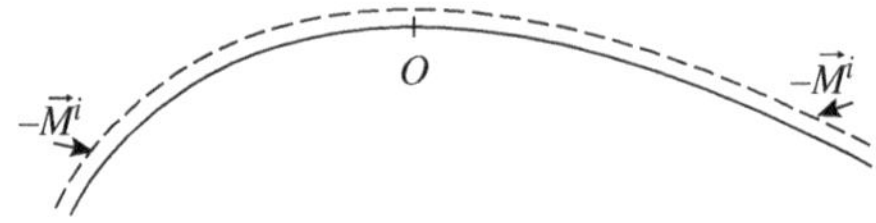

Figure 3.14 Application of the induction theorem to problem (i)

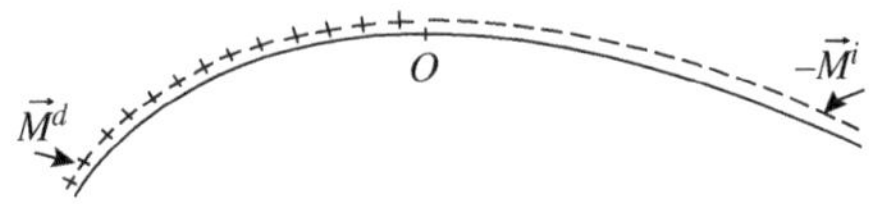

Figure 3.15 Replacement of the scattering problem of a curved sheet by a radiation problem

corresponds to Ufimtsev's correction in Michaeli's solution. When the direction of observation is above the shadow boundary of the reflected field or if the incident direction is below the tangent to the sheet at O, the reflection point lies on the fictitious surface, hence the integral over the current sheet $\vec{M}^i$ has a stationary point contribution in addition to the end-point contribution. In this case, the solution of problem (i) is calculated by applying the induction theorem which permits its replacement by the radiation problem of a magnetic current sheet $-\vec{M}^i$ as shown in Figure 3.14.

By adding this current sheet to the current sheet $\vec{M}^i$ of problem (ii), we see that the original problem is replaced by that of the radiation of two sheets of magnetic currents: $\vec{M}^d$ on the fictitious surface and $-\vec{M}^i$ on the real surface, as shown in Figure 3.15.

As in the preceding case, the field radiated by the sheet of current $-\vec{M}^i$ can be expressed in terms of new transition functions. Although the latter have a similar nature, they are not identical to those corresponding to the preceding case.

In fact, in Michaeli's procedure, they correspond to the sum of a surface diffraction contribution in the lit shadow transition region, on the fictitious surface which involves Fresnel and Pekeris functions and an end-point contribution which involves Michaeli's transition functions. The surface diffraction term on the fictitious surface in Michaeli's procedure comes from the second integral in (3.54) and must be added to the edge diffracted field (Figures 3.16(a) and (b)).

The asymptotic solution for the curved screen, in which the Keller diffraction coefficients due to the radiation of the current sheet $\vec{M}^d$ are adapted to a wedge with an exterior angle $n\pi$, is used in Liang's method for stating a general analytic form (or Ansatz) of the solution for the 2D curved wedge which recovers the solution for the curved screen for $n = 2$. Liang has verified that the solution constructed in this way is continuous and tends to the UTD solution when the incident and reflected shadow boundaries are far away from the faces.

Michaeli's solution is identical to Liang's solution in the zone where the transition regions overlap. However, since this solution has been constructed by using local plane or cylindrical approximations close to the edge, it does not tend continuously to the hybrid solutions described in Section 3.3 when the observation point moves

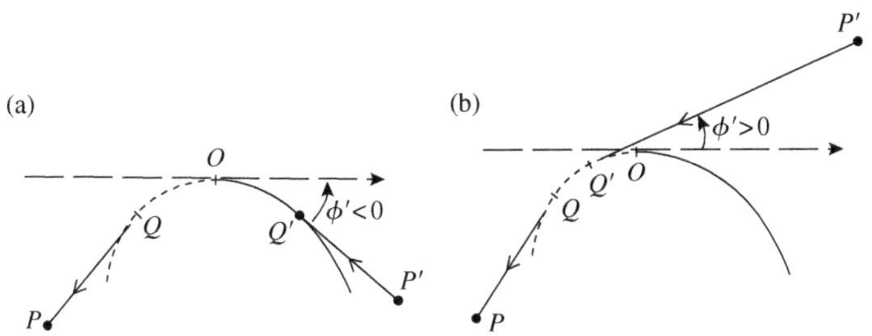

Figure 3.16 Pseudo-creeping ray P′Q′QP: (a) with the edge in the shadow region and (b) with the edge in the lit region

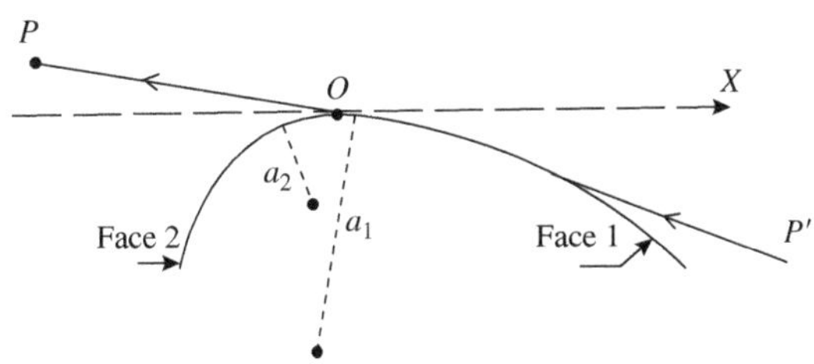

Figure 3.17 Geometry of the scattering by a curvature discontinuity

outside this zone. However, by means of a modification of the arguments of the transition functions and the introduction of a multiplication coefficient which tends to one on the transition surface, Michaeli was later able to construct a uniform solution (see Section 5.6.3 of Reference 25).

3.5.2 *Three-dimensional wedge*

A totally uniform solution valid at grazing incidence, for the diffraction of a plane wave by a perfectly conducting 3D wedge has been derived by Molinet [26] from Michaeli's uniform solution for a 2D curved wedge. The technique used is similar to that presented in Section 5.6.3 of Reference 25 and will not be detailed here.

3.5.3 *Curvature discontinuity*

Michaeli's procedure in Section 3.5.1 can be extended to a curvature discontinuity between two convex smooth surfaces. We consider again a source P' located below the tangent to the surface at O and an observation point located above that tangent (Figure 3.17).

At a point M on face 1, the total field for both polarisations s or h, may be written in the form

$$u_1(M) = u_1^{\infty}(M) + u_1^{f}(M), \tag{3.66}$$

where $u_1^{\infty}(M)$ is the PO field and $u_1^{f}(M)$ is the fringe current field.

The difference with the wedge problem is that the field on face 2 is not zero since in addition to the fringe current field, there also exists a creeping wave transmitted from face 1 to face 2 so that at a point M on face 2, the field may be written as

$$u_2(M) = u_2^{\infty}(M) + u_2^f(M), \tag{3.67}$$

where $u_2^{\infty}(M)$ is the field of the transmitted creeping wave and $u_2^f(M)$ is the fringe current field on face 2.

The total field radiated at P is given by the radiation integral (3.53) extended to both faces. Again, the fringe current contribution may be calculated by the spectral synthesis technique.

The field radiated by the PO field on face 1 can be calculated by extending face 1 beyond the curvature discontinuity, through continuity of the second-order derivatives and applying (3.54), which leads to an edge diffracted field given by an expression identical to (3.62).

The integral on face 2 of the surface field u_2^{∞} also gives an end-point effect similar to the first integral in (3.54) which also has to be attributed to the diffraction by the second-order edge.

In order to evaluate $u_2^{\infty}(M)$ and its normal derivative we use the spectral decomposition of $u_1^{\infty}(M)$ on face 1, at an observation point M close to O which is given by formulae (3.6)–(3.8). Applying the tangent plane approximation of the surface in the vicinity of O to this field we can write the transmitted spectrum in a form similar to (3.6), but with the amplitudes calculated on face 1 and the phase calculated on face 2

$$\begin{aligned} C_{\xi}^{i,r}(l,n) &= \exp\left[-ikl\left(1+\frac{\xi}{2m_2^2}\right) \mp ikn\frac{i\sqrt{\xi}}{m_2}\right], \\ A_s(\xi) &= u^i(Q')\frac{\exp[il_0'(k+\xi(m_1/a_1))]}{2\sqrt{\pi\xi}\,w_1'(\xi)}, \\ A_h(\xi) &= u^i(Q')\frac{\exp[il_0'(k+\xi(m_1/a_1))]}{2\sqrt{\pi}\,w_1'(\xi)}, \end{aligned} \tag{3.68}$$

where a_1 and a_2 are the radii of curvature at O of face 1 and face 2, respectively, and where

$$m_1 = \left(\frac{ka_1}{2}\right)^{1/3}, \qquad m_2 = \left(\frac{ka_2}{2}\right)^{1/3}.$$

Inserting (3.68) into the radiation integral of face 2 and using the Debye approximation for the Hankel function and a cylindrical approximation for face 2 close to O, we get by following the same steps as in (3.55)–(3.61) the following result:

$$u_{2,s,h}^e(P) = -u^i(Q')e^{ikl_0'}S_{s,h}(\sigma_{01}', \sigma_{02})\Pi(r) \tag{3.69}$$

with

$$S_{s,h}(\sigma_{01}', \sigma_{02}) = -4m_2\sqrt{\pi}e^{i\sigma_{02}^3/3}M_{s,h}(\sigma_{01}' - \sigma_{02}, \sigma_{02}), \tag{3.70}$$

where $M_{s,h}(X, Y)$ are Michaeli's transition functions given by (3.64) and (3.65), and where $r = OP$.

By adding to (3.69) the contribution from face 1, we finally obtain

$$u^e_{s,h}(P) = +u^i(Q')e^{ikl'_0}4\sqrt{\pi}[m_1 e^{-i\sigma^3_{01}/3} M_{s,h}(\sigma'_{01} - \sigma_{01}, \sigma_{01}) - m_2 e^{-i\sigma^3_{02}/3} M_{s,h}(\sigma'_{01} - \sigma_{02}, \sigma_{02})]\Pi(r). \tag{3.71}$$

The total diffracted field by the second-order edge, at grazing incidence, in the configuration shown in Figure 3.17 is given by

$$u^{te}_{s,h}(P) = u^e_{s,h}(p) + u^u_{s,h}(P), \tag{3.72}$$

where $u^u_{s,h}(P)$ is Ufimtsev's correction term calculated by the spectral synthesis technique.

3.6 Coated surfaces

The techniques presented in Sections 3.2–3.5 and the asymptotic solutions obtained by these techniques can be extended to a 2D curved wedge with coated faces or to a 2D curvature discontinuity on a coated surface.

3.6.1 Spectral representation of the Fock field on a smooth coated surface

We consider a 2D smooth convex surface coated with a thin layer of dielectric or absorbing material defined by its complex permittivity ε_1 and permeability μ_1. We suppose that the thickness of the coating is small compared to the wavelength λ_1 in the material. Under these conditions the total field along the exterior surface of the coating, excited by an electric or magnetic line source parallel to the generatrices of the surface, is given by (see Chapter 1, Equation 1.81)

$$u(\sigma, \nu) = u^i(Q')\frac{e^{iks}}{\sqrt{\pi}} \int_{-\infty}^{+\infty} \left[v(\xi - \nu) - \frac{v'(\xi) - im\zeta(\xi)v(\xi)}{w'_1(\xi) - im\zeta(\xi)w_1(\xi)} \times w_1(\xi - \nu) \right] e^{i\sigma\xi}\, d\xi, \tag{3.73}$$

where the notations are those defined in Figure 3.3 and where $u(\sigma, \nu) = E_z$ and $\zeta(\xi) = \zeta_E(\xi)$ for an electric line source and $u(\sigma, \nu) = H_z$ and $\zeta(\xi) = \zeta_H(\xi)$ for a magnetic line source.

The functions $\zeta_{E,H}(\xi)$ are related to the modal admittance (or impedance) $\zeta_{E,H}(p)$ by the transformation:

$$p = k\rho + m\xi, \tag{3.74}$$

where p is the complex modal index resulting from Watson's transformation of the modal solution for the coated circular cylinder and

$$\zeta_H(p) = i\frac{k_1}{k_0}\frac{\varepsilon_0}{\varepsilon_1}\frac{J'_p(k_1a)H_p^{(1)'}(k_1b) - J'_p(k_1b)H_p^{(1)'}(k_1a)}{J_p(k_1a)H_p^{(1)'}(k_1b) - J'_p(k_1b)H_p^{(1)}(k_1a)},$$
$$\zeta_E(p) = i\frac{k_1}{k_0}\frac{\mu_0}{\mu_1}\frac{J'_p(k_1a)H_p^{(1)}(k_1b) - J_p(k_1b)H_p^{(1)'}(k_1a)}{J_p(k_1a)H_p^{(1)}(k_1b) - J_p(k_1b)H_p^{(1)}(k_1a)}. \tag{3.75}$$

In (3.75), $a = a(s)$ and $b = b(s)$ are, respectively, the local radii of curvature of the exterior surface of the coating and of the interior perfectly conducting surface and ε_0, μ_0 are, respectively, the permittivity and the permeability of vacuum. The other quantities are

$$k_0 = \omega\sqrt{\varepsilon_0\mu_0}, \quad k_1 = \omega\sqrt{\varepsilon_1\mu_1}, \qquad \omega = 2\pi f,$$

where f is the frequency.

When $|p - k_1a| > |p|^{1/3}$ it is possible to replace the Bessel and Hankel functions in (3.75) by their Debye asymptotic expansions (see Reference 27, p. 710) and reduce these expressions to the following simpler formulae:

$$\zeta_H(p) = i\frac{k_1}{k}\frac{\varepsilon_0}{\varepsilon_1}\psi(x_1, p)\tan[k_1 d\psi(x_1, p)],$$
$$\zeta_E(p) = i\frac{k_1}{k}\frac{\mu_0}{\mu_1}\psi(x_1, p)\cot[k_1 d\psi(x_1, p)], \tag{3.76}$$

where

$$\psi(x_1, p) = \left(1 - \frac{p^2}{x_1^2}\right)^{1/2}, \quad x_1 = k_1a, \qquad d = a - b.$$

When the radius of curvature becomes very large compared to the wavelength, the function ψ tends to unity and formulae (3.76) tend to the impedance and admittance of a coated perfectly conducting planar surface.

Proceeding as in Section 3.2 and expanding $v(\xi - \nu)$ and $w_1(\xi - \nu)$ in Taylor series in the parameter ν, we obtain from (3.73)

$$u(l, \nu) = u^i(Q')\int_{-\infty}^{+\infty} e^{i(l'_0 - l)(k + \xi(m/a))}\frac{[1 + \nu^2(\xi/2) - im\zeta(\xi)(\nu + \nu^2\xi/6)]}{\sqrt{\pi}[w'_1(\xi) - im\zeta(\xi)w_1(\xi)]}\,d\xi, \tag{3.77}$$

where s has been replaced by $l'_0 - l$ (see Figure 3.3) and where a is the radius of curvature of the exterior surface of the coating at the observation point M which we suppose close to the origin O so that $a(s) \simeq a(O)$.

Finally, by using the same approximations as in Section 3.2

$$u(l,n) = \int_{-\infty}^{+\infty} \{A(\xi)[C_\xi^i(l,n) + C_\xi^r(l,n)] + B(\xi)[C_\xi^i(l,n) + C_\xi^r(l,n)]\}\, d\xi, \tag{3.78}$$

where $C_\xi^{i,r}(l,n)$ are given by (3.7) and where

$$\begin{aligned} A(\xi) &= u^i(Q')\frac{\exp[il_0'(k+\xi(m/a))]}{2\sqrt{\pi}[w_1'(\xi) - im\zeta(\xi)w_1(\xi)]}, \\ B(\xi) &= u^i(Q')\frac{-im\zeta(\xi)\exp[il_0'(k+\xi(m/a))]}{2\sqrt{\pi}\xi[w_1'(\xi) - im\zeta(\xi)w_1(\xi)]}. \end{aligned} \tag{3.79}$$

In (3.79) a stands for $a(O)$.

Again, by using (3.9) and (3.10) and the tangent plane approximation we see that (3.78) is a spectrum of inhomogeneous plane waves. The amplitude of the incident wave of the spectrum is given by the coefficient of C_ξ^i which we write in the form

$$A(\xi) + B(\xi) = \frac{u^i(Q')\exp[il_0'(k+\xi(m/a))]}{2\sqrt{\pi}[w_1'(\xi) - im\zeta(\xi)w_1(\xi)]}\frac{2}{1+R(\theta)}, \tag{3.80}$$

where θ is defined by (3.9) and where $R(\theta)$ is the reflection coefficient given by

$$R(\theta) = \frac{\sin\theta - \zeta}{\sin\theta + \zeta}. \tag{3.81}$$

3.6.2 Hybrid diffraction coefficients for a coated 2D wedge

By applying the spectral synthesis technique, we can directly write down the formulae giving the hybrid diffraction coefficients for a 2D coated wedge which generalise formulae (3.16) and (3.21) of Section 3.3. The result is

$$u(P) = \frac{1}{2}u^i(Q')e^{ikl_0'}\left(\frac{1}{\sqrt{\pi}}\int_{-\infty}^{+\infty} e^{i\sigma_0'\xi}\frac{D(\phi,\theta;\zeta)}{w_1'(\xi) - im\zeta(\xi)w_1(\xi)} \times\frac{2}{1+R(\xi)}\,d\xi\right)\Pi(r), \tag{3.82}$$

where $D(\phi,\theta;\zeta)$ is Maliuzhinets diffraction coefficient for an impedance wedge and where $u(P) = E_z(P), \zeta = \zeta_E$ for an electric line source and $u(P) = H_z(P), \zeta = \zeta_H$ for a magnetic line source.

Since

$$\left[\frac{\sin\theta + \zeta(\theta)}{\sin\theta}\right] D(\phi,\theta;\zeta) \cong \zeta(\theta)\left[\frac{\partial}{\partial\theta}D(\phi,\theta;\zeta)\right]_{\theta=0}, \tag{3.83}$$

Equation (3.82) may also be written in the form:

$$
\begin{aligned}
E_z(P) &= \frac{1}{2}E_z^i(Q')\frac{i}{m} f_E(\sigma_0')e^{ikl_0'}\left[\frac{\partial}{\partial\theta}D(\phi,\theta;\zeta_E)\right]_{\theta=0}\Pi(r),\\
H_z(P) &= \frac{1}{2}H_z^i(Q')\frac{i}{m} f_H(\sigma_0')e^{ikl_0'}\left[\frac{\partial}{\partial\theta}D(\phi,\theta;\zeta_H)\right]_{\theta=0}\Pi(r),
\end{aligned}
\tag{3.84}
$$

where

$$
f_{E,H}(X) = \frac{1}{\sqrt{\pi}}\int_\Gamma e^{iX\xi}\frac{-im\zeta_{E,H}(\xi)}{w_1'(\xi)-im\zeta_{E,H}(\xi)w_1(\xi)}\,d\xi. \tag{3.85}
$$

Formulae (3.84) and (3.85) are valid for ζ_H different from zero and not too small. When ζ_H is small we first take the limit $\zeta_H \to 0$ and then the limit $\theta \to 0$ in (3.83). Then we get

$$
\left[\frac{\sin\theta+\zeta_H(\theta)}{\sin\theta}\right]D(\phi,\theta;\zeta_H) \cong D_h(\phi,\theta), \tag{3.86}
$$

where $D_h(\phi,\theta)$ is Keller's diffraction coefficient for the hard boundary condition divided by the source factor S defined by Equation (3.13).

When $\zeta_E \to \infty$, (3.83) remains valid and

$$
\left[\frac{\partial}{\partial\,\theta}D(\phi,\theta;\zeta_E)\right]_{\theta=0} \to \frac{\partial}{\partial\theta}D_s(\phi,\theta). \tag{3.87}
$$

Hence, for $\zeta_H < 1/m$ and $\zeta_E \gg 1/m$, (3.84) takes the form

$$
\begin{aligned}
E_z(P) &= \frac{1}{2}E_z^i(Q')\frac{i}{m} f_E(\sigma_0')\,e^{ikl_0'}\left[\frac{\partial}{\partial\theta}D_s(\phi,\theta)\right]_{\theta=0}\Pi(r),\\
H_z(P) &= \frac{1}{2}H_z^i(Q')\frac{i}{m}\tilde{g}_H(\sigma_0')\,e^{ikl_0'}\,D_h(\phi,0)\Pi(r),
\end{aligned}
\tag{3.88}
$$

where

$$
\tilde{g}_H(X) = \frac{1}{\sqrt{\pi}}\int_\Gamma \frac{e^{iX\xi}}{w_1'(\xi)-im\zeta_H\,w_1(\xi)}\,d\xi. \tag{3.89}
$$

The generalisations of (3.21) follow the same steps and lead to the following solution, valid in the illuminated part of the transition region:

$$
\begin{aligned}
E_z(P) &= \frac{1}{2}E_z^i(Q')\frac{i}{m} f_E(\sigma_0')\,e^{i\sigma_0'^3/3}\left[\frac{\partial}{\partial\theta}D(\phi,\psi+\theta;\zeta_E)\right]_{\theta=0}\Pi(r),\\
H_z(P) &= \frac{1}{2}H_z^i(Q')\frac{i}{m} f_H(\sigma_0')e^{i\sigma_0'^3/3}\left[\frac{\partial}{\partial\theta}D\,(\phi,\psi+\theta;\zeta_H)\right]_{\theta=0}\Pi(r),
\end{aligned}
\tag{3.90}
$$

where $\sigma_0' = ml_0'/a$ with $l_0' < 0$.

When ζ_H is small, (3.90) may be simplified by using (3.86) and (3.87).

3.6.3 Grazing incidence and observation on a coated 2D wedge

Michaeli's solution presented in Section 3.5.1 can be extended to a coated wedge. The radiation integral (3.53) of the PO current is now replaced by

$$u(P) = \frac{1}{4}\int_0^\infty \Big[\Omega_1 H_0^{(1)}(k|\vec{r}-\vec{\rho}|)\Omega_2 u^\infty(\ell,0) - \Omega_2 H_0^{(1)}(k|\vec{r}-\vec{\rho}|)\Omega_1 u^\infty(\ell,0)\Big]\, d\ell, \tag{3.91}$$

where $\Omega_1 = \partial/\partial n$, $\Omega_2 = 1$ and where $u(P) = E_z(P)$, $u^\infty(\ell,0) = E_z(\ell,0)$ for an electric line source and $u(P) = H_z(P), u^\infty(\ell,0) = H_z(\ell,0)$ for a magnetic line source. By using the spectral representation of the fields $u^\infty(\ell,\nu)$ given by (3.77) we can calculate the normal derivatives of E_z and H_z on the surface. Inserting the result into (3.91) and proceeding as in Section 3.5.1, we obtain the following expression for the diffracted field $u^e(P)$ where $u^e = E_z$ or H_z:

$$u^e(P) = u^i(Q')\, e^{ikl'_0}\tilde{S}(\sigma'_0,\sigma_0)\Pi(r) \tag{3.92}$$

with

$$\tilde{S}(\sigma'_0,\sigma_0) = 4m\sqrt{\pi}\, e^{i\sigma_0^3/3}\tilde{M}(\sigma'_0-\sigma_0,\sigma_0) \tag{3.93}$$

and

$$\tilde{M}(X,Y) = \frac{1}{2\pi}\int_\Gamma e^{iX\xi}\frac{(\partial/\partial\xi)I_1(\xi,Y) - im\zeta(\xi)I_1(\xi,Y)}{w'_1(\xi) - im\zeta(\xi)w_1(\xi)}\, d\xi, \tag{3.94}$$

where the incomplete Airy function is defined by (3.61). We see that formulae for the edge diffracted field remain the same, only the transition functions $M(X,Y)$ are changed in $\tilde{M}(X,Y)$. In order to get the total edge diffracted field, we have to add to (3.92) Ufimtsev's correction calculated with the hybrid diffraction coefficients (3.84) and (3.90) for a coated wedge in which Maliuzhinets diffraction coefficient has to be replaced by Ufimtsev's diffraction coefficient, defined as in (3.51) by the difference between Maliuzhinets coefficient and the PO coefficient for an edge in a coated surface.

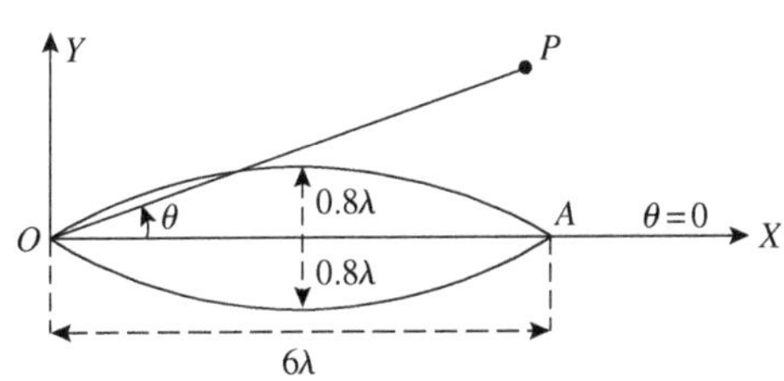

Figure 3.18 Geometrical parameters of the cross-section of an ogival cylinder

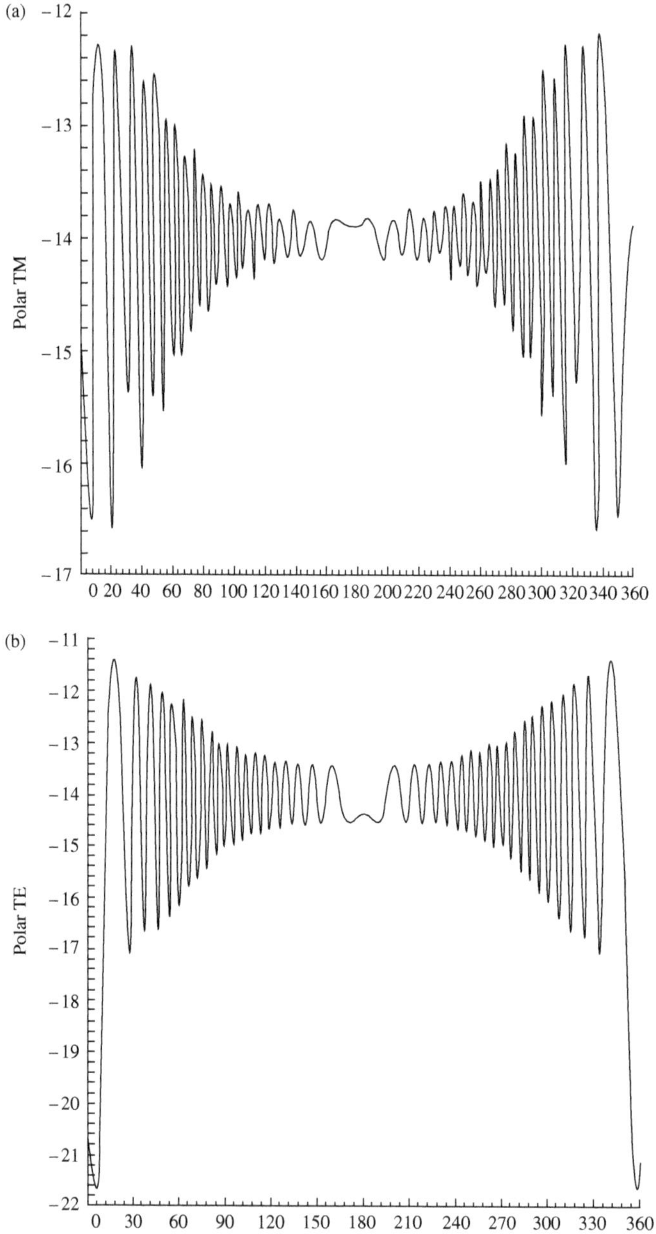

Figure 3.19 *Radiation pattern of a line source radiating in the presence of an ogival cylinder. Source location* $S_x = -10\lambda, S_y = 0\lambda$*, observation angle* θ *variable from 0 to 360°: ———, GTD results; – – – –, MM results. The dashed curve is exactly superposed to the solid curve and is therefore not visible on the diagram*

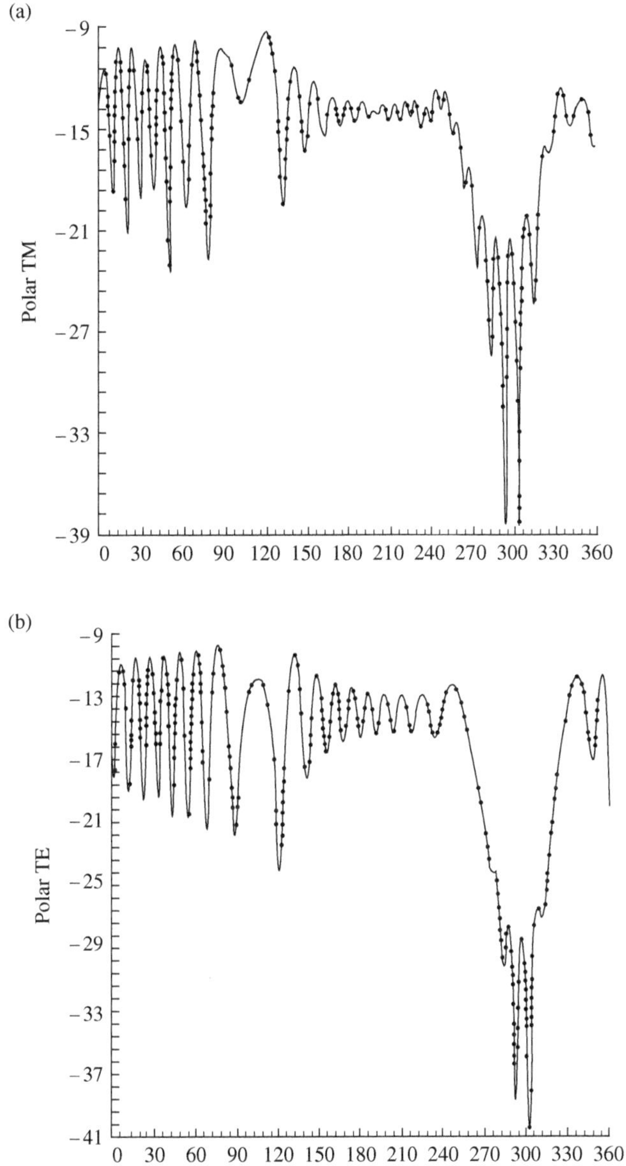

Figure 3.20 *Radiation pattern of a line source radiating in the presence of an ogival cylinder. Source location $S_x = 0\lambda$, $S_y = 5\lambda$, observation angle θ variable from 0 to 360°: ———, GTD results; ...... , MM results*

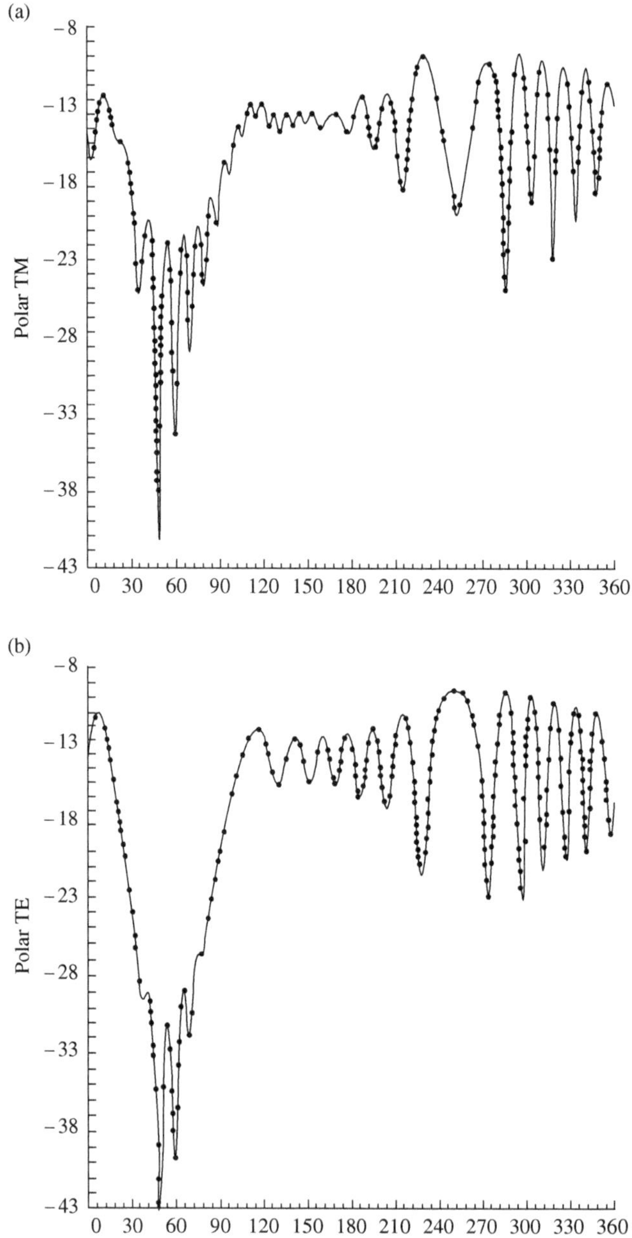

Figure 3.21 Radiation pattern of a line source radiating in the presence of an ogival cylinder. Source location $S_x = 0\lambda$, $S_y = -3.5\lambda$, observation angle θ variable from 0 to 360°: ________, GTD results; ......, MM results

3.7 Numerical results

Some numerical results corresponding to a line source radiating in the presence of a perfectly conducting ogival cylinder are shown. The geometrical parameters are defined in Figure 3.18 in units of wavelength λ.

The location of the source S is defined by its coordinates S_x and S_y.

For a source located on the negative Ox axis at $S_x = -10\lambda$, $S_y = 0\lambda$, the amplitude of the total radiated field in the presence of the ogival cylinder expressed in decibels (dB), is given in Figure 3.19 for both directions of polarisation TM and TE as a function

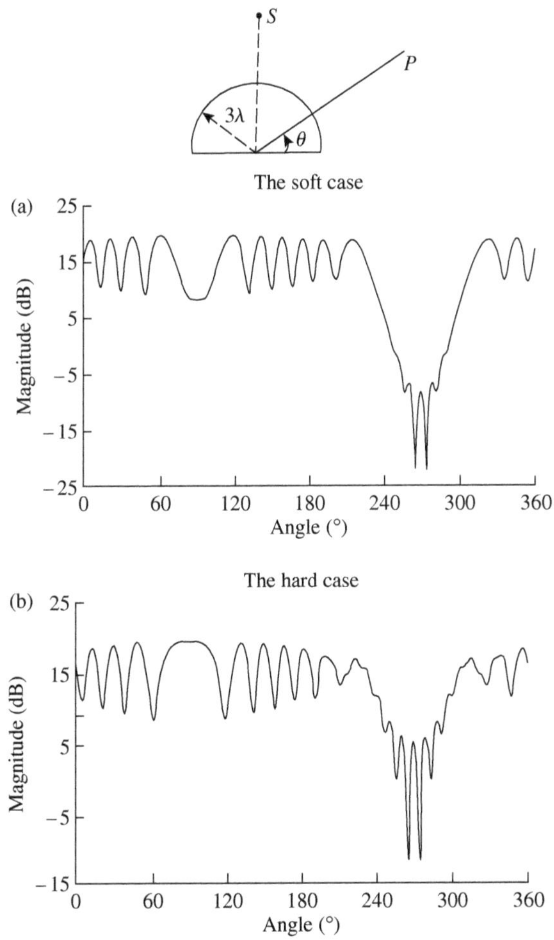

Figure 3.22 Radiation pattern of a line source radiating in the presence of a half-cylinder. Source location $\rho' = 6\lambda, \phi' = 90°$, observation angle θ variable from 0 to 360°: ———, MM results; – – – –, GTD results

of the observation angle θ. The GTD results are superposed to the results obtained by the method of moments (MM). We see that the agreement is excellent.

The curves in Figure 3.20 show the variation of the amplitude of the total radiated field (in dB) for a source located on the positive Oy axis at $S_x = 0\lambda$, $S_y = 5\lambda$. Again, the GTD results are in excellent agreement with the MM results.

Finally, for a source on the negative Oy axis at $S_x = 0\lambda$, $S_y = -3.5\lambda$, we obtain the curves shown in Figure 3.21. In this case, the observation direction $\theta = 30°$ corresponds to grazing incidence on the curved wedge at A (see Figure 3.18). We see that the GTD results are also in excellent agreement with the MM results in the zone of overlapping transition regions.

In Figure 3.22, we also show results obtained by Liang *et al.* [28] for the total field of a line source radiating in the presence of a half-cylinder with radius $a = 3\lambda$. Perfect agreement between the GTD results and the MM results is evident.

3.8 References

1 KOUYOUMJIAN, R. G., and PATHAK, P. H.: ‘A geometrical theory of diffraction for an edge in a perfectly conducting surface’, *Proceedings of the IEEE*, 1974, **62**, pp. 1448–61

2 PATHAK, P. H., and KOUYOUMJIAN, R. G.: ‘On the diffraction of edge excited surface rays’. Paper presented at the *USNC/URSI Meeting*, Standford University, Standford, CA, June 22–24, 1977

3 ALBERTSEN, N. C.: ‘Diffraction of creeping waves’. Report LD24, Electromagnetic Institute, Technical University of Denmark, 1974

4 MOLINET, F.: ‘Diffraction d’une onde rampante par une ligne de discontinuité du plan tangent’, *Annales des Télécommunications*, 1977, **32** (5–6), pp. 197–8

5 ALBERTSEN, N. C., and CHRISTIANSEN, P. L.: ‘Hybrid diffraction coefficients for first and second order discontinuities of two-dimensional scatterers’, *SIAM Journal of Applied Mathematics*, 1978, **34**, pp. 398–414

6 FONTANA, T. P., and DESCHAMPS, G. A.: ‘Uniform solution for the field diffracted by a curvature discontinuity in a cylinder’. EM Tech. Report, No. 77-14, University of Illinois, June 1977

7 MICHAELI, A.: ‘Transition functions for high-frequency diffraction by a curved perfectly conducting wedge, Part II: a partially uniform solution for a general wedge angle’, *IEEE Transactions on Antennas and Propagation*, 1989, **37**, pp. 1080–5

8 RHAMAT-SAMII, Y., and MITTRA, R.: ‘A spectral domain interpretation of high-frequency diffraction phenomena’, *IEEE Transactions on Antennas and Propagation*, 1977, **AP-25**, pp. 676–87

9 IDEMEN, M., and FELSEN, L. B.: ‘Diffraction of a whispering gallery mode by the edge of a thin cylindrically curved surface’, *IEEE Transactions on Antennas and Propagation*, 1981, **AP-29** (4), pp. 571–9

10 PATHAK, P. H.: ‘An asymptotic analysis of the scattering of plane waves by a smooth convex cylinder’, *Radio Science*, 1979, **14** (3), pp. 419–35

11 PATHAK, P. H., and KOUYOUMJIAN, R. G.: 'The dyadic diffraction coefficient for a perfectly conducting wedge'. Report 2183-4, ESL, Ohio State University, Columbus, Ohio, June 5, 1970
12 HILL, K. C., and PATHAK, P. H.: 'A UTD analysis of the excitation of surface rays by an edge in an otherwise smooth perfectly conducting convex surface'. URSI Radio Science Meeting, Blacksburg, Virginia, 1987
13 KAMINETSKY, L., and KELLER, J.: 'Diffraction coefficients for higher-order edges and vertices', *SIAM Journal of Applied Mathematics*, 1972, **22** (1), pp. 109–34
14 WESTON, V. H.: 'Effect of a discontinuity of curvature in high-frequency scattering, Part II', *IEEE Transactions on Antennas and Propagation*, 1965, **13** (4), pp. 611–13
15 SENIOR, T. B. A.: 'The diffraction matrix for a discontinuity in curvature', *IEEE Transactons on Antennas and Propagation*, 1972, **AP-20** (3), pp. 326–33
16 MICHAELI, A.: 'Diffraction by a discontinuity in curvature including the effect of the creeping wave', *IEEE Transactions on Antennas and Propagation*, 1990, **AP-38** (6), pp. 921–3
17 MICHAELI, A., and MOLINET, F.: 'An asymptotic solution for the diffraction by a discontinuity in curvature coupled with surface diffraction', *Annales des Télécommunications*, 1993, **48** (5–6), pp. 330–2
18 MICHAELI, A.: 'Transition functions for high-frequency diffraction by a curved perfectly conducting wedge, Part I: canonical solution for a curved sheet', *IEEE Transactions on Antennas and Propagation*, 1989, **AP-37**, pp. 1073–9
19 MICHAELI, A.: 'Transition functions for high-frequency diffraction by a curved perfectly conducting wedge, Part II: extension to overlapping transition regions', *IEEE Transactions on Antennas and Propagation*, 1989, **AP-37**, pp. 1086–92
20 UFIMTSEV, P. YA.: 'Method of edge waves in the physical theory of diffraction'. Air Force System Command, Foreign Tech. Div., Document ID No. FTD-HC-23-259-71, 1971 (Translated from the Russian version published by Soviet Publication House, Moscow, 1962)
21 KNOTT, E. K., and SENIOR, T. B.: 'Comparison of three high-frequency diffraction techniques', *Proceedings of the IEEE*, 1974, **62**, pp. 1468–74
22 UFIMTSEV, P. YA.: 'Elementary edge waves and the physical theory of diffraction', *Electromagnetics*, 1991, **11**, pp. 125–60
23 LIANG, M. C.: 'A generalized uniform GTD solution for the diffraction by a perfectly conducting wedge with convex faces'. PhD Thesis, Ohio State University, 1988
24 CHUANG, C. W., and LIANG, M. C.: 'A uniform asymptotic analysis of the diffraction by an edge in a curved screen', *Radio Science*, 1988, **23** (5), pp. 781–90
25 BOUCHE, D., MOLINET, F., and MITTRA, R.: 'Asymptotic methods in electromagnetics' (Springer-Verlag, Berlin, 1997)
26 MOLINET, F.: 'Uniform asymptotic solution for the diffraction of a plane wave by a perfectly conducting three-dimensional curved wedge', in GURAN, A., MITTRA, R., and MOSER, P. J. (Eds): 'Electromagnetic wave interactions',

Series on Stability, Vibration and Control of Systems, Series B, vol. 12 (World Scientific Publishing Company, Singapore, 1996) pp. 285–311

27 FELSEN, L. B., and MARCUVITZ, N.: 'Radiation and scattering of waves' (Prentice-Hall, Inc., New Jersey, 1973)

28 LIANG, M. C., PATHAK, P. H., and CHUANG, C. W.: 'A generalized uniform GTD ray solution for the diffraction by a wedge with convex faces'. URSI Radio Science Symposium, Prague, August 1990

Chapter 4

Asymptotic currents

4.1 Introduction

The geometrical theory of diffraction (GTD) and its uniform versions present some drawbacks for the calculation of the radar cross-section of a complex target, the most important of which are:

1 They do not allow us to calculate in general, the field on the caustics of the reflected rays.
2 They need sophisticated ray searching techniques which are very sensitive to the defects of the geometrical description when the surface is defined by computer aided geometrical design tools (B-splines etc.).

The success of the physical optics (PO) and the physical theory of diffraction (PTD) is mainly due to the abovementioned drawbacks of GTD.

The method of asymptotic currents is a natural extension of PO and PTD, which are founded on the radiation of the geometrical optics (GO) currents on the illuminated part of a target augmented by the radiation of the fringe currents close to the edges. Compared to PO, the asymptotic current method takes into account the currents on the shadow side of the target by introducing the creeping waves on convex surfaces and the whispering gallery modes on concave surfaces. Furthermore, in the illuminated part, close to the shadow boundary, the PO currents are replaced by more accurate transition zone solutions. Another difference is that the radiation of the fringe current is not calculated by introducing a diffraction coefficient referred to as Ufimtsev's coefficient D^u given by

$$D^u = D^{\mathrm{GTD}} - D^{\mathrm{OP}}$$

or by the radiation of an equivalent line source, but by using approximate expressions for the fringe currents in the radiation integral.

The asymptotic current method is well adapted for RCS computation of targets with large dimensions in wavelengths and low reflectivity for which PO and PTD fail to give an accurate result. This method is also of interest for the calculation of the coupling between antennas and nearby structures and for the development of hybrid numerical–asymptotic methods.

We consider two-dimensional (2D) and three-dimensional (3D) geometries which are either smooth or composed of regular convex or concave parts separated by edge type singularities forming curved wedges.

This chapter is divided into four parts.

The first part concerns a convex smooth cylindrical surface with arbitrary cross-section either perfectly conducting or coated with an absorbing material.

The diffraction by a smooth convex body is analysed by the boundary-layer method in Chapter 1, Section 1.2, for a scalar wave and in Chapter 2 for an electromagnetic wave. In each case general asymptotic solutions of the diffracted field are derived and discussed. In Section 4.2 of the present chapter, we focus our analysis on the asymptotic solutions giving the currents on the surface. Closed-form expressions for the currents on a 2D smooth convex surface are directly derived from the general formulae established before and extended to a coated surface.

The second part is devoted to a convex cylindrical perfectly conducting or coated surface delimited by sharp edges. The corresponding formulae are presented in Section 4.3.

In the third part (Section 4.4) we consider a concave perfectly conducting or coated surface with edges. A model problem solution is first derived by solving the problem of a line source in front of the concave side of a truncated circular cylindrical shell. Then the diffraction of the whispering gallery mode field with the edge of the truncated shell is treated by the spectral synthesis technique.

In the last part (Section 4.5), a general 3D solution is constructed for a convex or concave perfectly conducting face of a curved wedge.

At the end of each section, numerical applications of the formulae for the asymptotic currents are shown and compared with the method of moments (MM) results.

4.2 Asymptotic currents on a 2D smooth convex surface

4.2.1 Perfectly conducting surface

We consider a perfectly conducting convex cylindrical body, the generatrix of which is parallel to the OZ axis. A plane wave (or a line current) illuminates this structure at normal incidence. If Q' is the point of the surface lying on the shadow boundary, the field at an arbitrary point M on the convex cylinder, for the two principal polarisations of the incident field, may be directly derived from formulae (3.1) and (3.2) of Chapter 3, Section 3.2 by letting ν tend to 0. Additional information may also be found in Fock [1], Pathak [2] and Bouche *et al.* [3].

The result for a $e^{-i\omega t}$ time dependence is given by
For $\vec{H}^i_{\|}$:

$$H_z(M) = H_z^i(M)e^{i\sigma_0^3/3}g(\sigma_0) \qquad \text{if } M \text{ is in the lit region,} \tag{4.1}$$

$$H_z(M) = H_z^i(Q')\left[\frac{\rho(Q')}{\rho(M)}\right]^{1/6} e^{ikl'_0}g(\sigma'_0) \qquad \text{if } M \text{ is in the shadow region,} \tag{4.2}$$

where $\rho(M)$ is the radius of curvature of the surface at M and where $g(x)$ is the hard Fock function given by

$$g(x) = \frac{1}{\sqrt{\pi}}\int_{\Gamma} \frac{e^{ix\xi}}{w'_1(\xi)}\,d\xi, \tag{4.3}$$

where Γ is the contour in the complex ξ plane which runs from $\infty e^{2i\pi/3}$ to $\infty + i\varepsilon$.
The arguments of the Fock function in the lit region are given by

$$\sigma_0 = -m(M)\sin\psi(M) = -m(M)\cos\theta^i(M), \qquad m(M) = \left[\frac{k\rho(M)}{2}\right]^{1/3}, \tag{4.4}$$

where ψ is the angle of the incident ray with the tangent to the surface at M and $\theta^i = \pi/2 - \psi$ is the angle of reflection.

In the shadow region $\sigma'_0 = ml'_0/a$ for a circular cylinder of radius a and for a general convex cylinder with variable curvature:

$$\sigma'_0 = \int_{Q'}^{M} \frac{m(\ell)}{\rho(\ell)}\,d\ell. \tag{4.5}$$

For $\vec{E}^i_{\|}$:

$$\frac{\partial E_z}{\partial n}(M) = E_z^i(M)e^{i\sigma_0^3/3}\frac{k}{m(M)}f(\sigma_0) \qquad \text{if } M \text{ is in the lit region,} \tag{4.6}$$

$$\frac{\partial E_z}{\partial n}(M) = E_z^i(Q')\left[\frac{\rho(Q)}{\rho(M)}\right]^{1/6} \times e^{ikl'_0}\frac{k}{m(M)}f(\sigma'_0) \quad \text{if } M \text{ is in the shadow region,} \tag{4.7}$$

where n is the coordinate along the normal to the surface at M and where $f(x)$ is the soft Fock function given by

$$f(x) = \frac{1}{\sqrt{\pi}}\int_{\Gamma} \frac{e^{ix\xi}}{w_1(\xi)}\,d\xi. \tag{4.8}$$

The quantities H_z and $\partial E_z/\partial n$ are proportional to the surface current. For $\vec{H}^i_\parallel$ we have $\vec{H} = \hat{z}H_z$ and hence $\vec{J}_s = \hat{t}H_z$, where $\hat{t}$ is the unit vector of the tangent to the surface in the direction of propagation of the creeping wave. For $\vec{E}^i_\parallel$, we have

$$\vec{H} = \frac{\hat{t}}{i\omega\mu}\frac{\partial E_z}{\partial n},$$

and consequently

$$\vec{J}_s = \frac{i}{Z_0}\frac{\hat{z}}{k}\frac{\partial E_z}{\partial n}.$$

4.2.2 Imperfectly conducting or coated surface

When the surface is perfectly conducting, only H_z and $\partial E_z/\partial n$ are different from zero whereas $\partial H_z/\partial n = 0$, $E_z = 0$ for $\vec{H}^i_\parallel$ and $\vec{E}^i_\parallel$, respectively.

For an imperfectly conducting or coated surface, we have to calculate for each polarisation the field U_z and its normal derivative $\partial U_z/\partial n$ where U_z stands for H_z and E_z. Starting with formulae (3.73) of Chapter 3, Section 3.6 and letting ν tend to zero in these formulae and in their derivatives with respect to ν, we obtain by noting that $\nu = kn/m$, the following expressions for the field and its normal derivative:

For $\vec{H}^i_\parallel$:

$$\begin{aligned} H_z(M) &= H^i_z(M)e^{i\sigma_0^3/3}\tilde{g}_H(\sigma_0), \\ \frac{\partial H_z(M)}{\partial n} &= H^i_z(M)e^{i\sigma_0^3/3}\frac{k}{m(M)}f_H(\sigma_0), \end{aligned} \tag{4.9}$$

if M is in the lit region and

$$\begin{aligned} H_z(M) &= H^i_z(Q')\left[\frac{\rho(Q')}{\rho(M)}\right]^{1/6} e^{ikl'_0}\tilde{g}_H(\sigma'_0), \\ \frac{\partial H_z}{\partial n}(M) &= H^i_z(Q')\left[\frac{\rho(Q')}{\rho(M)}\right]^{1/6} e^{ikl'_0}\frac{k}{m(M)}f_H(\sigma'_0), \end{aligned} \tag{4.10}$$

if M is in the shadow region.

In these formulae there appear two new Fock functions introduced in Chapter 3, Section 3.6, defined by

$$\tilde{g}_H(x) = \frac{1}{\sqrt{\pi}}\int_\Gamma \frac{e^{ix\xi}}{w'_1(\xi) - im\zeta_H(\xi)w_1(\xi)}\,d\xi, \tag{4.11}$$

$$f_H(x) = \frac{1}{\sqrt{\pi}}\int_\Gamma e^{ix\xi}\frac{-im\zeta_H(\xi)}{w'_1(\xi) - im\zeta_H(\xi)w_1(\xi)}\,d\xi. \tag{4.12}$$

The arguments σ_0 and σ'_0 are given by (4.4) and (4.5) and l'_0 is the arc length between Q' and M along the exterior surface of the coating.

For $\vec{E}^i_{\|}$:

$$E_z(M) = E^i_z(M)e^{i\sigma_0^3/3}\tilde{f}_E(\sigma_0),$$
$$\frac{\partial E_z}{\partial n}(M) = E^i_z(M)e^{i\sigma_0^3/3}\frac{k}{m(M)}f_E(\sigma_0), \tag{4.13}$$

if M is in the lit region and

$$E^i_z(M) = E^i_z(Q')\left[\frac{\rho(Q')}{\rho(M)}\right]^{1/6} e^{ikl'_0}\tilde{f}_E(\sigma'_0),$$
$$\frac{\partial E^i_z}{\partial n}(M) = E^i_z(Q')\left[\frac{\rho(Q')}{\rho(M)}\right]^{1/6} e^{ikl'_0}\frac{k}{m(M)}f_E(\sigma'_0), \tag{4.14}$$

if M is in the shadow region.

Two other Fock functions appear in these formulae. They are defined by

$$\tilde{f}_E(x) = \frac{1}{\sqrt{\pi}}\int_\Gamma \frac{e^{ix\xi}}{w'_1(\xi) - im\zeta_E(\xi)w_1(\xi)}d\xi, \tag{4.15}$$

$$f_E(x) = \frac{1}{\sqrt{\pi}}\int_\Gamma e^{ix\xi}\frac{-im\zeta_E(\xi)}{w'_1(\xi) - im\zeta_E(\xi)w_1(\xi)}d\xi. \tag{4.16}$$

For a coated surface, the functions $\zeta_H(\xi)$ and $\zeta_E(\xi)$ are related to the modal impedance $\zeta_H(p)$ and admittance $\zeta_E(p)$ given by Equation (3.75) of Chapter 3, Section 3.6, by the transformation

$$p = k\rho + m\xi.$$

For an imperfectly conducting surface defined locally by a surface impedance Z we get the same formulae with ζ_H and ζ_E replaced respectively by Z and $1/Z$.

When the surface impedance tends to zero, we see that $\tilde{g}_H(x)$ tends to $g(x)$, $f_H(x)$ and $\tilde{f}_E(x)$ tend to zero and $f_E(x)$ tends to $f(x)$ and formulae (4.9), (4.10) and (4.13), (4.14) reduce to the corresponding formulae for a perfectly conducting surface given by (4.1), (4.2) and (4.6), (4.7), respectively.

4.2.3 *Numerical calculation of the Fock functions*

When the observation point is in the shadow region, the Fock functions (4.11), (4.12), (4.15) and (4.16) are calculated by the residue series. Very close to the shadow boundary in the shadow region, for $x < 0.1$, the number of residues which are necessary becomes prohibitive. Then the Fock functions are calculated by numerical integration. The latter is also used in the lit region where $x < 0$.

In this case, the exponential function $\exp(ix\xi)$ tends to infinity on the infinite branches of the integration contour, since $\operatorname{Im}\xi > 0$ on these branches. But since $w_1(\xi)$ increases like $\exp(\frac{2}{3}|\xi|^{3/2})$, the integral converges. However, for $|x| \gg 1$, the integrands of the Fock functions oscillate rapidly along the path of integration so that the numerical integration becomes inaccurate and finally diverges for a fixed sampling of the path, when $|x|$ continues to augment. In this case, the integrals can be evaluated by the stationary phase method after having replaced the Airy function and its derivative by their asymptotic expansions.

When $x < 0$, a stationary point $\xi_s < 0$ exists, given by

$$(-\xi_s)^{1/2} + x = 0.$$

Then the stationary phase evaluation leads to the following approximations:

$$\begin{aligned}
\tilde{g}_H(x) &= \frac{2x}{x - m\zeta_H} e^{-ix^3/3} = \frac{2\cos\theta^i}{\cos\theta^i + \zeta_H} e^{-ix^3/3} = (1 + R_H)\, e^{-ix^3/3}, \\
\tilde{f}_E(x) &= \frac{2x}{x - m\zeta_E} e^{-ix^3/3} = \frac{2\cos\theta^i}{\cos\theta^i + \zeta_E} e^{-ix^3/3} = (1 + R_E)\, e^{-ix^3/3}, \\
f_H(x) &= -im\zeta_H(1 + R_H)\, e^{-ix^3/3}, \\
f_E(x) &= -im\zeta_E(1 + R_E)\, e^{-ix^3/3}.
\end{aligned} \tag{4.17}$$

These formulae, which correspond to the GO approximation, are applied when $|x| > 3$.

4.2.4 *Numerical results*

Some numerical results for the surface currents and the radiated field are shown for a circular and an elliptic cylinder illuminated by a plane wave in Figures 1.25–1.32 of Chapter 1, Section 1.3. Additional results are given in Figure 4.1 for a perfectly conducting elliptic cylinder defined by $a = 1$ m, $b = 0.75$ m and in Figure 4.2 for a perfectly conducting elliptic cylinder defined by $a = 1$ m, $b = 0.5$ m (see Figure 1.24 of Chapter 1, Section 1.3). The asymptotic currents at 1 GHz on the surface for an incident magnetic field parallel to the generatrix ($H_{\parallel}$) and an incident electric field parallel to the generatrix ($E_{\parallel}$) are plotted as a function of the polar angle of an observation point on the surface and compared to the currents obtained by the MM. It can be seen that the correspondence with the MM result is excellent.

4.3 Asymptotic currents on a 2D convex surface delimited by sharp edges

4.3.1 *Perfectly conducting convex surface delimited by sharp edges*

A typical geometry for this problem is an ogival cylinder, the cross-section of which is represented in Figure 4.3.

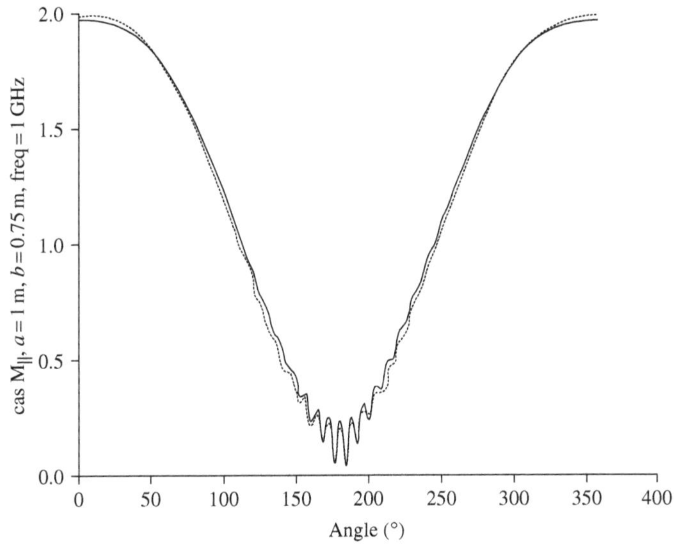

Figure 4.1 *Modulus of the current on a perfectly conducting elliptic cylinder as a function of the polar angle, $a = 1\,m$, $b = 0.75\,m$, $F = 1\,GHz$. Comparison between asymptotic currents and the currents obtained by MM, for $H_{||}$ polarisation: ——, asymptotic; ……, MM*

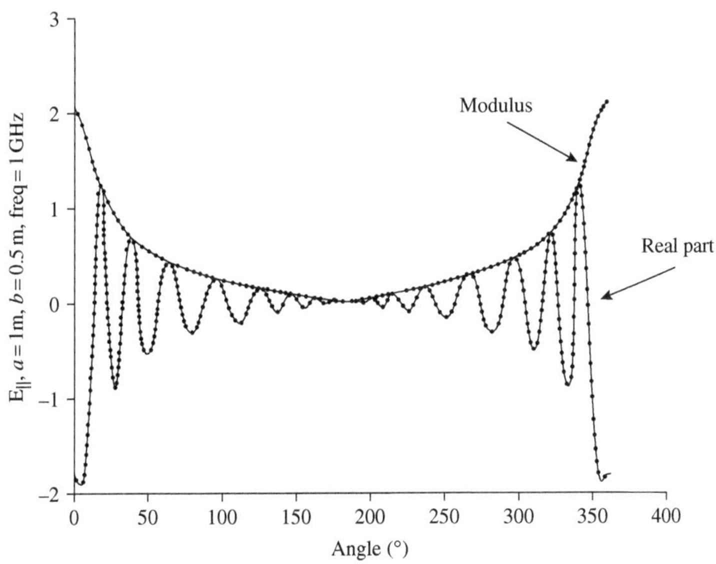

Figure 4.2 *Real part and modulus of the current on a perfectly conducting elliptic cylinder as a function of the polar angle, $a = 1\,m$, $b = 0.5\,m$, $F = 1\,GHz$: ——, asymptotic; ……, MM*

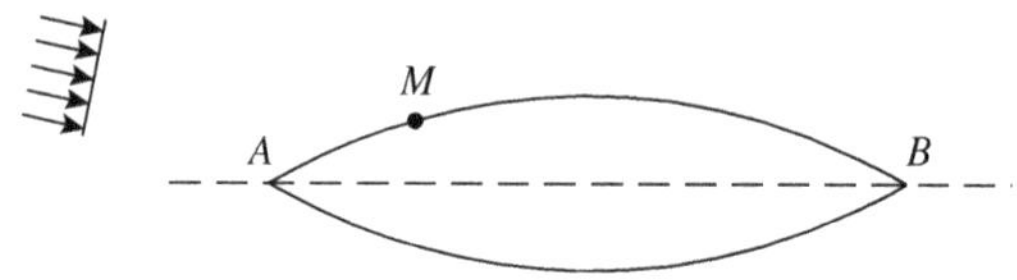

Figure 4.3 Typical geometry

We consider an incident plane wave or a line current illuminating the ogival cylinder. The field at an arbitrary point M on the convex surface is due to different diffraction phenomena:

(a) the diffraction by the smooth part of the surface which we have already treated in Section 4.2
(b) the single diffraction by the edge at A or B
(c) the double diffraction by the edges A and B
(d) the multiple diffractions (more than two), by the edges A and B.

Multiple diffractions by two edges give generally a small contribution to the field at M, at high frequencies. However, this remark is not essential for the formulation of the problem since all the diffraction phenomena (b)–(d) may be reduced to two fundamental problems:

1 the diffraction of a space wave into a creeping wave and vice versa
2 the diffraction of a creeping wave into another creeping wave.

The solution to problem (1) may be obtained by treating either the diffraction of a plane wave (or a local plane wave when the source is a line current) by a wedge followed by the excitation of a creeping wave propagating along the surface from the edge to the observation point M or the radiation of a line current located at M, which excites a creeping wave propagating along the surface to the edge where it is diffracted in a space wave. By applying the reciprocity theorem it is a simple matter to write down the solution of one of these problems when the solution to the reciprocal problem is known. We will consider the case of a creeping wave diffracted by an edge which allows us to use the spectral decomposition technique of the Fock functions introduced in Chapter 3. But a direct approach is also possible and has been developed by Michaeli [4] starting with a spectral decomposition in inhomogeneous plane waves of the Sommerfeld integral arising in the exact solution of the diffraction of a plane wave by a straight wedge and applying to each spectral component the analytical extension of the solution to the diffraction problem of a plane wave by a smooth surface.

4.3.1.1 Edge excited currents on an illuminated edge

As mentioned before, we first treat the reciprocal problem of a line source at M exciting a creeping wave which is diffracted by the edge into a space wave. In order to solve this problem, we start with the general solution of a line source located close to a circular cylinder of radius a. This canonical problem has an exact solution given by the following expression of its Green's function (see Reference 5, p. 287):

$$G(r', \phi'; r, \phi; k) = \frac{i}{8} \int_C e^{ip|\phi-\phi'|} H_p^{(1)}(kr_>)$$
$$\times \left[H_p^{(2)}(kr_<) - \frac{\Omega H_p^{(2)}(ka)}{\Omega H_p^{(1)}(ka)} H_p^{(1)}(kr_<) \right] dp, \tag{4.18}$$

where (r', ϕ') and (r, ϕ) are the polar coordinates of the source and of the observation point, respectively, $r_>$ and $r_<$ are the greater and lesser of the coordinates r', r and $\Omega = 1$ for an electric line source and $\Omega = \partial/\partial r$ for a magnetic line source. The contour C runs from $-\infty + \varepsilon$ to $+\infty + \varepsilon$ where ε is an arbitrary small positive number.

We suppose that ka is large and that both the source and the observation point lie in the surface boundary layer. Then in the Fock region, the Hankel functions can be replaced by their asymptotic expansion of the Olver type (see Reference 6, p. 715).

Retaining only the first terms, we have

$$H_p^{(1,2)}(ka) = \mp i (m\sqrt{\pi})^{-1} w_{1,2}(\xi), \tag{4.19}$$

where $w_{1,2}$ are the Airy functions defined in Chapter 1, Section 1.2.4 and where the transformation

$$p = ka + m\xi, \qquad m = \left(\frac{ka}{2}\right)^{1/3}$$

has been used.

Replacing the Hankel functions in (4.18) by their asymptotic expansions (4.19), one obtains for the components of the magnetic and electric fields, parallel to the generatrix, the following expressions:

$$\begin{aligned} H_z(S) &= 2ikZ_0^{-1} M_z F_h(\sigma, \nu', \nu)\Pi(t), \\ E_z(S) &= 2ikZ_0 J_z \frac{-im^2}{kt} F_s(\sigma, \nu', \nu)\Pi(t), \end{aligned} \tag{4.20}$$

where

$$F_h(\sigma, \nu', \nu) = \frac{e^{i\pi/4}\sigma^{1/2}}{4\sqrt{\pi}} \int_\Gamma w_1(\xi - \nu') \left[w_2(\xi - \nu) - \frac{w_2'(\xi)}{w_1'(\xi)} w_1(\xi - \nu) \right] e^{i\sigma\xi}\, d\xi,$$

$$F_s(\sigma, \nu', \nu) = -\frac{e^{-i\pi/4}\sigma^{3/2}}{2\sqrt{\pi}} \int_\Gamma w_1(\xi - \nu') \left[w_2(\xi - \nu) - \frac{w_2(\xi)}{w_1(\xi)} w_1(\xi - \nu) \right] e^{i\sigma\xi} d\xi$$
$$\tag{4.21}$$

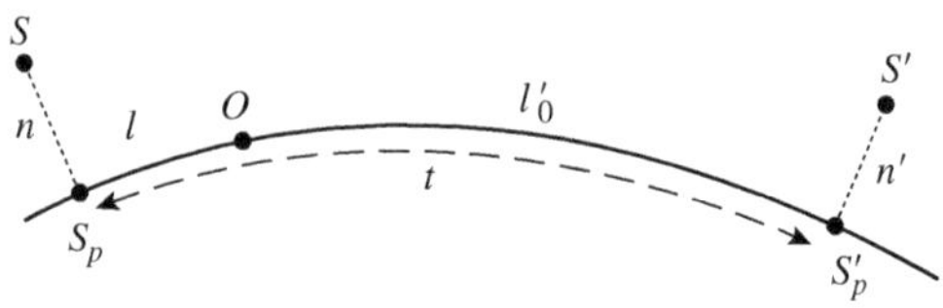

Figure 4.4 Geometrical parameters of the problem

and

$$\nu' = \frac{kn'}{m}, \quad \nu = \frac{kn}{m}, \quad t = \text{arc } S'_P S_p, \quad \sigma = \frac{mt}{a},$$
$$m = \left[\frac{ka}{2}\right]^{1/3}, \quad \Pi(t) = \frac{e^{i\pi/4}}{\sqrt{8\pi k}} \frac{e^{ikt}}{\sqrt{t}} \tag{4.22}$$

and where M_z and J_z are, respectively, the intensities of the magnetic and electric line currents and Z_0 is the characteristic impedance of the vacuum.

The geometrical parameters t, n, n' are defined in Figure 4.4.

Proceeding as in Chapter 3, Section 3.2 by expanding (4.21) in Taylor series with respect to ν which is supposedly small, and making use of the Wronskian $[w_1, w_2] = 2i$, the integrals in (4.21) may be written as

$$\pm 2i\, e^{-ikt} \int_\Gamma A_{s,h}(\xi)\big[C^i_\xi(l,n) \mp C^r_\xi(l,n)\big]\, d\xi, \tag{4.23}$$

where

$$C^{i,r}_\xi(l,n) \cong \exp\left[-ikl\left(1+\frac{\xi}{2m^2}\right) \mp ikn\frac{i\sqrt{\pi}}{m}\right]$$
$$\cong \exp[-ik(l\cos\theta \pm n\sin\theta)],$$
$$A_s(\xi) = \frac{1}{2}\frac{w_1(\xi-\nu')}{w_1(\xi)\sqrt{\xi}} \exp\left[il'_0\left(k+\xi\frac{m}{a}\right)\right], \tag{4.24}$$
$$A_h(\xi) = \frac{1}{2}\frac{w_1(\xi-\nu')}{w'_1(\xi)} \exp\left[il'_0\left(k+\xi\frac{m}{a}\right)\right]$$

and where

$$l = \text{arc } OS_p < 0 \quad \text{and} \quad l'_0 = \text{arc } OS'_p > 0, \quad \theta = i\sqrt{\xi}m^{-1}.$$

One recognises in (4.23) and (4.24) a spectral decomposition of the integrals in (4.21) in a spectrum of quasi-plane waves. By spectral synthesis, this spectrum is transformed by diffraction at the edge A in a diffracted space ray, the field of which is given at large distances from the edge by

$$\begin{aligned} H_z^d(P) &= ikZ_0^{-1}M_z\Pi(l_0') \\ &\quad\times\left(\frac{e^{-i\pi/4}(\sigma_0')^{1/2}}{2\sqrt{\pi}}\int\limits_\Gamma \frac{w_1(\xi-\nu')}{w_1'(\xi)}e^{i\sigma_0'\xi}D_h(\phi,\theta)\,d\xi\right)\Pi(r), \\ E_z^d(M) &= ikZ_0J_z\frac{m}{k}\frac{\Pi(l_0')}{l_0'} \\ &\quad\times\left(\frac{e^{-3i\pi/4}(\sigma_0')^{3/2}}{\sqrt{\pi}}\int\limits_\Gamma \frac{w_1(\xi-\nu')}{w_1(\xi)}e^{i\sigma_0'\xi}\frac{D_s(\phi,\theta)}{\theta}\,d\xi\right)\Pi(r), \end{aligned} \tag{4.25}$$

where $l_0' = \text{arc}\,MA$ and $\sigma_0' = ml_0'/a$.

The coefficients D_s, D_h are, respectively, Keller's diffraction coefficients for soft and hard boundary conditions, divided by the factor $(8\pi k)^{-1/2}\exp(i\pi/4)$. Since D_s and D_h vary slowly with respect to the integration variable ξ, we can replace them, outside the transition regions, by

$$D_h(\phi,\theta) \cong D_h(\phi,0), \qquad D_s(\phi,\theta) \cong \theta\left[\frac{\partial}{\partial\theta}D_s(\phi,\theta)\right]_{\theta=0}. \tag{4.26}$$

By taking into account (4.26) in (4.25), we obtain the final result

$$\begin{aligned} H_z^d(P) &= ikZ_0^{-1}M_z\Pi(l_0')D_h(\phi,\theta)\Pi(r)V(\sigma_0',\nu'), \\ E_z^d(P) &= ikZ_0J_z\frac{\Pi(l_0')}{l_0'}\left[\frac{\partial}{\partial\theta}D_s(\phi,\theta)\right]_{\theta=0}U(\sigma_0',\nu')\Pi(r), \end{aligned} \tag{4.27}$$

where

$$\begin{aligned} V(\sigma_0',\nu') &= \frac{1}{2}e^{-i\pi/4}\left(\frac{\sigma_0'}{\pi}\right)^{1/2}\int\limits_\Gamma \frac{w_1(\xi-\nu')}{w_1'(\xi)}e^{i\sigma_0'\xi}\,d\xi, \\ U(\sigma_0',\nu') &= \frac{1}{2}e^{-3i\pi/4}\left(\frac{\sigma_0'^3}{\pi}\right)^{1/2}\int\limits_\Gamma \frac{w_1(\xi-\nu')}{w_1(\xi)}e^{i\sigma_0'\xi}\,d\xi. \end{aligned} \tag{4.28}$$

In order to obtain the results for the reciprocal problem, we have to replace ikM_z by $Z_0H_z^i(A)$ and ikJ_z by $Z_0^{-1}E_z^i(A)$ and remove in formulae (4.27) the factor $\Pi(r)$. Consequently, the field at S' due to an incident plane wave (or local plane wave) diffracted by the edge A is given by (see Figure 4.5),

$$\begin{aligned} H_z^d(S') &= H_z^i(A)\Pi(l_0')D_h(\phi,0)V(\sigma_0',\nu'), \\ E_z^d(S') &= E_z^i(A)\frac{m'}{k}\frac{\Pi(l_0')}{l_0'}\left[\frac{\partial}{\partial\theta}D_s(\phi,\theta)\right]_{\theta=0}U(\sigma_0',\nu'). \end{aligned} \tag{4.29}$$

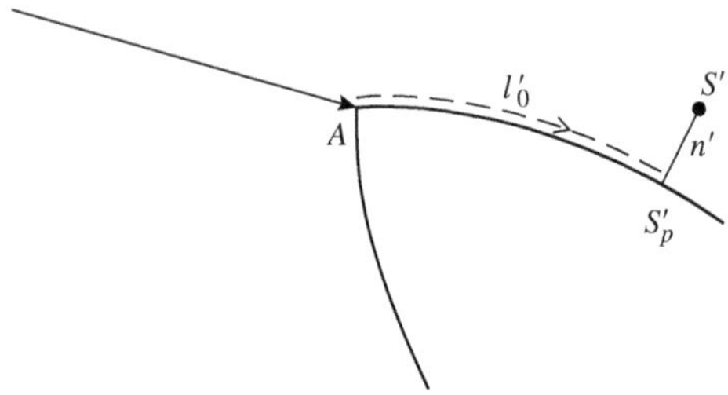

Figure 4.5 *Reciprocal problem*

From (4.29) we can calculate $\partial H_z/\partial n'$ and $\partial E_z/\partial n'$. For an observation point M on the surface ($\nu' = 0$), we obtain:

$$H_z^d(M) = H_z^i(A)\Pi(l_0')D_h(\phi, 0)N_h(\sigma_0'),$$
$$\frac{\partial E_z^d}{\partial n'}(M) = -E_z^i(A)\frac{\Pi(l_0')}{l_0'}\left[\frac{\partial}{\partial\theta}D_s(\phi,\theta)\right]_{\theta=0} N_s(\sigma_0'), \tag{4.30}$$
$$\frac{\partial H_z^d}{\partial n'}(M) = E_z^d(M) = 0,$$

where $N_{s,h}$ are the Nicholson functions for the soft and hard boundary conditions defined by

$$N_s(X) = \frac{e^{-3i\pi/4}X^{3/2}}{\sqrt{\pi}}\int_\Gamma \frac{w_1'(\xi)}{w_1(\xi)}e^{iX\xi}\,d\xi,$$
$$N_h(X) = \frac{e^{-i\pi/4}X^{1/2}}{2\sqrt{\pi}}\int_\Gamma \frac{w_1(\xi)}{w_1'(\xi)}e^{iX\xi}\,d\xi. \tag{4.31}$$

For a surface of variable curvature (see Chapter 1, Section 1.2.5), we must replace in (4.30) the parameter σ_0' by

$$\sigma_0' = \int_0^{l_0'} \frac{m\,ds}{\rho}$$

and introduce the factor $[\rho(A)/\rho(M)]^{1/6}$ not present in (4.30).

4.3.1.2 Edge excited currents on a shadowed edge

When the edge is in the shadow region of the incident field, its contribution to the currents at a point M on the surface is due to a creeping wave diffracted by the edge in another creeping wave which reaches the point M (see Figure 4.6).

In order to construct the solution for $\vec{E}_\parallel^i$ polarisation we need to calculate the field at a point S slightly removed from the surface at a height n and then take the derivative of the result with respect to n, since for a perfect conductor, only the normal derivative of the electric field on the surface is different from zero.

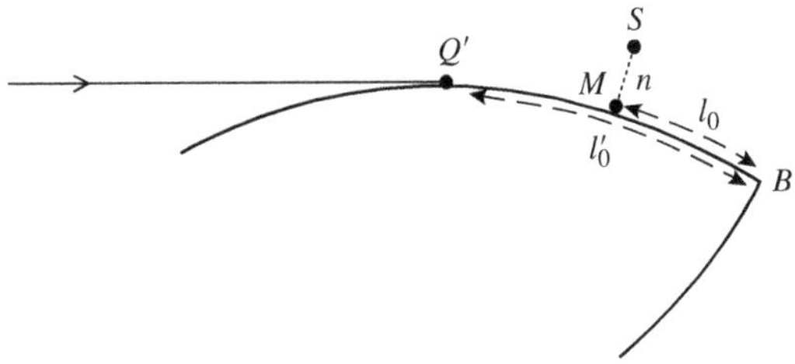

Figure 4.6 Ray paths for a shadowed edge

For both polarisations, the technique for constructing the solution consists in decomposing the field of the creeping wave reaching B in a spectrum of inhomogeneous plane waves. To each component of this spectrum, we then apply the solution presented in Section 4.3.1.1 and given by equations (4.30). Performing these operations and synthesising the resulting spectrum lead to the following expressions of the field diffracted by the edge at a point M on the surface:

$$H_z^d(M) = \frac{1}{2} H_z^i(Q') \left[\frac{\rho(Q')}{\rho(M)} \right]^{1/6} e^{ikl'_0} g(\sigma'_0) \Pi(l_0) D_h(\phi, 0) N_h(\sigma_0), \tag{4.32}$$

$$\frac{\partial E_z^d}{\partial n}(M) = -\frac{1}{2} E_z^i(Q') \left[\frac{\rho(Q')}{\rho(M)} \right]^{1/6} \frac{i}{m(M)} e^{ikl'_0} f(\sigma'_0) \frac{\Pi(l_0)}{l_0} \times \left[\frac{\partial^2}{\partial\theta\,\partial\theta'} D_s(\theta, \theta') \right]_{\theta=\theta'=0} N_s(\sigma_0), \tag{4.33}$$

where $l'_0 = \text{arc}\, Q'B$, $l_0 = \text{arc}\, BM$ and

$$\sigma'_0 = \int_{Q'}^{B} \frac{m\, ds}{\rho}, \qquad \sigma_0 = \int_{B}^{M} \frac{m\, ds}{\rho}.$$

4.3.1.3 Grazing incidence

When the incident ray is tangent to one of the two faces of the wedge at the edge, the transition regions of the direct and reflected waves and the transition region of the edge diffracted wave overlap (see Figures 4.7(a) and (b)).

In order to treat this problem, we consider the reciprocal problem of a line source situated on or close to one face (say S_1), with an observation point in the direction defined by the tangent to the same face at the edge (Figure 4.8).

Using Michaeli's procedure presented in Chapter 3, Section 3.5.1, we replace the diffraction problem by a radiation problem of PO type currents and fringe currents. The radiation of the PO currents is calculated by extending the real surface S_1 continuously beyond the edge into a fictitious surface S'_1 and by writing that

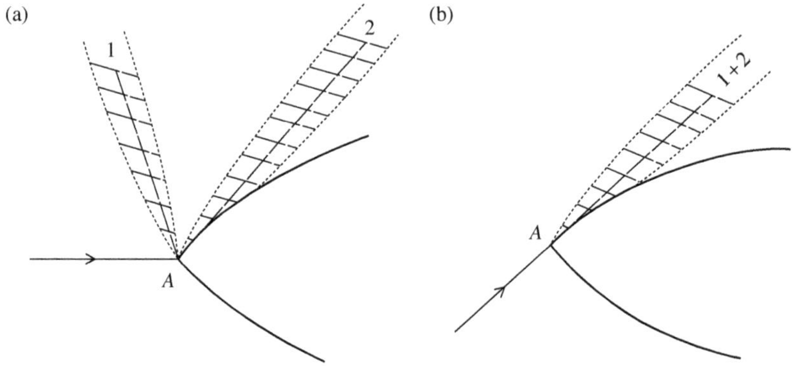

Figure 4.7 Transition regions for the reflected and diffracted rays: (a) separated transition regions; (b) overlapping transition regions

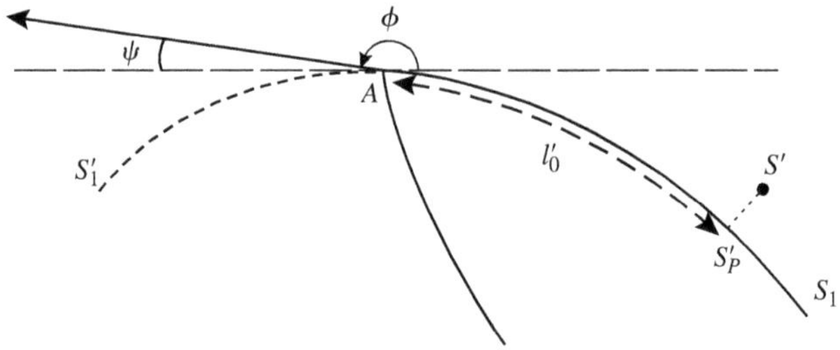

Figure 4.8 Reciprocal problem for grazing incidence

the integral of the PO currents over S_1 is equal to the integral of the PO currents over the extended surface $S_1 + S_1'$ minus the integral of the PO currents over S_1'. This last integral gives the contribution of the edge to the radiated field, in the configuration shown in Figure 4.8, to which we must add the contribution of the fringe currents. The latter is calculated by the spectral synthesis technique described in Section 4.3.1.1 and leads to the result given by Equation (4.30) in which Keller's diffraction coefficients D_s and D_h have to be replaced by Ufimtsev's diffraction coefficients D_h^u and D_s^u defined by

$$D_{s,h}^{u} = D_{s,h} - D_{s,h}^{OP},$$

where all the coefficients have been divided by $(8\pi k)^{-1/2}\exp(i\pi/4)$.

The PO contribution to the edge diffracted field is calculated by using the spectral decomposition (4.23), (4.24) of the creeping wave excited by the source S' on the smooth surface $(S_1 + S_1')$ and replacing S_1' close to the edge by its circular cylindrical approximation. The details of the reduction to a simpler form of the radiation integral over S_1' are identical to those presented in Chapter 3, Section 3.5.1 and will not be reproduced here. The final result is given by

$$H_z^e(P) = ikZ_0^{-1}M_z e^{-i\sigma_0'^3/3} P_h(\sigma_0' - \sigma_0, \sigma_0) \left[\frac{\rho(M)}{\rho(A)}\right]^{1/6} e^{ikl_0'}\Pi(r),$$

$$\frac{\partial E_z^e}{\partial n}(P) = -ikZ_0 J_z e^{-i\sigma_0'^3/3} \frac{k}{m(M)} P_s(\sigma_0' - \sigma_0, \sigma_0) \left[\frac{\rho(M)}{\rho(A)}\right]^{1/6} e^{ikl_0'}\Pi(r), \tag{4.34}$$

where $l_0' = \text{arc}\, MA$, $\sigma_0' = \int_M^A m\, ds/\rho$, $\sigma_0 = m(A)\psi$, and where the new transition functions are modified Nicholson functions defined by

$$P_h(X, Y) = \frac{1}{2\pi} \int_\Gamma \frac{w_1(\xi)}{w_1'(\xi)} \frac{\partial I_1}{\partial \xi}(\xi, Y) e^{iX\xi}\, d\xi,$$

$$P_s(X, Y) = \frac{1}{2\pi} \int_\Gamma \frac{w_1'(\xi)}{w_1(\xi)} I_1(\xi, Y) e^{iX\xi}\, d\xi \tag{4.35}$$

in the integrand of which appears the incomplete Airy function or its derivative.

Proceeding as in Sections 4.3.1.1 and Section 4.3.1.2, we obtain the result for the reciprocal problem by replacing ikM_z by $Z_0 H_z^i(A)$ and ikJ_z by $Z_0^{-1}E_z^i(A)$ and suppressing the factor $\Pi(r)$. This yields

$$H_z^e(M) = H_z^i(A) e^{-i\sigma_0'^3/3} P_h(\sigma_0' - \sigma_0, \sigma_0) \left[\frac{\rho(M)}{\rho(A)}\right]^{1/6} e^{ikl_0'},$$

$$\frac{\partial E_z^e}{\partial n}(M) = -E_z^i(A) e^{-i\sigma_0'^3/3} \frac{k}{m(M)} P_s\left(\sigma_0' - \sigma_0, \sigma_0\right) \left[\frac{\rho(M)}{\rho(A)}\right]^{1/6} e^{ikl_0'}. \tag{4.36}$$

When the observation direction is below the tangent to face S_1 in Figure 4.8, a creeping wave contribution from the fictitious surface S_1' has to be added to formulae (4.36) since this contribution belongs also to the edge diffracted field. Moreover, as mentioned at the beginning of Section 4.3.1.3, the complete solution also comprises Ufimtsev's correction term of the PO contribution.

4.3.2 *Imperfectly conducting or coated wedge with convex faces*

For an imperfectly conducting convex surface, the dominant term of the asymptotic expansion of the field radiated by a line source S' located close to the surface, at an observation point S also close to the surface (both S' and S are in the surface boundary layer) is still given by (4.20) but the Fock functions are replaced by the following modified Fock functions:

$$F_H(\sigma,\nu',\nu) = e^{i\pi/4}\frac{\sigma^{1/2}}{4\sqrt{\pi}}\int_\Gamma w_1(\xi-\nu')$$
$$\times\left[w_2(\xi-\nu) - \frac{w_2'(\xi) - im\zeta_H w_2(\xi)}{w_1'(\xi) - im\zeta_H w_1(\xi)}w_1(\xi-\nu)\right]e^{i\sigma\xi}\,d\xi,$$
$$F_E(\sigma,\nu',\nu) = -\frac{e^{-i\pi/4}\sigma^{3/2}}{2\sqrt{\pi}}\int_\Gamma w_1(\xi-\nu')$$
$$\times\left[w_2(\xi-\nu) - \frac{w_2'(\xi) - im\zeta_E w_2(\xi)}{w_1'(\xi) - im\zeta_E w_1(\xi)}w_1(\xi-\nu)\right]e^{i\sigma\xi}\,d\xi. \tag{4.37}$$

A spectral representation of these functions in a spectrum of inhomogeneous plane waves is readily obtained by expanding $w_2(\xi-\nu)$ and $w_1(\xi-\nu)$ in Taylor series as in Section 4.3.1.1. Then applying the same technique, we obtain the following formulae.

For an illuminated wedge:

$$H_z^d(M) = H_z^i(A)\Pi(l_0')\left[\frac{\rho(A)}{\rho(M)}\right]^{1/6}\frac{i}{m(M)}N_H(\sigma_0')\left[\frac{\partial}{\partial\theta}D(\phi,\theta;\zeta_H)\right]_{\theta=0},$$

$$\frac{\partial H_z^d}{\partial n}(M) = -H_z^i(A)\Pi(l_0')\frac{k}{m(M)}\left[\frac{\rho(A)}{\rho(M)}\right]^{1/6}\frac{i}{m(M)}\tilde{N}_H(\sigma_0')$$
$$\times\left[\frac{\partial}{\partial\theta}D(\phi,\theta;\zeta_H)\right]_{\theta=0},$$

$$E_z^d(M) = E_z^i(A)\frac{\Pi(l_0')}{l_0'}\left[\frac{\rho(A)}{\rho(M)}\right]^{1/6}\frac{m(M)}{k}\tilde{N}_E(\sigma_0')\left[\frac{\partial}{\partial\theta}D(\phi,\theta;\zeta_E)\right]_{\theta=0},$$

$$\frac{\partial E_z^d}{\partial n}(M) = -E_z^i(A)\frac{\Pi(l_0')}{l_0'}\left[\frac{\rho(A)}{\rho(M)}\right]^{1/6}N_E(\sigma_0')\left[\frac{\partial}{\partial\theta}D(\phi,\theta;\zeta_E)\right]_{\theta=0}, \tag{4.38}$$

where $D(\phi,\theta;\zeta)$ is Maliuzhinet's diffraction coefficient for a straight wedge divided by $(8\pi k)^{-1/2}\exp(i\pi/4)$ and

$$\sigma_0' = \int_A^M \frac{m\,ds}{\rho}, \qquad l_0' = \text{arc}\,AM$$

and where the Fock–Nicholson functions are defined by

$$N_H(X) = \frac{e^{-i\pi/4}}{2}\left(\frac{X}{\pi}\right)^{1/2}\int_\Gamma \frac{-im\zeta_H(\xi)w_1(\xi)}{w_1'(\xi) - im\zeta_H(\xi)w_1(\xi)}e^{iX\xi}\,d\xi,$$

$$\tilde{N}_H(X) = \frac{e^{-i\pi/4}}{2}\left(\frac{X}{\pi}\right)^{1/2}\int_\Gamma \frac{-im\zeta_H(\xi)w_1'(\xi)}{w_1'(\xi) - im\zeta_H(\xi)w_1(\xi)}e^{iX\xi}\,d\xi,$$

$$N_E(X) = e^{-3i\pi/4}\left(\frac{X^3}{\pi}\right)^{1/2}\int_\Gamma \frac{-im\zeta_E(\xi)w_1'(\xi)}{w_1'(\xi) - im\zeta_E(\xi)w_1(\xi)}e^{iX\xi}\,d\xi,$$

$$\tilde{N}_E(X) = e^{-3i\pi/4}\left(\frac{X^3}{\pi}\right)^{1/2}\int_\Gamma \frac{-im\zeta_E(\xi)w_1(\xi)}{w_1'(\xi) - im\zeta_E(\xi)w_1(\xi)}e^{iX\xi}\,d\xi.$$

For a shadowed wedge (see Figure 4.6 for the notations):

$$\begin{aligned}
H_z^d(M) &= \frac{1}{2}H_z^i(Q')\left[\frac{\rho(Q')}{\rho(M)}\right]^{1/6} e^{ikl_0'}\frac{i}{m(M)}f_H(\sigma_0')\Pi(l_0)\frac{i}{m(M)}N_H(\sigma_0)\\
&\quad\times\left[\frac{\partial^2}{\partial\theta\partial\theta'}D(\theta,\theta';\zeta_H)\right]_{\theta=\theta'=0},\\
\frac{\partial H_z^d}{\partial n}(M) &= -\frac{1}{2}H_z^i(Q')\left[\frac{\rho(Q')}{\rho(M)}\right]^{1/6} e^{ikl_0'}\frac{k}{m(M)}f_H(\sigma_0')\Pi(l_0)\\
&\quad\times\left(\frac{i}{m(M)}\right)^2\tilde{N}_H(\sigma_0)\left[\frac{\partial^2}{\partial\theta\partial\theta'}D(\theta,\theta';\zeta_H)\right]_{\theta=\theta'=0},\\
E_z^d(M) &= \frac{1}{2}E_z^i(Q')\left[\frac{\rho(Q')}{\rho(M)}\right]^{1/6} e^{ikl_0'}\frac{m(M)}{k}f_E(\sigma_0')\frac{\Pi(l_0)}{l_0}\frac{i}{m(M)}\tilde{N}_E(\sigma_0)\\
&\quad\times\left[\frac{\partial^2}{\partial\theta\partial\theta'}D(\theta,\theta';\zeta_E)\right]_{\theta=\theta'=0},\\
\frac{\partial E_z^d}{\partial n}(M) &= -\frac{1}{2}E_z^i(Q')\left[\frac{\rho(Q')}{\rho(M)}\right]^{1/6} e^{ikl_0'}f_E(\sigma_0')\frac{\Pi(l_0)}{l_0}\frac{i}{m(M)}N_E(\sigma_0)\\
&\quad\times\left[\frac{\partial^2}{\partial\theta\partial\theta'}D(\theta,\theta';\zeta_E)\right]_{\theta=\theta'=0}.
\end{aligned}\tag{4.39}$$

For grazing incidence we obtain a similar result to that in (4.36), but with N_h and N_s, respectively, replaced by P_H and P_E given by

$$P_H(X,Y) = \frac{1}{2\pi}\int_\Gamma w_1(\xi)\frac{(\partial I_1/\partial\xi)(\xi,Y) - im\zeta_H(\xi)I_1(\xi,Y)}{w_1'(\xi) - im\zeta_H(\xi)w_1(\xi)}e^{iX\xi}\,d\xi, \tag{4.40}$$

$$P_E(X,Y) = \frac{1}{2\pi}\int_\Gamma w_1'(\xi)\frac{(\partial I_1/\partial\xi)(\xi,Y) - im\zeta_E(\xi)I_1(\xi,Y)}{w_1'(\xi) - im\zeta_E(\xi)w_1(\xi)}e^{iX\xi}\,d\xi. \tag{4.41}$$

Two other functions $\tilde{P}_H$ and $\tilde{P}_E$ are needed to express $\partial H_z^d/\partial n$ and E_z^d. $\tilde{P}_H$ is obtained by replacing the factor $w_1(\xi)$ in the integrand of (4.40) by $w_1'(\xi)$. Similarly $\tilde{P}_E$ is obtained by replacing the factor $w_1'(\xi)$ in the integrand of (4.41) by $w_1(\xi)$.

4.3.3 *Improvement of the asymptotic currents close to the edge for a perfectly conducting wedge*

The asymptotic solutions (4.20) are not valid when the observation point S is very close to the source S', both being in the boundary layer of the curved surface. As a consequence, the asymptotic currents given by (4.30) are not valid when σ_0' tends to zero. This leads to an inaccuracy in the radiated field which is sensitive for $\vec{E}_\|^i$ polarisation.

In order to improve the asymptotic currents close to the edge, for the $\vec{E}_\|^i$ polarisation, we use formulae (4.30) with the Nicholson's function for $kl_0' \geq 2$. For $0.1 < kl_0' < 2$, an improved solution is given by the uniform asymptotic solution for a straight wedge in which the GO term is replaced by the PO term of a curved wedge. More precisely, for a straight wedge illuminated by a plane wave the UTD solution (derived from formulae (106) and (107) of Chapter 3.3 of Reference 7) at an observation point M situated on the illuminated face is given by

$$\begin{aligned}\frac{\partial E_z}{\partial n}(M) &= -2ik\sin\phi e^{-ik\rho\cos\phi}\\ &\quad - 4ik\sin\frac{\phi}{2}\frac{e^{i\pi/4}}{\sqrt{8\pi k}}B(\phi)\left[1 - F\left(2k\rho\cos^2\frac{\phi}{2}\right)\right]\frac{e^{ik\rho}}{\sqrt{\rho}}\\ &\quad - \frac{e^{i\pi/4}}{\sqrt{8\pi k}}F\left(2k\rho\cos^2\frac{\phi}{2}\right)C(\phi)\frac{e^{ik\rho}}{\rho\sqrt{\rho}},\end{aligned} \tag{4.42}$$

where $N\pi$ is the exterior wedge angle, ϕ is the angle between the reference face and the direction of the source and ρ is the radial distance of the observation point M to the edge.

The first term on the right-hand side of (4.42) corresponds to the GO term whereas the other two terms give the fringe current term. In (4.42), the coefficients B and C are given by

$$\begin{aligned}B(\phi) &= \frac{1}{N}\frac{\cos\phi/2}{\cos^2(\phi/2N) - \cos^2(\pi/2N)},\\ C(\phi) &= \frac{1}{N}\left\{\frac{\tan\phi/2}{\cos^2(\phi/2N) - \cos^2(\pi/2N)}\right.\\ &\qquad \left. - \frac{1}{N}\frac{\sin\phi/N}{[\cos^2(\phi/2N) - \cos^2(\pi/2N)]^2}\right\},\end{aligned} \tag{4.43}$$

and the function F is the Kouyoumjian and Pathak transition function for a wedge.

For $N = 2$, we have $B(\phi) = 1$ and $C(\phi) = 0$. Hence the solution (4.42) reduces to the exact solution of the half plane. The behaviour of the fringe current term when

ρ tends to zero is

$$\left(\frac{\partial E_z}{\partial n}\right)_f \cong \frac{2e^{i\pi/4}}{\sqrt{2\pi}} \sqrt{k} \frac{e^{ik\rho}}{\sqrt{\rho}} \sin\frac{\phi}{2},$$

which corresponds to the first term of the modal series solution.

For $N < 2$, we have $C(\phi) \neq 0$ and the solution behaves like $1/\rho$ when ρ tends to zero whereas the first term of the exact modal series solution behaves like $(\rho^{1-1/N})^{-1}$. It is therefore necessary to modify the solution (4.42) when $k\rho$ tends to zero. In our numerical results presented in Section 4.3.4, we have used the first term of the exact modal series solution for $k\ell'_0 \leq 0.1$. We have not found in the literature a uniform solution of the fringe current of a wedge for $\vec{E}^i_\parallel$ polarisation. Such a solution exists however for $\vec{H}^i_\parallel$ polarisation [8], but in this case it brings no sensitive effect on the radiation integral which is accurate with the representation of the fringe currents given by formulae (4.30).

4.3.4 *Numerical results*

The asymptotic solutions established in Chapter 3 have been tested on an ogival cylinder, by comparison of the results on the currents and on the radiated field with the MM results. The geometry of the ogival cylinder is shown in Figure 4.9.

The contour is defined by two arcs of a circle having the same radius, which intersect at A and B and form an ogive which is symmetric with respect to the main axis AB. Two different sizes have been considered $OA = 3\lambda$, $OC = 0.8\lambda$ and $OA = 1.5\lambda$, $OC = 0.4\lambda$, where λ is the wavelength. In both cases, the half wedge angle is nearly equal to 30°. As in the case of the elliptic cylinder, the direction of the incident field is fixed (θ^i = constant) and the observation angle θ varies around the centre of the ogive from 0 to 360°. For a point M in the shadow region, the asymptotic currents have been calculated by adding the contributions of the creeping wave launched at the shadow boundary, the creeping wave excited at the edge A and the two creeping waves launched at the shadow boundaries and diffracted at edge B in two other creeping waves reaching the point M. Double diffraction by edges has been neglected.

Typical results are shown in Figures 4.10–4.16. The results in Figures 4.10 and 4.11 correspond to a perfectly conducting ogival cylinder with half axes $a = 3\lambda$,

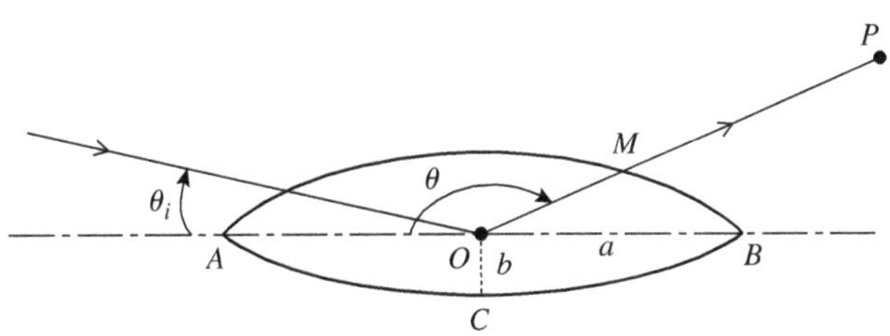

Figure 4.9 Ogival cylinder

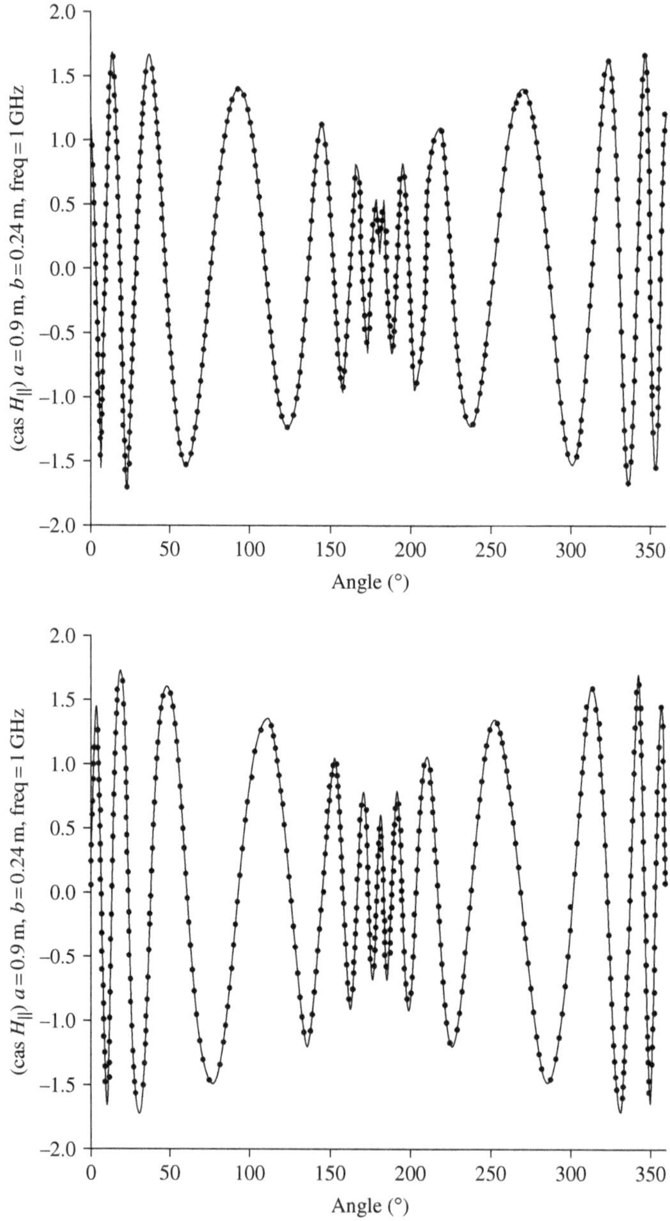

Figure 4.10 *Perfectly conducting ogival cylinder, $a = 3\lambda$, $b = 0.8\lambda$, incidence $\theta^i = 0$, $F = 1$ GHz. Comparison of the real and imaginary parts of the asymptotic currents with the results obtained by MM, $H_{\|}$ polarisation: ——, asymptotic; ……, MM*

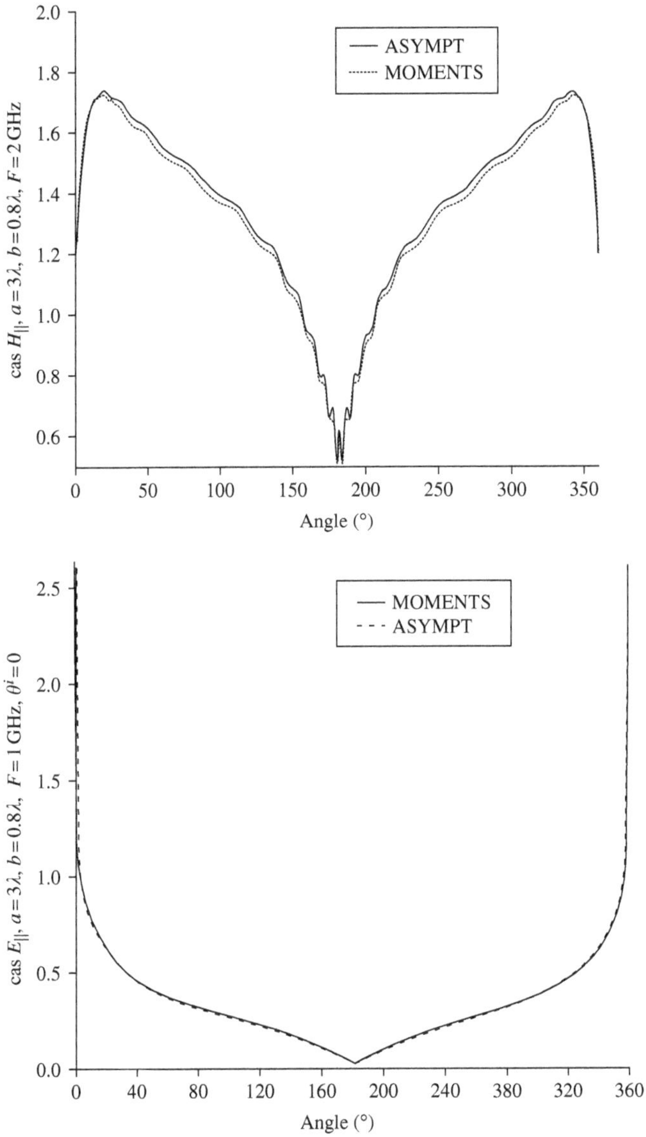

Figure 4.11 *Perfectly conducting ogival cylinder, $a = 3\lambda$, $b = 0.8\lambda$, incidence $\theta^i = 0$, $F = 1$ GHz. Comparison of the modulus of the asymptotic currents with the results obtained by MM, $H_{\|}$ and $E_{\|}$ polarisations*

$b = 0.8\lambda$, illuminated by a plane wave incident along the main axis of the ogive ($\theta^i = 0$). The agreement with the MM result is excellent, even for $\vec{E}^i_{\|}$ polarisation where the field close to the edges has been approximated by matching three different solutions as described in Section 4.3.3.

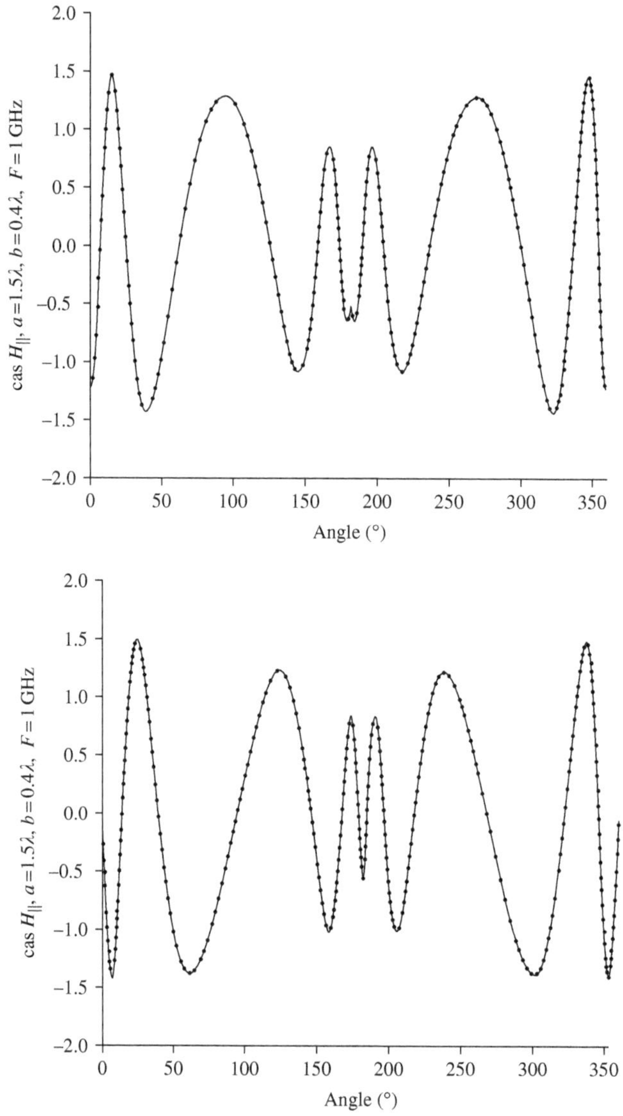

Figure 4.12 *Perfectly conducting ogival cylinder, $a = 1.5\lambda$, $b = 0.4\lambda$, incidence $\theta^i = 0$. Comparison of the real and imaginary parts of the asymptotic currents with the MM results, for $H_{\|}$ polarisation: ——, asymptotic; ……, MM*

For a smaller perfectly conducting ogival cylinder $a = 1.5\lambda$, $b = 0.4\lambda$, the same excellent agreement with the MM results is obtained as can be seen in the two diagrams of Figure 4.12. The diagrams of Figures 4.13–4.15 correspond to a perfectly conducting ogival cylinder $a = 3\lambda$, $b = 0.8\lambda$, illuminated at $\theta^i = 30°$. In this case,

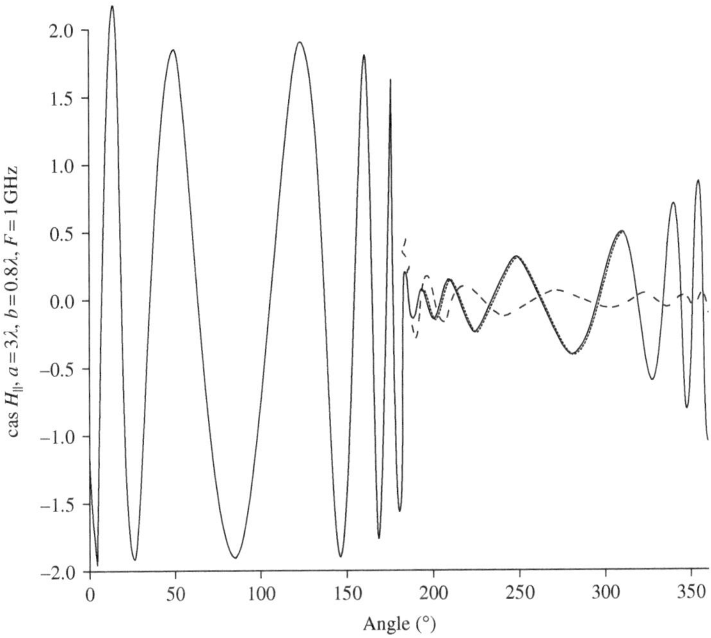

Figure 4.13 Perfectly conducting ogival cylinder, $a = 3\lambda$, $b = 0.8\lambda$, grazing incidence $\theta^i = 30°$, $H_{\|}$ polarisation. Real part of the currents: ......, asymptotic; ——, MM; – – – –, UTD

the direction of propagation of the incident field is tangent at A to the lower face of the corresponding wedge. When the proper solution for overlapping penumbra regions is used (dotted line), an excellent agreement with the MM results (solid line) is obtained for the real and imaginary parts and the modulus of the current. When the ordinary UTD solution valid outside the overlapping transition regions is used, the results shown in Figures 4.13–4.15 (dashed line) for the real part, imaginary part and modulus of the current on the lower face ($180° < \theta < 360°$) are completely wrong.

The results for a coated ogival cylinder with half-axes $a = 1.5\lambda$, $b = 0.4\lambda$ are given in Figure 4.16. The coating is a dielectric material with $\varepsilon_r = 3$, $\mu_r = 1$ and thickness $\lambda/60$. Good agreement with the MM results is obtained for the H_z component of the surface field. The results for $\vec{E}_z^i$ polarisation (not shown) are also in good agreement with the MM results.

4.4 Asymptotic currents on a 2D concave surface delimited by sharp edges

4.4.1 Introduction

It has been shown in Chapter 1, Section 1.2 that the curvilinear boundary of acoustic or electromagnetic objects carries specific waves. These are creeping waves on convex

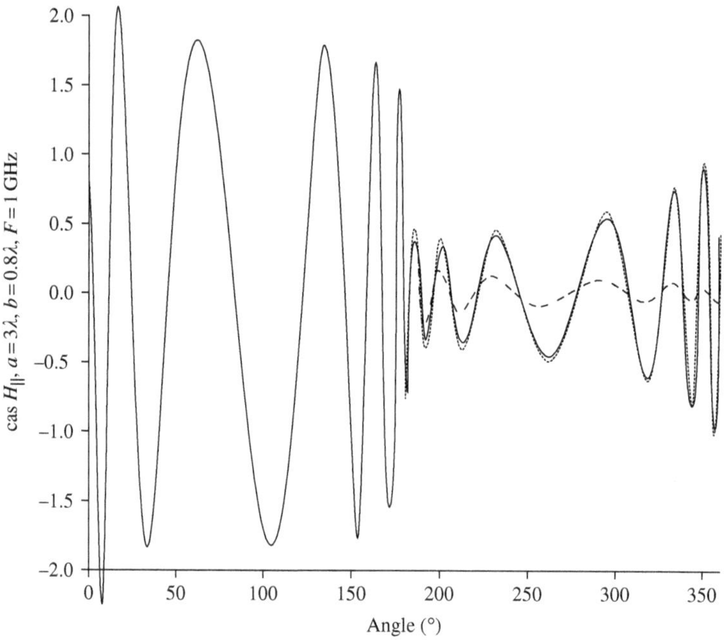

Figure 4.14 *Perfectly conducting ogival cylinder,* $a = 3\lambda$, $b = 0.8\lambda$, *grazing incidence* $\theta^i = 30°$, $H_{\|}$ *polarisation. Imaginary part of the currents: ……, Asymptotic; ——, MM; – – – –, UTD*

surfaces and whispering gallery waves on concave surfaces. If the normal coordinate n on the concave side is stretched as

$$\nu = \frac{2m^2 n}{\rho}, \qquad n < 0$$

then the boundary-layer method gives a solution on the concave side of the form (amplitude term in the general asymptotic form stated in the boundary-layer method):

$$u_0(s, \nu) = A(s)v(\xi - \nu), \tag{4.44}$$

where $v(x)$ is the Airy function defined by

$$v(x) = \frac{w_1(x) - w_2(x)}{2i}$$

and where $A(s)$ is a slowly varying function of the curvilinear coordinate s (Figure 4.17). The parameter ξ is a constant depending on the boundary conditions.

The solution (4.44) is a solution of the source free equations. In the presence of a line source, a matching procedure must be performed at the periphery of the caustic boundary layer of the rays emanating radially from the source.

The surface wave is generated by the interference of waves with a large number of reflections propagating inside the boundary layer. Such a solution cannot be

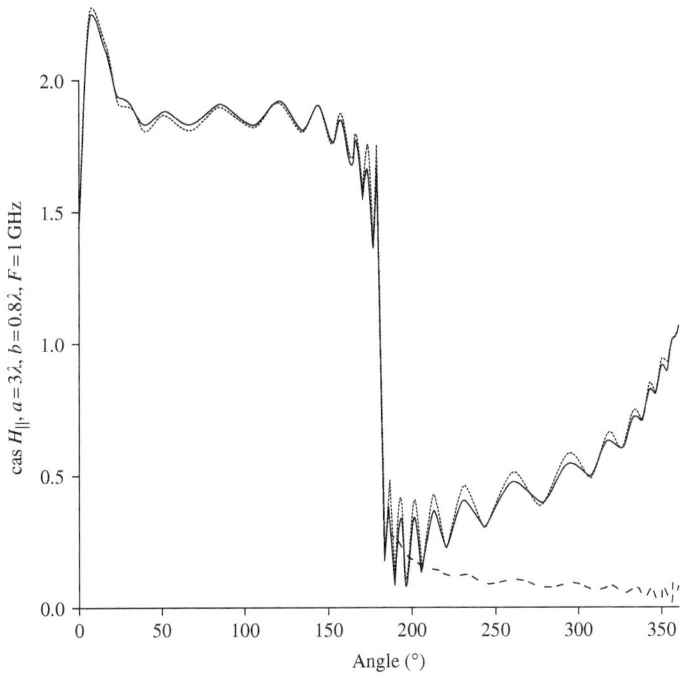

Figure 4.15 *Perfectly conducting ogival cylinder, $a = 3\lambda$, $b = 0.8\lambda$, grazing incidence $\theta^i = 30°$, $H_{\|}$ polarisation. Modulus of the currents: ......, asymptotic; ______, MM; – – – –, UTD*

constructed by the ray method. To describe the surface wave, we must use the solutions of the homogeneous Helmholtz equation near the surface S. Then we will be able to construct a superposition of these solutions that describes a surface wave propagating from the line source.

In order to find the way to solve this problem we turn, first of all, to the simplest etalon or model problem and then extend the solution of this problem to the general case. The simplest model problem is a line source near a circular boundary, where the variables are separable. We will show that a representation can be constructed for the surface wave in the form of a contour integral. This integral will show us the pattern of the desired superposition which describes the surface wave in the general case.

Once this problem has been solved, we use again the spectral theory of diffraction (STD) for calculating the whispering gallery wave field diffracted at a singularity in the surface of the sharp edge type.

4.4.2 *Solution of the canonical problem of a line source parallel to the generatrix of a concave circular cylinder*

A solution to this problem has been obtained separately by Wasylkiwskyj [9] and Babich and Buldyrev [5]. A complete analysis of the domain of validity of different

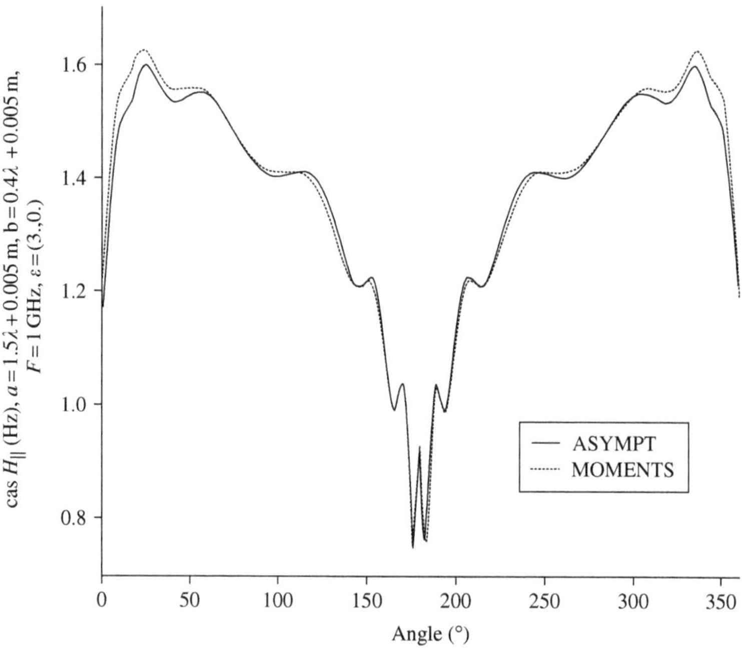

Figure 4.16 *Coated ogival cylinder, outer ellipse $a = 1.5\lambda$, $b = 0.4\lambda$, characteristics of the coating: $\varepsilon_r = 3$, $\mu_r = 1$, thickness $\lambda/60$. Comparison of the modulus of the asymptotic currents with the homologous MM results, for $H_{||}$ polarisation*

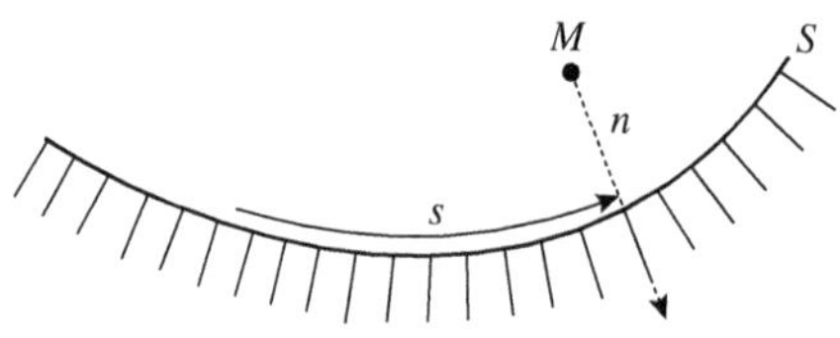

Figure 4.17 *Coordinate system (s, n)*

representations such as rays, whispering gallery modes, integral representations and their hybridisation is given in Reference 10.

The method used by all these authors consists in constructing the Green function in a domain delimited by the circular cylinder and two perfectly absorbing planes OP_1 and OP_2 passing through the axis O of the cylinder (Figure 4.18).

The perfectly absorbing planes make an interior angle slightly smaller than π in order to suppress the periodicity with respect to the azimuth angle ϕ, which can be extended to an infinite angular space $-\infty < \phi < +\infty$. However, the absorbing planes introduce a parasitic diffraction at the origin. But it will be shown that its

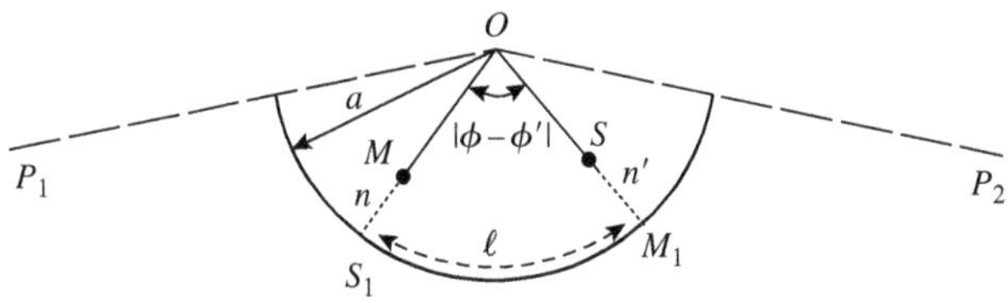

Figure 4.18 Canonical problem for a concave surface

contribution is of order $(ka)^{-1}$ whereas the principal contribution is of order $(ka)^{-1/2}$, where a is the radius of the cylinder. This parasitic effect can therefore be neglected.

The solution obtained by Wasylkiwskyj [9] is given by

$$G(\vec{\rho},\vec{\rho}') = \frac{1}{2\pi}\int_C e^{i\nu|\phi-\phi'|}\tilde{g}(\rho,\rho';\nu)\,d\nu, \tag{4.45}$$

where (ρ',ϕ') and (ρ,ϕ) are, respectively, the polar coordinates of the source S^S and of the observation point M and where ν is the spectral variable associated to the angular variable $\phi-\phi'$. The function $\tilde{g}$ is defined on the Riemann sheet $\operatorname{Re}\nu>0$:

$$\tilde{g}(\rho,\rho';\nu) = \begin{cases} g(\rho,\rho';\nu), & \operatorname{Re}\nu>0 \\ g(\rho,\rho';-\nu), & \operatorname{Re}\nu<0 \end{cases} \tag{4.46}$$

and $g=g_h$ for a magnetic line source and $g=g_s$ for an electric line source with

$$g_h(\rho,\rho';\nu) = \frac{i\pi}{2}J_\nu(k\rho_<)\left[H_\nu^{(1)}(k\rho_>) - \frac{H_\nu^{(1)'}(ka)}{J_\nu'(ka)}J_\nu(k\rho_>)\right], \tag{4.47}$$

$$g_s(\rho,\rho';\nu) = \frac{i\pi}{2}J_\nu(k\rho_<)\left[H_\nu^{(1)}(k\rho_>) - \frac{H_\nu^{(1)}(ka)}{J_\nu(ka)}J_\nu(k\rho_>)\right], \tag{4.48}$$

where $\rho_{\gtrless} = {}^{\max}_{\min}\{\rho,\rho'\}$.

The integration path C in the complex ν plane is defined in Figure 4.19.

To satisfy (4.46), a branch cut is introduced on the imaginary axis. The functions g_h and g_s have poles given respectively by the zeros ν_n' and ν_n of $J_\nu'(ka)$ and $J_\nu(ka)$, with $\operatorname{Re}\nu_n'>0$ and $\operatorname{Re}\nu_n>0$. These poles are of finite number. They are situated on the real axis and are less than ka. For $\operatorname{Re}\nu_n'>ka$ or $\operatorname{Re}\nu_n>ka$, $J_\nu'(ka)$ (respectively $J_\nu(ka)$), decreases monotonously and tend to zero when $\operatorname{Re}\nu_n'$ (respectively $\operatorname{Re}\nu_n$) tend to infinity.

By deforming the contour C into the contour C_1 (see Figure 4.19), we get

$$\begin{aligned} G_h(\vec{\rho},\vec{\rho}') = {} & \frac{\pi}{2}\sum_n e^{i\nu_n|\phi-\phi'|}\frac{J_{\nu_n'}(k\rho_<)}{(\partial/\partial\nu)J_\nu'(ka)|_{\nu=\nu_n'}}J_{\nu'_n}(k\rho_>)H_{\nu_n'}^{(1)'}(ka) \\ & + \frac{1}{2\pi}\int_{C_1} e^{i\nu|\phi-\phi'|}\tilde{g}_h(\rho,\rho';\nu)\,d\nu, \end{aligned} \tag{4.49}$$

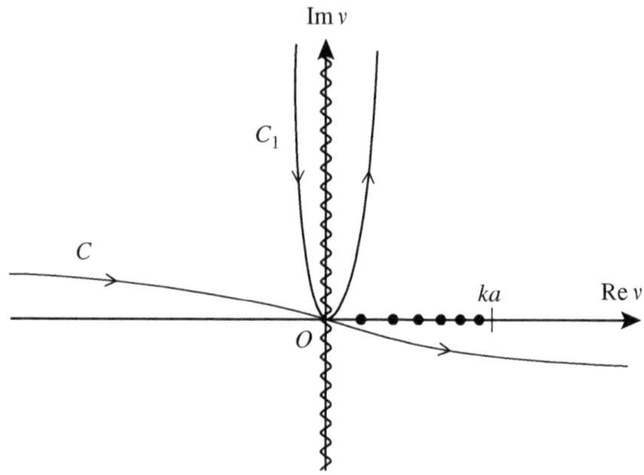

Figure 4.19 *Integration path and poles corresponding to the whispering gallery modes*

$$G_s(\vec{\rho},\vec{\rho}') = \frac{\pi}{2}\sum_n e^{i\nu_n|\phi-\phi'|}\frac{J_{\nu_n}(k\rho_<)}{(\partial/\partial\nu)J_\nu(ka)|_{\nu=\nu_n}}J_{\nu_n}(k\rho_>)H^{(1)}_{\nu_n}(ka)$$
$$+\frac{1}{2\pi}\int_{C_1} e^{i\nu|\phi-\phi'|}\tilde{g}_s(\rho,\rho';\nu)\,d\nu. \tag{4.50}$$

Since the exponential function in the residual integrals in (4.49) and (4.50) decreases rapidly along the imaginary ν axis when $|\phi-\phi'|$ is large, these integrals describe the behaviour of the solution when $|\phi-\phi'|$ is small which corresponds to an observation point close to the source, when both the source and the observation point are near the surface.

When the magnetic line source is on the surface, the whispering gallery modes which propagate from the source in the positive sense of the angle φ have the general analytical form

$$U_{\nu_n'}(\rho,\varphi) = Ae^{i\nu_n'\varphi}J_{\nu_n'}(k\rho), \qquad \varphi=\phi-\phi', \tag{4.51}$$

where A is a constant depending on ν_n'.

A similar analytical form is verified by the derivative with respect to the source coordinates of the whispering gallery mode excited by an electric line source.

Now, if in (4.51) $J_\nu(x)$ is replaced by

$$J_\nu(x) = \tfrac{1}{2}[H^{(1)}_\nu(x)+H^{(2)}_\nu(x)], \tag{4.52}$$

and if the Hankel functions in (4.52) are replaced by their Debye asymptotic expansions for $x \gg 1$, we can write (4.51) as the sum of two waves, one propagating towards the surface and the other propagating away from the surface. The phase of the wave propagating towards the surface comes from the asymptotic expansion of

$H_\nu^{(1)}(k\rho)$ which is given by

$$H_\nu^{(1)}(k\rho) \cong \sqrt{\frac{2}{\pi\rho\sin\sigma}}\,e^{ik(\rho\sin\sigma-(\nu/k)\sigma)-i\pi/4}, \qquad \sigma = \arc\cos\frac{\nu}{k\rho},$$
$$0 \le \operatorname{Re}\sigma < \pi. \tag{4.53}$$

Inserting (4.53) into (4.51), we get for the phase $S_{\nu'_n}(\rho,\varphi)$ of the modal rays propagating towards the surface S:

$$S_{\nu'_n}(\rho,\varphi) = \frac{\nu'_n}{k}\varphi + \rho\sin\sigma'_n - \frac{\nu'_n}{k}\arc\cos\frac{\nu'_n}{k\rho}, \qquad \sigma'_n = \arc\cos\frac{\nu'_n}{k\rho}, \tag{4.54}$$

and the gradient of the phase is given by

$$\begin{aligned}\overrightarrow{\nabla}S_{\nu'_n}(\rho,\varphi) &= \frac{\nu'_n}{k\rho}\hat{\phi} + \hat{\rho}\left(\sin\sigma'_n + \rho\cos\sigma'_n\frac{d\sigma'_n}{d\rho} - \frac{\nu'_n}{k}\frac{d\sigma'_n}{d\rho}\right)\\ &= \cos\sigma'_n\hat{\phi} + \sin\sigma'_n\hat{\rho}.\end{aligned} \tag{4.55}$$

Equation (4.55) shows that the modal rays which are directed as $\overrightarrow{\nabla}S$, make a constant angle with the circle of radius ρ, given by $\sigma'_n = \arc\cos(\nu'_n/k\rho)$. They also make a constant angle with the circular cylinder of radius a, given by $\psi'_n = \arc\cos(\nu'_n/ka)$ (see Figure 4.20).

These rays are tangential to a caustic which is also a circular cylinder, the radius ρ_c of which is given by

$$\arc\cos\frac{\nu'_n}{k\rho_c} = 0$$

and therefore

$$\rho_c = \frac{\nu'_n}{k} = a\cos\psi'_n. \tag{4.56}$$

If we treat in the same way the Hankel function $H_{\nu'_n}^{(2)}(k\rho)$ in (4.49), we obtain a congruence of rays leaving the surface S and having the same caustic. These rays

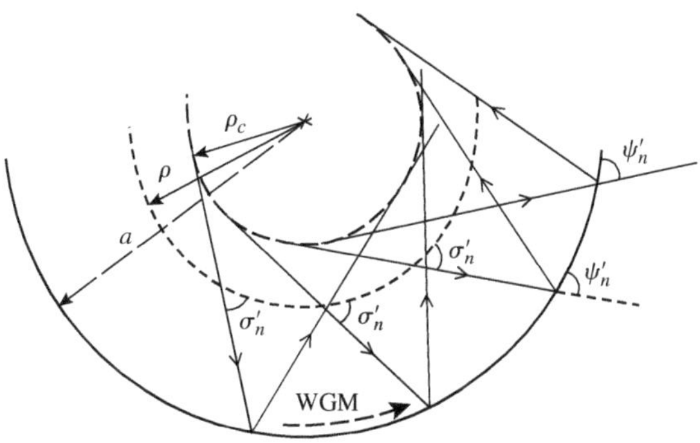

Figure 4.20 Modal rays and caustic

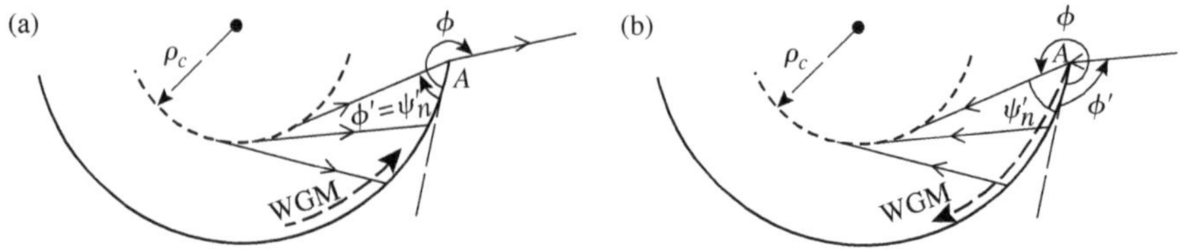

Figure 4.21 *Modal wave representations of the whispering gallery mode propagating towards or away from the wedge*

correspond to the modal rays reflected by the concave surface. It is possible to treat the diffraction by the edge of a truncated cylinder by using these modal rays. A whispering gallery mode propagating towards the edge A is equivalent to a divergent modal wave. The incident ray makes an angle ψ'_n with the plane tangent to the cylinder at the edge (Figure 4.21(a)). Hence, the corresponding diffracted ray is directly given by the law of edge diffraction. If the whispering gallery mode is excited by the diffraction of a plane wave by the edge of a truncated circular cylinder, the corresponding diffracted wave is a convergent wave and the angle of diffraction depends on the mode and is given by $2\pi - \psi'_n$ (Figure 4.21(b)).

The solution constructed in this way is in accordance with the exact solution established in Reference 11.

Now, if we want to apply the same technique to the field defined by (4.49) and (4.50), we are confronted with two problems. The first problem is the residual integral which can be neglected only if the source is not too close to the edge. This difficulty, which in the asymptotic current method has to be considered in the reciprocal situation when the observation point is close to the edge, can be solved by replacing the original problem by its tangent plane approximation close to the edge. The second problem comes from the fact that the modes with ν'_n close to ka, have a caustic surface which, according to (4.56), is close to the surface S of the circular cylinder. In this case, the Debye approximation of the Hankel functions is not valid and hence, the modal wave is not a local plane wave. The modal wave representation of the field excited on the concave side of the surface by a line source is therefore not applicable for treating the diffraction by an edge.

We will keep therefore the integral form of the solution which we will transform before applying the STD.

4.4.3 Transformation of the integral form of the solution

We start with the identities

$$\begin{aligned}
\frac{1}{J'_\nu(ka)} &= \frac{2}{H'^{(2)}_\nu(ka)}\frac{1}{1-R_h}, &\qquad R_h &= -\frac{H'^{(1)}_\nu(ka)}{H'^{(2)}_\nu(ka)},\\
\frac{1}{J_\nu(ka)} &= \frac{2}{H^{(2)}_\nu(ka)}\frac{1}{1-R_s}, &\qquad R_s &= -\frac{H^{(1)}_\nu(ka)}{H^{(2)}_\nu(ka)}.
\end{aligned} \tag{4.57}$$

Then, we replace in (4.49) and (4.50) the Bessel functions by Hankel functions, using (4.52), and taking into account (4.57), we can write

$$g_h(\rho,\rho';\nu) = i\frac{\pi}{2}H_\nu'^{(2)}(ka)\frac{F_\nu'(\rho_>,a)J_\nu(k\rho_<)}{2J_\nu'(ka)} = i\frac{\pi}{2}\frac{F_\nu'(\rho_>,a)J_\nu(k\rho_<)}{1-R_h}, \tag{4.58}$$

where

$$F_\nu'(\rho,a) = H_\nu^{(1)}(k\rho) + R_h H_\nu^{(2)}(k\rho). \tag{4.59}$$

By using the identity

$$2J_\nu(k\rho_<) = F_\nu'(\rho_<,a) + (1-R_h)H_\nu^{(2)}(k\rho_<), \tag{4.60}$$

it follows that

$$g_h(\rho,\rho';\nu) = i\frac{\pi}{4}\frac{F_\nu'(\rho_>,a)F_\nu'(\rho_<,a)}{1-R_h} + i\frac{\pi}{4}F_\nu'(\rho_>,a)H_\nu^{(2)}(k\rho_<), \tag{4.61}$$

with a similar equation for g_s

$$g_s(\rho,\rho';\nu) = i\frac{\pi}{4}\frac{F_\nu(\rho_>,a)F_\nu(\rho_<,a)}{1-R_s} + i\frac{\pi}{4}F_\nu(\rho_>,a)H_\nu^{(2)}(k\rho_<), \tag{4.62}$$

where

$$F_\nu(\rho,a) = H_\nu^{(1)}(k\rho) + R_s H_\nu^{(2)}(k\rho). \tag{4.63}$$

Equations (4.61) and (4.62) are the analytical forms of the solution used by Babich and Buldyrev [5].

Now, in (4.45) we deform the integration path C in the complex ν plane into the path C_0 which for $\mathrm{Re}\,\nu < 0$ follows the positive imaginary axis and for $\mathrm{Re}\,\nu > 0$ is identical to C (Figure 4.22).

Then, the original integral (4.45) is written as the sum of three terms which for a magnetic line source are:

$$\begin{aligned} G_h(\vec{\rho},\vec{\rho}') = {} & \frac{i}{8}\int_0^\infty \frac{F_\nu'(\rho_<,a)F_\nu'(\rho_>,a)}{1-R_h} e^{i\nu|\phi-\phi'|}\,d\nu \\ & + \frac{i}{8}\int_{C_0} H_\nu^{(2)}(k\rho_<)F_\nu'(\rho_>,a)e^{i\nu|\phi-\phi'|}\,d\nu \\ & + \frac{1}{8}\int_{i\infty}^{0} e^{2i\pi\nu}\frac{F_\nu'(\rho_<,a)F_\nu'(\rho_>,a)}{1-e^{2i\pi\nu}R_h} e^{i\nu|\phi-\phi'|}\,d\nu, \end{aligned} \tag{4.64}$$

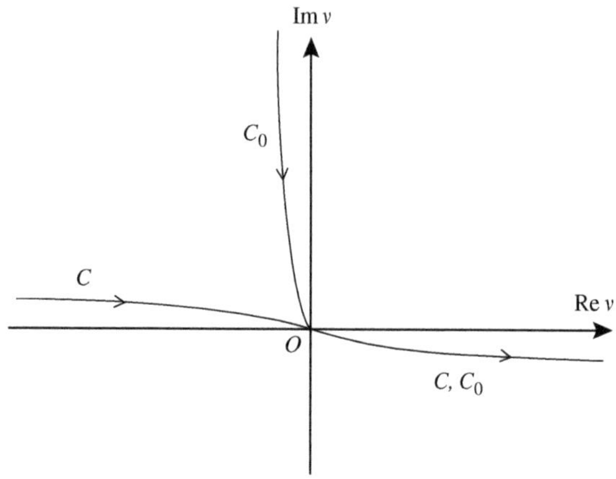

Figure 4.22 Deformation of the integration path C into C_0

where we have used the relation

$$H_{-\nu}^{(1,2)}(x) = e^{\pm i\nu\pi} H_{\nu}^{(1,2)}(x). \tag{4.65}$$

If in the last integral on the right-hand side of (4.64), we replace the Hankel functions by their Debye asymptotic approximations, we can see that the integrand has no stationary point. The dominant term of its asymptotic expansion is given by the end-point contribution at $\nu = 0$ which can be calculated by integration by parts and gives a term of order k^{-1}.

We do not modify this result by replacing this integral by

$$\frac{i}{8}\int_{-i\infty}^{0} \frac{F_{\nu}'(\rho_<, a)F_{\nu}'(\rho_>, a)}{1 - R_h} e^{i\nu|\phi-\phi'|}\, d\nu, \tag{4.66}$$

since the sum

$$\frac{1}{8}\int_{i\infty}^{0} e^{2i\pi\nu} \frac{F_{\nu}'(\rho_<, a)F_{\nu}'(\rho_>, a)}{1 - e^{2i\pi\nu} R_h} e^{i\nu|\phi-\phi'|}\, d\nu + \int_{0}^{-i\infty} \frac{F_{\nu}'(\rho_<, a)F_{\nu}'(\rho_>, a)}{1 - R_h} e^{i\nu|\phi-\phi'|}\, d\nu \tag{4.67}$$

has no end-point contribution.

In the same way, by using the Debye asymptotic approximation for the Hankel functions in the integrand of the second integral on the right-hand side of (4.64), it can be shown that this integrand has two stationary points situated on the real axis which correspond to a direct and a reflected ray (Figure 4.23).

By deforming the path of integration into steepest descent paths, passing through these points, we obtain by the steepest descent method (SDM) the corresponding contributions to the field. The existence of the stationary points is submitted to the

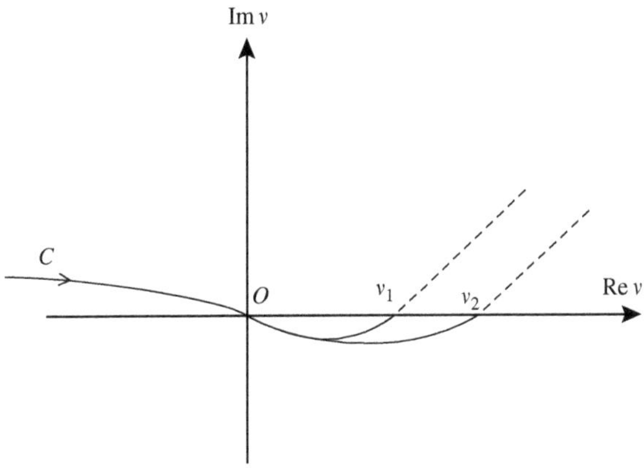

Figure 4.23 Integration path deformation through the stationary points

conditions:

$$0 \leq |\phi - \phi'| \leq \frac{\pi}{2} - \arcsin\left(\frac{\rho_<}{\rho_>}\right): \qquad \text{for the direct ray,}$$

$$0 \leq |\phi - \phi'| \leq \tau_1, \quad \tau_1 = \frac{\pi}{2} + \arcsin\left(\frac{\rho_<}{\rho_>}\right) - 2\arcsin\left(\frac{\rho_<}{a}\right)$$

for the reflected ray.

If the source S and the observation point M are both on the surface, these rays exist only when $|\phi - \phi'| = 0$, so that S and M coalesce, and their contributions are strictly zero. When S and M are in the boundary layer, the value of τ_1 is close to zero since ρ' and ρ differ from a by a small quantity of the order of m/k where m is the Fock parameter $(ka/2)^{1/3}$ so that the rays exist only if M is close to S. In this case it can be shown by SDM that their contributions are negligibly small [9].

We can therefore neglect the contribution of the second integral on the right-hand side of (4.64).

In addition, if we neglect the end-point contribution in (4.66) which can be attributed to the parasitic effect introduced by the absorbing planes delimiting the domain in which we have solved the Helmholtz equation (Figure 4.18), we obtain the final result for a magnetic line source excitation:

$$G_h(\vec{\rho}, \vec{\rho}') = \frac{i}{8} \int_{C_1} \frac{F'_\nu(\rho_<, a) F'_\nu(\rho_>, a)}{1 - R_h} e^{i\nu|\phi-\phi'|}\, d\nu, \tag{4.68}$$

where the integration path C_1 is shown in Figure 4.24.

Similarly, for an electric line source excitation, we get:

$$G_s(\vec{\rho}, \vec{\rho}') = \frac{i}{8} \int_{C_1} \frac{F_\nu(\rho_<, a) F_\nu(\rho_>, a)}{1 - R_s} e^{i\nu|\phi-\phi'|}\, d\nu. \tag{4.69}$$

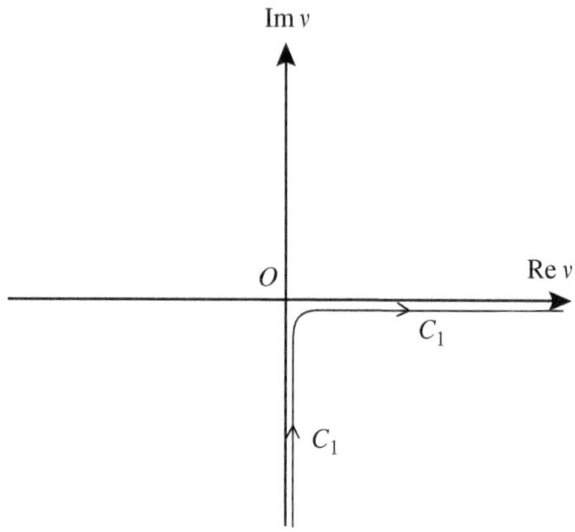

Figure 4.24 Integration path C_1

When in (4.68) and (4.69) the Hankel functions are approximated by the Debye asymptotic expansion, we see that the dominant contribution to the integral comes from ν close to ka provided that $|\phi - \phi'|$ remains small compared to π. Since for $ka \gg 1$, the Debye expansion of $H_\nu^{(1,2)}(ka)$ remains valid very close to $\nu = ka$, it follows that the most important contribution to the integrals (4.68) and (4.69) comes from the neighbourhood of $\nu = ka$. This argument leads to the following change of variables:

$$\nu = ka + m\xi, \qquad m = \left(\frac{ka}{2}\right)^{1/3}. \tag{4.70}$$

For $ka \gg 1$ and $\xi \leq 1$, the Bessel and Hankel functions may be replaced by their asymptotic expansions:

$$\begin{aligned} J_\nu(ka) &\cong (m\sqrt{\pi})^{-1} v(\xi), \\ H_\nu^{(1,2)}(ka) &\cong \mp i\left(m\sqrt{\pi}\right)^{-1} w_{1,2}(\xi), \end{aligned} \tag{4.71}$$

and since

$$\begin{aligned} \nu &= k(a+n) + m\left(\xi - \frac{kn}{m}\right) = k\rho + m(\xi - h), \qquad h = \frac{kn}{m}, \\ &= k(a+n') + m\left(\xi - \frac{kn'}{m}\right) = k\rho' + m(\xi - h'), \qquad h' = \frac{kn'}{m}, \end{aligned} \tag{4.72}$$

we have also, for $h \leq 1$, $h' \leq 1$:

$$\begin{aligned} &J_\nu(k\rho) \cong (m\sqrt{\pi})^{-1} v(\xi - h), \qquad J_\nu(k\rho') \cong \left(m\sqrt{\pi}\right)^{-1} v(\xi - h'), \\ &H_\nu^{(1,2)}(k\rho) \cong \mp i(m\sqrt{\pi})^{-1} w_{1,2}(\xi - h), \\ &H_\nu^{(1,2)}(k\rho') \cong \mp i(m\sqrt{\pi})^{-1} w_{1,2}(\xi - h'). \end{aligned} \tag{4.73}$$

We obtain by inserting (4.71)–(4.73) into (4.68) and (4.69):

$$G_{h,s}(\vec{\rho},\vec{\rho}\,') = \frac{e^{ik\ell}}{4\pi m} S_{h,s}^{(0)}(\sigma, h, h'), \tag{4.74}$$

where ℓ is the distance along the concave surface between the projections S_1 and M_1 of the source S and the observation point M on the surface and $\sigma = m\ell/a$. The Fock-type functions $S_{h,s}^{(0)}$, where the superscript (0) means that no GO contribution has been extracted, are given by

$$\begin{aligned} S_h^{(0)}(\sigma, h, h') &= \frac{1}{4}\int_{C_\xi} e^{i\sigma\xi} F'_\xi(h) F'_\xi\left(h'\right) \frac{w_2'(\xi)}{v'(\xi)}\, d\xi, \\ S_s^{(0)}(\sigma, h, h') &= \frac{1}{4}\int_{C_\xi} e^{i\sigma\xi} F_\xi(h) F_\xi(h') \frac{w_2(\xi)}{v(\xi)}\, d\xi, \end{aligned} \tag{4.75}$$

where

$$\begin{aligned} F_\xi(X) &= w_1(\xi - X) - \frac{w_1(\xi)}{w_2(\xi)} w_2(\xi - X), \\ F'_\xi(X) &= w_1(\xi - X) - \frac{w_1'(\xi)}{w_2'(\xi)} w_2(\xi - X). \end{aligned} \tag{4.76}$$

Since $v(\xi)$ has its fastest growth on the straight line $\arg\xi = -2\pi/3$ and decays only algebraically when $\arg\xi = \pi/3$ whereas its fastest decay is on the real axis and is exponential (see Reference 5, p. 408 or Reference 3, p. 514), it is advantageous to deform the integration path in the straight line $\exp(i4\pi/3), \exp(i\pi/3)$ passing through the origin (Figure 4.25). This path is designated by C_ξ.

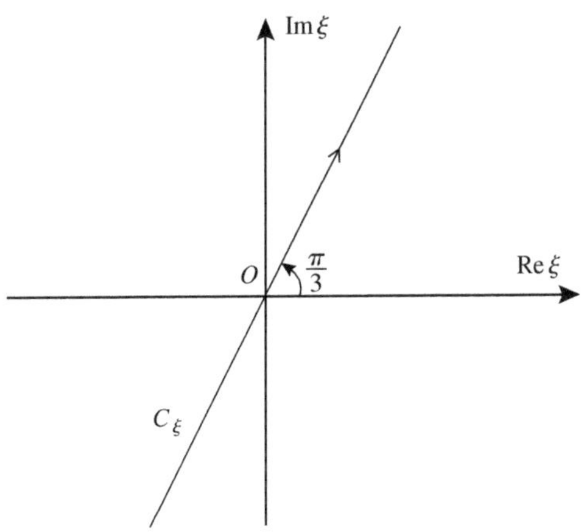

Figure 4.25 Integration path C_ξ

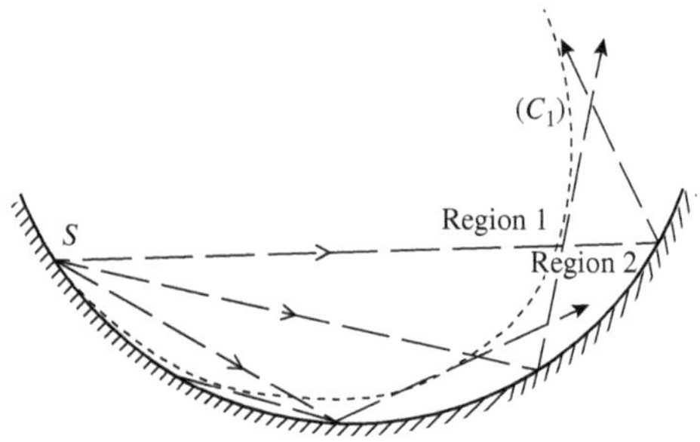

Figure 4.26 Caustic of simply reflected rays

Formulae (4.75) are valid when the angular distance between the source and the observation point is not too large. In order to specify this rule more carefully, we consider the rays emanating from the source which we suppose is located on the concave surface (Figure 4.26) and which undergoes one reflection on the surface. These rays have an envelope or caustic C_1 which is real and which divides the space above the concave surface in two domains denoted by regions 1 and 2.

The field in the neighbourhood of the caustic, in region 2, expressed in the curvilinear coordinates s_c, n_c, where s_c is the abscissa along the curve C_1 and n_c the distance of the observation point M to the caustic, is given by

$$u(M) = A_0(s_c)e^{iks_c}v(-\nu_c), \tag{4.77}$$

where $v(x)$ is the Airy function and where $A_0(s_c)$ is a slowly varying function of the radius ρ_c of the caustic at the abscissa s_c and $\nu_c = kn_c/m_c$ where $m_c = (k\rho_c/2)^{1/3}$ (see Chapter 1, Section 1.2.8).

The caustic boundary is defined by $\nu \leq 1$, hence,

$$n_c \leq \frac{m_c}{k}. \tag{4.78}$$

The condition (4.78) defines the maximum distance along the concave surface where formulae (4.75) are valid. This distance corresponds approximatively to $\sigma = 3.5$ with $\sigma = ml/a$. For a larger distance, the reflected field at a point on the surface can be calculated by the laws of GO which result from Debye's asymptotic approximations of the Hankel functions. In this case the asymptotic expansions (4.71) and (4.73) are no longer valid. However, we cannot replace them throughout by Debye's asymptotic expansion. The correct procedure consists in extracting from (4.68) and (4.69) the direct field or the direct and the reflected field and in treating the remainder integral by using the expansions (4.71) and (4.73). This procedure can be repeated when the distance from M to S augments by extracting the rays which have 2 or more, say N reflections on the surface.

The GO terms are easily extracted by writing

$$\frac{1}{1 - R_{s,h}} = \sum_{n=1}^{N} R_{s,h}^{n} + \frac{R_{s,h}^{N+1}}{1 - R_{s,h}}, \tag{4.79}$$

and by approximating the Hankel functions in the first term on the right-hand side of (4.79) by the Debye asymptotic expansion and evaluating the integrals by the stationary phase method. The second term which we call the remainder is calculated by replacing in the integrals (4.68) and (4.69) the Bessel and Hankel functions by their expansions (4.71) and (4.73). We do not perform these operations which are rather lengthy and which can be found in References 5, 8 and 10.

4.4.4 Coated concave surface

For a coated concave surface, an extension of the solutions (4.75) to the canonical problem of Figure 4.18 where the perfectly conducting concave surface is covered by an absorbing material of constant thickness characterised by a complex permittivity and permeability, leads to a Green's function for a magnetic (H) and electric (E) line source, given by

$$G_{H,E}(\sigma,n,n') = \frac{e^{ik\ell}}{4\pi m} S^{(0)}_{H,E}(\sigma,n,n'), \tag{4.80}$$

where the distances n' and n are measured from the surface of the coating to the source or the observation point. The Fock-type functions $S^{(0)}_{H,E}$ are defined by

$$S^{(0)}_{H,E}(\sigma,n,n') = \frac{1}{4}\int_{C_\xi} F_\xi(n, Z_{H,E})F_\xi(n', Z_{H,E})\frac{\Omega_{H,E}w_2(\xi)}{\Omega_{H,E}v(\xi)}\,d\xi, \tag{4.81}$$

where for $x = n$ or n', $Z = Z_H$ or Z_E and $\Omega = \Omega_H$ or Ω_E.

$$F_\xi(x, Z) = w_1(\xi - \nu) - \frac{\Omega w_1(\xi)}{\Omega w_2(\xi)} w_2(\xi - X), \quad \nu = \frac{kx}{m} \tag{4.82}$$

and where Ω is the operator

$$\Omega = \frac{\partial}{\partial\xi} - imZ. \tag{4.83}$$

4.4.5 Edge-excited currents on a perfectly conducting concave surface

By expanding $F_\xi(n)$ and $F'_\xi(n)$ in a Taylor series with respect to n it is possible to represent $S^{(0)}_h$ and $S^{(0)}_s$ in the form of a spectrum of local inhomogeneous plane waves. Applying STD to each component of this spectrum leads to the solution of the diffraction problem of a whispering gallery type field radiated by a line source, with an edge in a concave surface and by reciprocity to the following expressions for the edge-diffracted field on the concave surface for an illuminated edge:

$\vec{H}^i_\parallel$ polarisation

$$\begin{aligned} &H^d_z(M) = H^i_z(A)\Pi(\ell)D_h(\phi,0)\bar{S}_h(\sigma), \\ &\frac{\partial H^d_z}{\partial n}(M) = 0. \end{aligned} \tag{4.84}$$

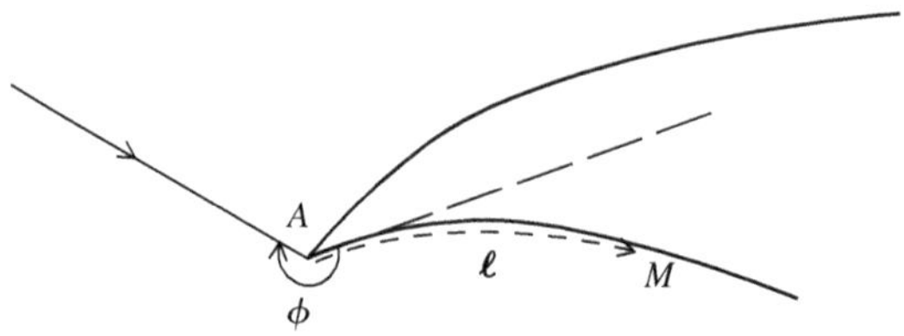

Figure 4.27 Wedge configuration and ray path with edge A in the lit region

$\vec{E}^i_{\parallel}$ polarisation

$$E_z^d(M) = 0,$$
$$\frac{\partial H_z^d}{\partial n}(M) = -E_z^i(A)\frac{\Pi(\ell)}{\ell}\left[\frac{\partial}{\partial\theta}D_s(\phi,\theta)\right]_{\theta=0}\bar{S}_s(\sigma), \tag{4.85}$$

where

$$\bar{S}_h(\sigma) = \frac{e^{-i\pi/4}}{2}\left(\frac{\sigma}{\pi}\right)^{1/2}\int_{C_\xi}\frac{e^{i\sigma\xi}}{v'(\xi)w_2'(\xi)}\,d\xi,$$
$$\bar{S}_s(\sigma) = e^{-3i\pi/4}\left(\frac{\sigma^3}{\pi}\right)^{1/2}\int_{C_\xi}\frac{e^{i\sigma\xi}}{v(\xi)w_2(\xi)}\,d\xi. \tag{4.86}$$

The parameter σ is defined by $\sigma = m\ell/a$ for a circular concave surface where $\ell = \text{arc } AM$ (Figure 4.27).

Formulae (4.84) and (4.85) have the same analytical form as (4.29) and (4.30). Only the Fock type functions are modified. They have been written for a circular concave surface. For a general concave surface with variable curvature they must be adapted with the same modifications as in Section 4.3.1.1.

For an edge situated in the shadow region of the incident GO field, the formulae for the z-components of the field, which completely define the currents, can also be directly written down by reproducing the analytical form of (4.32) and (3.16):

$\vec{H}^i_{\parallel}$ polarisation

$$H_z(M) = \tfrac{1}{2}H^i(Q')e^{ik\ell'}g(\sigma')\Pi(\ell)D_h(0,N\pi)\bar{S}_h(\sigma),$$
$$\frac{\partial E_z^d}{\partial n}(M) = -\frac{1}{2}E_z^i(Q')\frac{i}{m}e^{ik\ell'}f(\sigma')\frac{\Pi(\ell)}{\ell} \times\left[\frac{\partial^2}{\partial\theta\partial\theta'}D_s(\theta,N\pi+\theta')\right]_{\theta=\theta'=0}\bar{S}_s(\sigma), \tag{4.87}$$

where the notations are defined in Figure 4.28.

Formulae (4.87) give the field excited at M by the diffraction of a creeping wave in a whispering gallery type wave. They have been written for a circular convex and

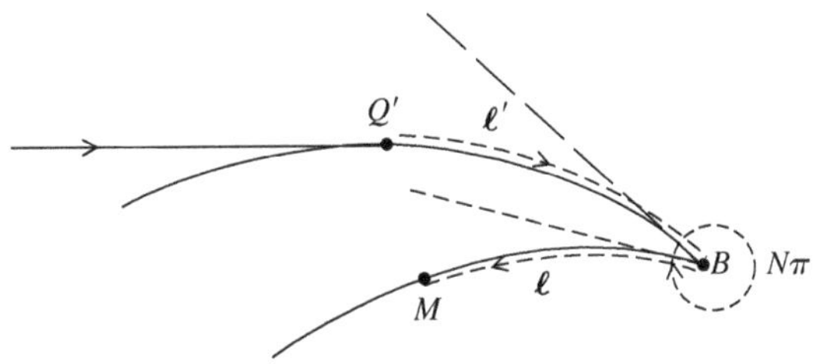

Figure 4.28 Wedge configuration and ray path with edge B in the shadow region

concave surface. For a general 2D wedge with a convex and a concave face, they must be adapted as in Section 4.3.1.2.

4.4.6 Edge-excited currents on a coated concave surface

Similarly, the formulae for edge-excited currents on a coated concave surface can be directly written down by following the analytical structure of (4.38) and (4.39). Only the special functions will be modified. They are given by

$$\begin{aligned}
\bar{S}_H(x) &= \frac{e^{-i\pi/4}}{2}\left(\frac{x}{\pi}\right)^{1/2}\int\limits_{C_\xi}\frac{-im\,\zeta_H(\xi)}{[\Omega_H w_2(\xi)][\Omega_H v(\xi)]}e^{ix\xi}\,d\xi,\\
\bar{\bar{S}}_H(x) &= e^{-i\pi/4}\left(\frac{x}{\pi}\right)^{1/2}\int\limits_{C_\xi}\frac{-im\,\zeta_H(\xi)}{[\Omega_H w_2(\xi)][\Omega_H v(\xi)]}e^{ix\xi}\,d\xi,\\
\bar{S}_E(x) &= e^{-3i\pi/4}\left(\frac{x^3}{\pi}\right)^{1/2}\int\limits_{C_\xi}\frac{-im\,\zeta_E(\xi)}{[\Omega_E w_2(\xi)][\Omega_E v(\xi)]}e^{ix\xi}\,d\xi,\\
\bar{\bar{S}}_E(x) &= e^{-3i\pi/4}\left(\frac{x^3}{\pi}\right)^{1/2}\int\limits_{C_\xi}\frac{-im\,\zeta_E(\xi)}{[\Omega_E w_2(\xi)][\Omega_E v(\xi)]}e^{ix\xi}\,d\xi.
\end{aligned} \tag{4.88}$$

4.5 Three-dimensional perfectly conducting convex–concave surface

The extension of the asymptotic solutions established in Section 4.3.3 for the surface currents on a 2D perfectly conducting convex or concave surface delimited by a sharp edge to a 3D geometry follows the same steps as those described in Section 3.3.2 of Chapter 3.

The first step consists in treating the oblique incidence on a 2D geometry in order to introduce the vectorial nature of the field.

A point M on the convex or concave face of a wedge is defined by its coordinates (ℓ, z) where ℓ is the curvilinear abscissa of M projected on the cross-section of the cylinder (Figures 4.29(a) and (b)).

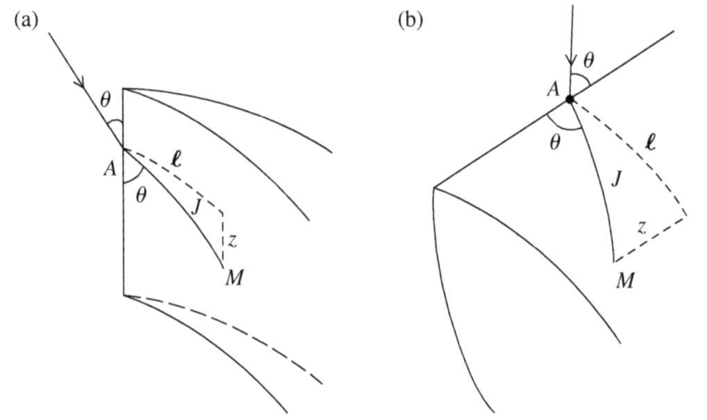

Figure 4.29 Curvilinear and ray-fixed coordinates of a point M on a concave face (a) and a convex face (b)

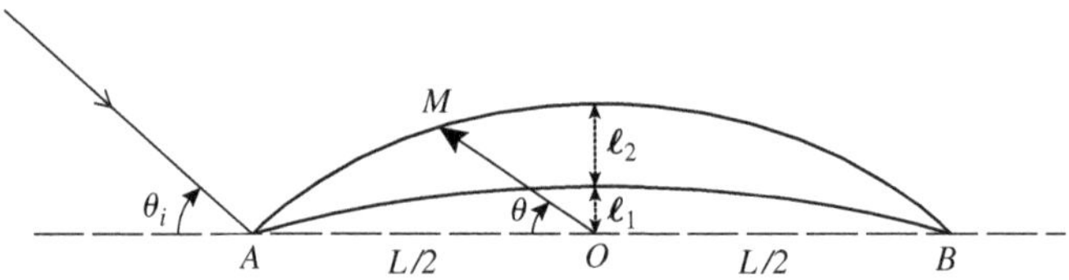

Figure 4.30 Crescent-shaped cylinder

We consider first an illuminated edge. The incident field is expressed in the spherical coordinates θ, ϕ:

$$\vec{E}_i = (E^i_\theta \hat{\theta} + E^i_\phi \hat{\phi}) e^{ikz\cos\theta} e^{ik_t(x\cos\theta + y\sin\theta)}, \tag{4.89}$$

where $k_t = k\sin\theta$ and

$$E^i_z = -E^i_\theta \sin\theta, \qquad H^i_z = Y_0 E^i_\phi \sin\theta, \tag{4.90}$$

Y_0 being the characteristic admittance of the vacuum.

The diffracted magnetic field on the surface at M is represented by its $\hat{z}$ and $\hat{\ell}$ components

$$\vec{H}^d = H^d_z \hat{z} + H^d_\ell \hat{\ell}. \tag{4.91}$$

Since $\vec{H}^d_\ell = \hat{\ell} \cdot \vec{H}^d_t$ where $\vec{H}^d_t$ is the transverse component of the diffracted magnetic field at M we obtain by using the formula giving the transverse component $\vec{H}^d_t$ as a function of the longitudinal components H^d_z and E^d_z

$$\vec{H}^d_t = \frac{1}{k_t^2}[ik_z \nabla_t H_z + i\omega\varepsilon \hat{z} \times \nabla_t E_z] \tag{4.92}$$

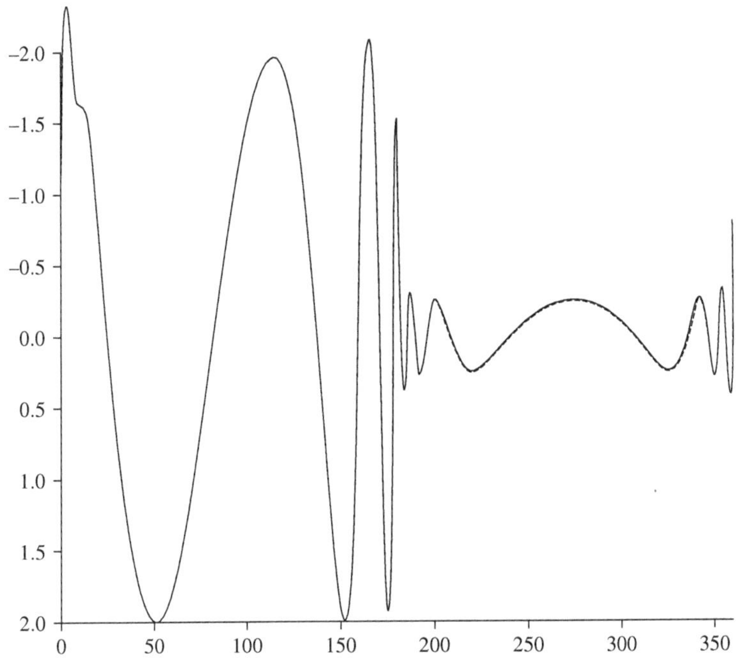

Figure 4.31 *Real part of the asymptotic currents on a cylinder with crescent-shaped cross-section compared to the MM results, for $H_{\|}$ polarisation. Parameters used in the numerical application: $\theta_i = 60°$, θ varies from 0 to 180° on the convex face and from 180 to 360° on the concave face, $L = 6\lambda$, $\ell_1 = \ell_2 = 0.4\lambda$, $F = 300$ MHz: ——, MM; ----, asymptotic*

and by stating $\hat{n} = \hat{z} \times \hat{\ell}$

$$H_\ell^d = \frac{i}{k \sin^2 \theta} \left(-Y_0 \frac{\partial E_z^d}{\partial n} + \cos \theta \frac{\partial H_z^d}{\partial \ell} \right). \tag{4.93}$$

The longitudinal components in (4.93) are given by the 2D solution in which k is replaced by k_t and the phase is multiplied by $\exp(ikz \cos \theta)$. Moreover, since the dominant term in the derivative of H_z^d with respect to ℓ comes from $e^{ik_t \ell}$, we have:

$$\frac{\partial H_z^d}{\partial \ell} \cong ik \sin \theta H_z^d. \tag{4.94}$$

Inserting (4.93) with (4.94) in (4.91) gives the vectorial expression of $\vec{H}^d$ at a point M on the surface.

Now we will express the diffracted magnetic field in the ray fixed coordinates $(\hat{n}', \hat{b}, \hat{t})$ already defined earlier and write

$$\vec{H}^d = H_b^d \hat{b} + H_t^d \hat{t}, \tag{4.95}$$

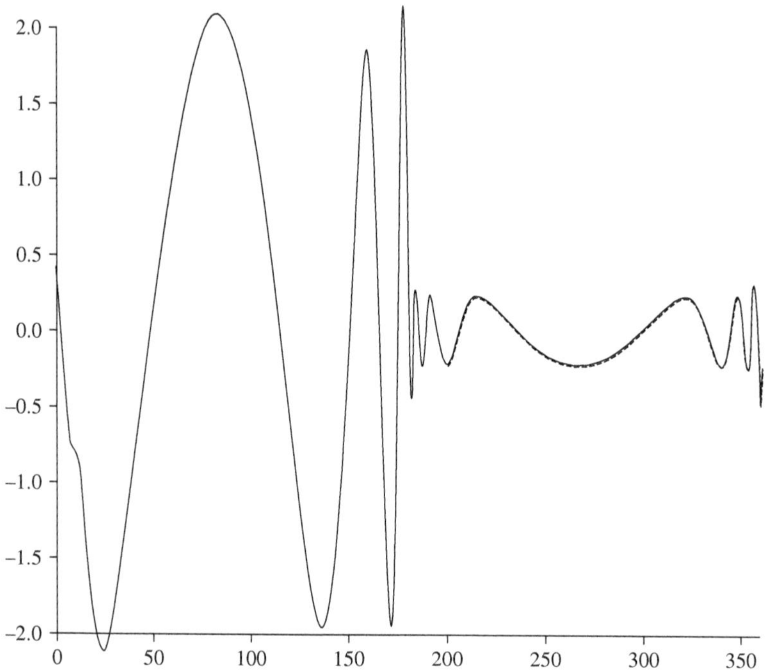

Figure 4.32 *Imaginary part of the asymptotic currents on a cylinder with crescent-shaped cross-section compared to the MM results, for $H_{\|}$ polarisation. Parameters used in the numerical application: $\theta_i = 60°$, θ varies from 0 to 180° on the convex face and from 180 to 360° on the concave face, $L = 6\lambda$, $\ell_1 = \ell_2 = 0.4\lambda$, $F = 300$ MHz: ——, MM; ----, asymptotic*

where

$$\begin{aligned} H_b^d &= \vec{H}^d \cdot \hat{b} = H_z^d \sin\theta - H_\ell^d \cos\theta, \\ H_t^d &= \vec{H}^d \cdot \hat{t} = H_z^d \cos\theta + H_\ell^d \sin\theta. \end{aligned} \tag{4.96}$$

Inserting (4.96) into (4.95) and taking into account (4.93) and (4.94) we obtain

$$\vec{H}^d = \frac{1}{\sin\theta}\left[H_z^d \hat{b} + \frac{i}{k} Y_0 \frac{\partial E_z^d}{\partial n}(\cot\theta \hat{b} - \hat{t})\right] \tag{4.97}$$

and by taking into account (4.90) and using the notations

$$H_z^d = H_z^i h_z^d, \qquad \frac{\partial E_z^d}{\partial n} = E_z^i \frac{\partial e_z^d}{\partial n} \tag{4.98}$$

we obtain finally for the fringe current

$$\vec{J}^f(M) = \hat{n}' \times \vec{H}^d = Y_0 \bar{\bar{D}} \cdot \vec{E}^i(A), \tag{4.99}$$

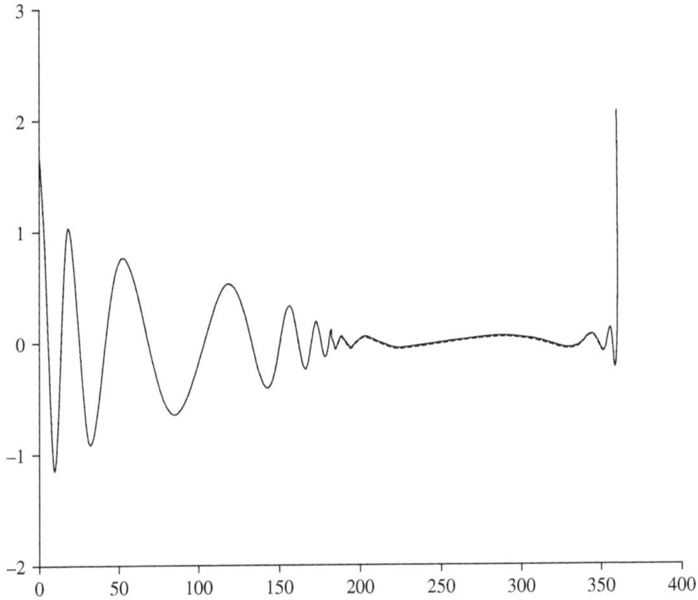

Figure 4.33 *Real part of the asymptotic currents on a cylinder with crescent-shaped cross-section compared to the MM results, for $H_{\|}$ polarisation. Parameters used in the numerical application: $\theta_i = 15°$, θ varies from 0 to 180° on the convex face and from 180 to 360° on the concave face, $L = 6\lambda$, $\ell_1 = \ell_2 = 0.4\lambda$, $F = 300\,MHz$: ——, MM; ----, asymptotic*

where $\bar{\bar{D}}$ is the dyad

$$\bar{\bar{D}} = \left\{\hat{t}\hat{\phi}h_z^d - (\mathrm{T}_0\hat{t}\hat{\theta} + \hat{b}\hat{\theta})\frac{i}{k}\frac{\partial e_z^d}{\partial n}\right\}(\hat{n}' \cdot \hat{n}) \tag{4.100}$$

and $\mathrm{T}_0 = \cot\theta$ is the torsion factor of the geodesic. For a convex surface $\hat{n}' = \hat{n}$ whereas for a concave surface $\hat{n}' = -\hat{n}$.

For a shadowed edge, formulae (4.91)–(4.97) are still valid, the angle θ now being the angle between the incident creeping ray or whispering gallery ray at the point of diffraction on the shadowed edge with the z axis.

In the general case of a curved edge, θ is the angle between the incident ray and the tangent to the edge at the point of diffraction. The formulae for H_z^d and $\partial E_z^d/\partial n$ have furthermore to be adapted by taking into account the divergence factor of the surface rays.

4.6 Numerical results

The asymptotic solutions established in Chapter 4 have been tested on a cylinder with a crescent-shaped cross-section having the same outer radius as the ogive analysed in

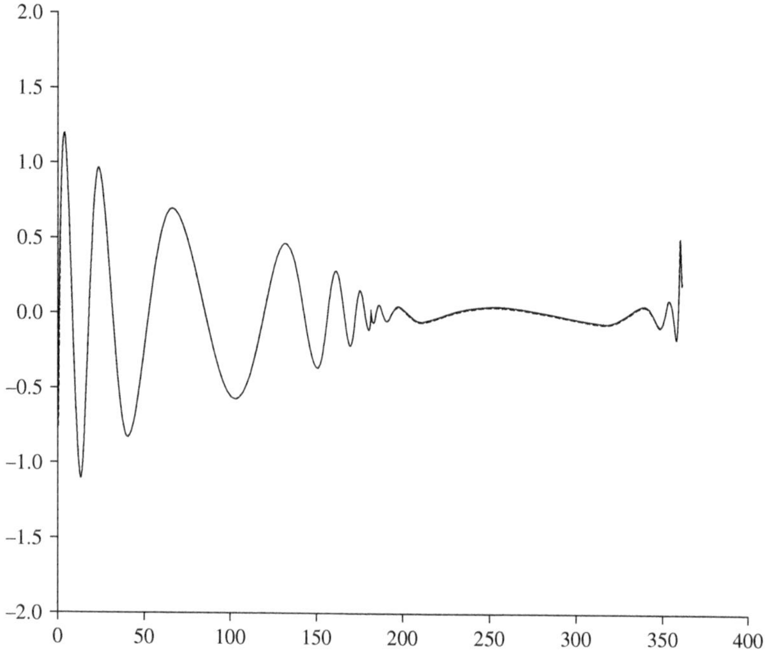

Figure 4.34 Imaginary part of the asymptotic currents on a cylinder with crescent-shaped cross-section compared to the MM results, for $H_{\parallel}$ polarisation. Parameters used in the numerical application: $\theta_i = 15°$, θ varies from 0 to 180° on the convex face and from 180 to 360° on the concave face, $L = 6\lambda$, $\ell_1 = \ell_2 = 0.4\lambda$, $F = 300$ MHz: ——, MM; ----, asymptotic

Section 4.3.4, a distance L between the edges of 6λ and a thickness $\ell_2 = \ell_1$ of 0.4λ (Figure 4.30).

The direction of propagation of the incident plane wave is fixed and the observation angle varies around the centre O of the segment AB from 0 to 360°.

Typical results for the surface currents, when a plane wave is incident on a perfectly conducting crescent-shaped cylinder, are presented in Figures 4.31–4.34 and compared with those given by MM.

The curves in Figures 4.31 and 4.32 correspond to an incident direction of 60° whereas the curves in Figures 4.33 and 4.34 correspond to an incident direction of 15°. For $\theta_i = 60°$, only the real and imaginary parts of the current for $\vec{H}^i_{\parallel}$ polarisation are shown. The currents for $\vec{E}^i_{\parallel}$ polarisation (not shown) are very small on the concave side. For both polarisations the agreement with the MM results is excellent.

For $\theta_i = 15°$ the current on the concave side takes higher values for $\vec{E}^i_{\parallel}$ polarisation as shown in Figures 4.33 and 4.34. Again a good agreement with the MM results is obtained.

In all the results shown, double diffractions on the two edges have been neglected.

4.7 References

1 FOCK, V. A.: 'Electromagnetic diffraction and propagation' (Pergamon, New York, 1965)

2 PATHAK, P. H.: 'An asymptotic analysis of the scattering of plane waves by a smooth convex cylinder', *Radio Science*, 1979, **14** (3), pp. 419–35

3 BOUCHE, D., MOLINET, F., and MITTRA, R.: 'Asymptotic methods in electromagnetics' (Springer-Verlag, Berlin, 1997)

4 MICHAELI, A.: 'Asymptotic analysis of edge-excited currents on a convex face of a perfectly conducting wedge under overlapping penumbra region conditions', *IEEE Transactions on Antennas and Propagation*, 1996, **44** (1), pp. 97–101

5 BABICH, V. M., and BULDYREV, V. S.: 'Short wavelength diffraction theory' (Springer-Verlag, Berlin, 1991) (First published in Russian by Nauka, Moscow, 1972)

6 FELSEN, L. B., and MARCUVITZ, N.: 'Radiation and scattering of waves' (Prentice-Hall Inc., Englewood Cliffs, NJ, 1973)

7 JAMES, G. L.: 'Geometrical theory of diffraction for electromagnetic waves', IEE Electromagnetic Waves Series, vol. 1 (Peter Peregrinus, Hertfordshire, England, 1976, 3rd edn revised 1986)

8 MURTHY, P. K., and THIELE, G.: 'Non uniform currents on a wedge illuminated by a TE plane wave', *IEEE Transactions on Antennas and Propagation*, 1986, **AP-34** (8), pp. 1038–45

9 WASYLKIWSKYJ, W.: 'Diffraction by a concave perfectly conducting circular cylinder', *IEEE Transactions on Antennas and Propagation*, 1975, **AP-23** (4), pp. 480–92

10 ISHIHARA, T., FELSEN, L. B., and GREEN, A.: 'High frequency fields excited by a line source located on a perfectly conducting concave cylindrical surface', *IEEE Transactions on Antennas and Propagation*, 1978, **AP-26** (6), pp. 757–67

11 IDEMEN, M., and FELSEN, L. B.: 'Diffraction of a whispering gallery mode by the edge of a thin cylindrically curved surface', *IEEE Transactions on Antennas and Propagation*, 1981, **AP-29** (4), pp. 571–9

Chapter 5
Hybrid methods

5.1 Introduction – state-of-the-art

Hybrid methods were introduced in the early 1970s for solving some difficulties encountered with asymptotic methods. A typical situation is the diffraction of an object which is large compared to the wavelength but which has locally in some specific areas a complexity in the geometry (fine details) or in the boundary conditions (thick penetrable material) which fall outside the domain of applicability of asymptotic methods. To this category of targets, we also need to add singularities of the surface for which no explicit expressions for the coefficient of diffraction exist.

The hybrid methods have been continuously developed since that time. Nowadays, the same types of problems are still of interest, but the position of the problem is changed since hybrid methods are now developed for augmenting the possibilities of numerical techniques by reducing the storage and the CPU time.

A short review on the different types of hybrid methods which have been developed or proposed in the literature is given in Reference 1. Several synthesis papers on specific applications of hybrid methods have also been published by Newman [2], Medgyesi-Mitschang and coworkers [3, 4] and Thiele [5].

The different methods which have been proposed in the literature are generally characterised as either field-based or current-based analysis. The field-based approach is particularly attractive for radiation problems where the radiator is in proximity of large surfaces for which GTD solutions through ray techniques apply. A typical example is a $\lambda/2$ wire antenna in front of a plate or a corner reflector. In the current-based formulation in which one attempts to determine the equivalent surface currents that represent an object, the analysis proceeds from an approximation of the currents on large parts of the object, obtained from physical optics (PO) or more refined asymptotic solutions. Both methods have also been combined in a mixed field-current-based approach.

In what follows, we classify the different hybrid methods proposed in the literature by specifying the technique that is used in order to reduce the complexity of the problem.

Four different techniques have been identified:

1. The partition of the surface in a GTD region where the currents are approximated by asymptotic solutions and a moment method (MM) region where the currents are calculated by solving an integral equation. The GTD region is usually chosen away from the singularities of the surface where simple asymptotic solutions, like PO, are applicable.
2. The reduction of the number of unknowns in the integral equation method by using a Green's function verifying the boundary conditions on a part of the surface of the object. This method is also called the hybrid MM/Green's function method [2]. The Green's function is either an approximate Green's function obtained by asymptotic techniques or an exact Green's function also calculated with the MM method. The last case is referred to as an MM/matrix Green's function solution which is of particular interest when a part of the scatterer remains fixed whereas the geometry of other parts changes. This is typically the case when an antenna has to be optimised in the presence of an obstacle (support).
3. The decomposition of the original problem into two simpler problems, one of them having an asymptotic solution when excited by the incident field and application of an iteration technique for solving the coupling between both parts. This method has been widely used when the structure is composed of two separate bodies with an initial solution given by the PO approximation on one of them.
4. The use of some *a priori* knowledge on the behaviour of the field on the surface of the scatterer in order to derive basis functions better adapted to the problem. Of special importance is the information concerning the phase of the field which when incorporated in the basis functions leads to a number of unknowns necessary to achieve accuracy which is virtually independent of the wave number k.

In the first category, the accuracy of the current-based hybrid method depends strongly on the validity of the asymptotic approximation for the equivalent currents in the GTD region. Numerous applications using the PO approximations on rather simple geometries have been presented in the literature. A hybrid formulation for arbitrary three-dimensional (3D) bodies was first presented by Medgyesi-Mitschang and Putnam [6]. These authors have used a PO approximation in which the electric field radiated by the PO current due to the impressed magnetic field in the GTD region is considered when enforcing the boundary conditions for the MM region. Later on, Jakobus and Landstorfer [7] proposed a PO approximation involving two coupling mechanisms. First, the MM region contributes to the incident magnetic field in the GTD region. In other words, each basis function in the MM region is acting as a source for the GTD region where the currents are calculated with the PO approximation. Second, the electric field radiated by these augmented PO currents is considered when enforcing the boundary conditions for the MM region. In the determination of the PO currents, the GO shadow condition on the GTD region must be imposed to the impressed field and also to the field radiated by each differential element of current of the MM region, which can be quite time consuming. However, for reflector antenna applications where no shadowing effects occur, this hybrid MM/PO method is very efficient. Nevertheless, the hybrid MM/PO method is not generally valid for an irregularly shaped surface because it neglects effects such as multiple reflections, diffraction and surface waves as well as higher-order interactions between them. Correction terms to PO have been introduced in the hybrid MM/PO method by Hodges and Rahmat-Samii [8] by using an iteration procedure on the magnetic field integral equation similar to that applied by Thiele and his coworkers in their iterative method for solving scattering problems [9, 10]. Since the magnetic field integral equation (MFIE) is a Fredholm integral equation of the second kind, it can be solved by successive approximations by starting with an estimate for the solution, such as the PO approximation. Therefore, by including additional higher-order terms in the MFIE, a higher accuracy of the PO approximation is obtained. The procedure consists in applying the electric field integral equation (EFIE) and MFIE to separate regions of an object. Two distinct solution methods are used to solve the coupled equations. The MM is used to solve for the currents in the EFIE region, while the iterative technique is applied in the MFIE region. A second iterative process is utilised to solve the coupled equations, starting with an estimate for the solution in the MFIE region such as the PO approximation limited to the impressed magnetic field. The second iterative process uses the Neumann series solution for currents in the MFIE region to compute an improved value for the current in the EFIE region and so on. The convergence of the method is accelerated when the zeroth-order solution in the MFIE region is chosen identical to the hybrid MM/PO solution described in Reference 7. Since the iterative hybrid method retains the MFIE terms which are neglected in Reference 7, the accuracy is maintained when PO alone is not sufficient at the expense of increased computational effort. The MM/PO hybrid method has also been improved in References 7 and 8 by the use of refined asymptotic currents close to the edge of a half plane and a wedge.

In all these methods, the continuity of the currents through the boundary between the MM and GTD regions is accomplished by the use of piecewise continuous subdomain basis functions which overlap the boundary. However, despite the improvements which have been mentioned, these methods are difficult to apply, in general, to complex 3D structures which are not constituted by physically separated objects.

The methods of categories (2) and (3) are parts of a more general technique which we call ‘regularisation procedure’. Since asymptotic methods apply only to objects the dimensions of which are large compared to the wavelength and are either smooth or have some singularities like sharp edges or curvature discontinuities, the basic idea is to regularise the surface of the object and adapt its geometry in order to fulfil these requirements. However, such a procedure modifies the solution of the problem. We must therefore find a complementary problem which is defined in such a way that its solution added to the solution of the regularised geometry equals the solution of the original problem. Many applications of iterative methods of category (2) which have been published in the past are particular examples of the regularisation procedure. Among them are wire antennas attached to a support which constitutes a canonical geometry (wedge, cylinder, sphere) which can be considered as a regularised geometry if we separate the antenna from its support. In the general case, the regularisation procedure may be performed by a smooth surface surrounding externally an irregular area, or passing through the irregularities. The first procedure is well adapted to cavities whereas the second one applies to protrusions which stand widely out of the surface. The regularisation procedure will be described in detail in Sections 5.3 and 5.4 of this chapter.

The methods announced in category (4) are new techniques, some of which are still under development. They can be classified into two groups:

- *a priori* phase determination
- analytically or asymptotically derived characteristic basis functions.

5.1.1 A priori *phase determination*

Numerical methods based on classical finite elements or finite difference techniques for wave phenomena become expensive at high frequency because, to follow the variation of the phase of the solution, a minimal number of elements per wavelength is required. This is true for volume as well as for surface finite element techniques. A frequently used rule of thumb is 10 points per wavelength. This of course can vary depending on the type (surface or volume), on the order of accuracy of the method, and on the precision required (low RCS objects demand higher accuracy). To circumvent this difficulty, a number of approaches have been proposed, where in some way, some information about the phase is included inside the finite element formulation, in order to use much larger elements. We first present an approach, initially proposed by Després, incorporating plane waves in the shape functions.

5.1.1.1 Plane waves in the shape functions

Usual shape functions for the finite element method are polynomials in coordinates local to the element, and cannot follow properly the oscillations of the solution, hence we are forced, with such elements, to use a minimal number of elements per wavelength (smaller of course for higher-order elements). A natural idea is therefore to incorporate an exponential with a phase satisfying the wave equation, namely a plane wave in the shape functions. To this end, Després and Cessenat [11, 12] derived the so-called ultraweak variational formulation for the mixed boundary value problem for the Helmholtz equation in homogeneous media. Their approach was generalised by Huttunen *et al.* [13] to inhomogeneous media. The domain V is partitioned into k subdomains V_k. In this formulation, the unknowns x_k are the traces of a linear combination of the solution u and its normal derivative on the boundaries ∂V_k, namely $x_k = (-\partial_{nk}u + i\omega u)\partial V_k$. As in a classical finite element technique, a variational formulation is established for the unknowns x_k, which verifies:

$$B(x_k, v_k) = b, \tag{5.1}$$

where B is a bilinear form, and v_k are test functions. Functions x_k and v_k are both in space $L^2(V_k)$, v_k are solutions of the Helmholtz equation. To obtain a discrete formulation, x_k and v_k in (5.1) are taken in a finite dimensional space, spanned by a basis of shape functions T_{kl}, with k the index of the subdomain and l the index of the shape function in the subdomain. These functions are derived from plane waves (note that other solutions of the wave equation can also be chosen) with wavevectors v_{kl} in V_k, namely $T_{kl}(r) = \exp(i\omega v_{kl} r)$ in V_k, 0 outside V_k. The choice of the unit vectors v_{kl} is not unique. Without any information on the solution, a natural choice is to take the vectors equidistributed in the plane, along p directions, as in Reference 12. As a result, the discrete form of the problem can be written as a matrix equation

$$(D - C)X = b, \tag{5.2}$$

where X is the vector of unknowns, with pk components, D is block diagonal with $(p \times p)$ blocks, C has also $p \times p$ block structure, with zero entries between non-adjacent elements, the right-hand side vector b incorporates the boundary conditions. We refer to Reference 13 where explicit formulae for the values of the D and C matrices, and b vector entries are given.

Some numerical examples are provided in the references. The inclusion of a phase inside the shape functions allows us to choose elements spanning several wavelengths, up to six in Reference 13. Note however that, without any selection of the direction using some information from an asymptotic solution, the number of degrees of freedom per element is equal to p, and can be large to get a precise result. From the physical point of view, this result is expected, because you require in the basis plane waves close to the directions of the most important rays. Moreover, the matrix equation turns out to be ill conditioned. In spite of these drawbacks, the reduction in the number of degrees of freedom allows us to reduce the computation time. Note, however, that the numerical examples provided in the references (circular or elliptic cylinders, spheres) are academic and that, to the best of our knowledge, the method has not been applied to complex 3D objects.

Other teams also implement plane-wave-derived shape functions, but prefer to start from the usual variational formulation of the Helmholtz equation, obtained by multiplying the equation with a test function, and integrating in the domain where the solution is searched. We briefly describe the work of Bettess and coworkers, dealing with acoustic wave scattering in two and three dimensions [14, 15]. The unknown function u within each n node finite element is expressed using a basis of shape functions that are products of usual polynomial nodal shape functions N_j (e.g., on a triangle, one can choose a linear basis function which is 1 at node I, and 0 at all other nodes), and plane waves. The distribution of directions is homogeneous in space as previously. The number of directions used is large (from 100 to 400, in some examples). The volume exterior to the diffracting object is meshed up to an artificial boundary with absorbing boundary condition. A numerical example of diffraction by a sphere is provided in the references. It turns out that the number of degrees of freedom can be reduced by a factor of between 30 and 40, as compared with the classical finite element approach, on this specific case. Plane wave elements for surface integral equations and their application to diffraction by a sphere are also presented in Reference 15, with the same uniform distribution as in free space.

The above methods appear promising, but their efficiency could be drastically improved if some selection of the directions chosen for the plane waves was performed, instead of using a uniform distribution. The possibility of clustering the plane wave in the vicinity of some directions is mentioned in the references, and some extra numerical examples, showing that the coefficient of plane waves is significant only around certain preferred directions, are provided, but no application with *a priori* selection is presented. The first evident advantage of such a strategy will be to reduce the number of degrees of freedom (DOF). A more subtle one will be to improve conditioning, as can be deduced from results presented in Reference 16, showing that the matrix becomes ill conditioned when the number of plane waves is increased. Such an approach of selection of directions from an asymptotic solution was studied by Giladi and Keller [17], using a volume finite element method, and by Kwon *et al.* [18] using a surface integral equation.

5.1.1.2 Phase selection

In Reference 17, the shape functions depend on the node indexed by i and an eikonal S_l indexed by l. They write:

$$T_{il}(r) = N_i(r)\exp(ikS_l(r)).$$

There are as many eikonals as rays determined using the GTD. For example, Reference 17 deals with scattering by a parabolic cylinder, with only a reflected ray and hence one eikonal, and a circular cylinder, with a reflected and a creeping ray, hence two eikonals for the scattered field. Otherwise, the variational formulation is the same as in References 14–16.

The mesh and therefore the number of DOF does not depend on the frequency. The reduction in number of DOF is therefore drastic (around 1/300 in some examples presented in Reference 17) as compared to the previous approaches. The precision is better than with a classical finite element method with much lower cost. However, the examples are very simple and no applications to complicated objects are presented, though the steps for the application of the method to such cases are outlined in the conclusion.

Kwon *et al.* [18] use a surface integral equation (the MFIE) to solve diffraction problems by 3D scatterers. They first perform a computation at moderate frequency f_l and extract, using a Fourier transform, the phase fronts: at each node i, n wave vectors k_{in} accounting for the propagation of the phase of the current are determined. This procedure is applied except in the close vicinity of point and line discontinuities, where a fine mesh is used. It is also possible to extract these wave vectors using the method of asymptotic currents. To compute at high frequency f_h, the current away from discontinuities is expanded on a basis of shape functions $\exp(i(f_h/f_l)k_{in})$.

This technique allows us to use a coarser mesh, with elements spanning half a wavelength. The limitation comes from a degradation of accuracy due to grating lobe scattering from the amplitude discontinuities. This result is a bit disappointing when compared to the much bigger elements used in the previous approaches. It should be improved by using a higher-order polynomial instead of a piecewise constant function, that would provide some form of tapering and limit the above phenomenon, as noted in Reference 18.

5.1.1.3 Other related approaches

The incorporation in the phase function of the PO phase was used, in association with the fast multipole method, by Darrigrand [19], following works by de la Bourdonnaye, Zhou, Abboud and Nédélec (microlocal discretisation), also with significant improvements in the size of the elements and in the number of DOF. However, this approach is restricted to convex obstacles.

5.1.1.4 Conclusions

A priori phase determination and its incorporation in the shape functions of a volume or surface finite element method appear promising, with drastic reduction of the amount of the computational effort required to compute the solutions of diffraction problems at high frequency. However, most applications presented in the literature are limited to simple shapes, except the work performed in Ohio State University [18], though it should be possible to apply such methods to more realistic and application oriented diffraction problems.

5.1.2 Analytically or asymptotically derived characteristic basis functions

A new technique called analytically derived 'characteristic basis functions' (CBF) has been reported [20]. It consists in defining large domain or macro basis functions using some *a priori* knowledge concerning the analytical form of the current.

Consider for instance a perfectly conducting strip of width a, large compared to the wavelength λ. The current density induced on this strip by an incident plane wave can be represented with a good approximation, as a sum of three terms:

- the PO current corresponding to the current induced by the incident plane wave on an infinite plate,
- the fringe currents associated with each edge of the strip.

The analytical expression of the PO current is well known and that of the fringe currents can be directly derived from the rigorous solution of the problem of diffraction by a half plane. The knowledge of the analytical form of these currents allows us to choose three entire domain basis functions defined respectively by the PO and the fringe current terms, to represent the unknown current on the strip and to construct a matrix equation by imposing the boundary conditions. Hence we need to invert only a 3×3 matrix to obtain the induced current density.

This method can be extended to a 2D-corner reflector which can be either concave or convex by introducing secondary characteristic basis functions describing the interaction between the two facets [21]. Each primary CBF defined on one facet generates a secondary CBF on the other facet so that a resulting matrix with dimensions 12×12 has to be inverted to obtain the induced current density. The secondary CBF can be determined by using the plane wave spectral decomposition technique and retaining only the PO term associated with each spectral component. Extension of the method to a 2D-faceted body is straightforward and leads to a drastic reduction of unknowns compared to the ordinary MM method for large body problems. However, its extension to scattering problems involving 3D-faceted bodies, although theoretically possible, has not yet been accomplished. The principal difficulty is to introduce general formulae for the corner diffraction.

For curved surfaces, it is possible to introduce asymptotically derived CBF and to proceed as before. It has been shown in Chapter 4 that analytical formulae derived by asymptotic techniques exist for convex and concave surfaces leading to a good approximation of the induced currents in terms of smooth body diffracted waves and edge excited creeping and whispering gallery waves. By choosing the analytical form of these elementary contributions as CBF in the integral equation verified by the currents, an accurate numerical solution may also be obtained for a 2D-faceted body with curved facets with a small number of unknowns.

Asymptotically derived basis functions have also been reported for the full wave analysis of large planar phased-array antennas [22]. The method is based on the MM solution of a fringe integral equation in which the unknown function is the difference between the exact solution of the finite array and that of the associated infinite array. The unknown currents can be interpreted as being produced by the field diffracted at the array edge which is excited by the Floquet waves pertinent to the infinite configuration. Following this physical interpretation efficient array–global basis functions have been derived, using uniform asymptotic techniques, for representing the unknown in the integral equation. A very small number of basis functions are needed to obtain an accurate solution.

In the next sections our analysis will be focused on the regularisation procedure. We will consider separately the two situations of a cavity and a protrusion which in a practical problem may be simultaneously present, and define in each case the procedure which allows us to decompose the problem. Since the procedure is founded on the equivalence theorem, we will present in Section 5.2 some consequences of this theorem which can be directly used to conceive a hybrid method.

In Section 5.3, they are applied respectively to a cavity in a perfectly conducting surface and to a protrusion standing out of such a surface.

The methods developed in this chapter are then generalised, in Section 5.4, to a coated metallic surface having local irregularities in the geometry and in the boundary conditions.

In Section 5.5, we present a brief review of asymptotic solutions, which are well adapted to the hybrid methods described in the previous Sections (5.2–5.4). Some numerical results are shown in Section 5.6.

5.2 Equivalence theorem and its consequences

The iterative methods which will be described in Sections 5.3 and 5.4 are founded on the equivalence theorem presented in the book by Harrington [23] and formulated in a more direct way by Chen [24] by using the integral equations verified by the equivalent currents.

We will present the basic problem without demonstration and focus our attention on its consequences.

Let $(\vec{E}^i, \vec{H}^i)$ be the electromagnetic field radiated in the vacuum by electric and magnetic sources occupying a finite volume in space, with volume densities $(\vec{J}^i, \vec{M}^i)$. Let S be a closed surface surrounding an arbitrary volume V containing the sources and let $\vec{E}^i(M), \vec{H}^i(M)$ be the field radiated by the sources at an arbitrary point M of S.

Let us further define fictitious current densities on S:

$$\vec{J}_s = \hat{n} \times \vec{H}^i, \qquad \vec{M}_s = \vec{E}^i \times \hat{n}, \tag{5.3}$$

where $\hat{n}$ is the unit vector of the normal to S, oriented towards the region exterior to V and let $\vec{E}^r(P), \vec{H}^r(P)$ be the electromagnetic field radiated by the fictitious currents at a point P exterior to V. It can be rigorously proved that

$$\vec{E}^r(P) = \vec{E}^i(P), \qquad \vec{H}^r(P) = \vec{H}^i(P). \tag{5.4}$$

In other words, Equation (5.4) means that we can replace the primary sources $(\vec{J}^i, \vec{M}^i)$ by the equivalent sources $(\vec{J}_s, \vec{M}_s)$ for calculating the field outside the volume V. Inside the volume V, the field radiated by the sources $(\vec{J}_s, \vec{M}_s)$ is identically zero since the total radiated field by all the sources must be equal to $(\vec{E}^i, \vec{H}^i)$.

The equivalence theorem that we have just stated, is illustrated schematically in Figure 5.1.

5.2.1 *Corollary 1*

If the sources are in the region exterior to V and if we change the sign of the normal unit vector $\hat{n}$ in the definition (5.3) of the current $(\vec{J}_s, \vec{M}_s)$ by replacing $\hat{n}$ by $\hat{n}' = -\hat{n}$, we find that the field radiated by these modified currents inside V is equal to $(\vec{E}^i, \vec{H}^i)$.

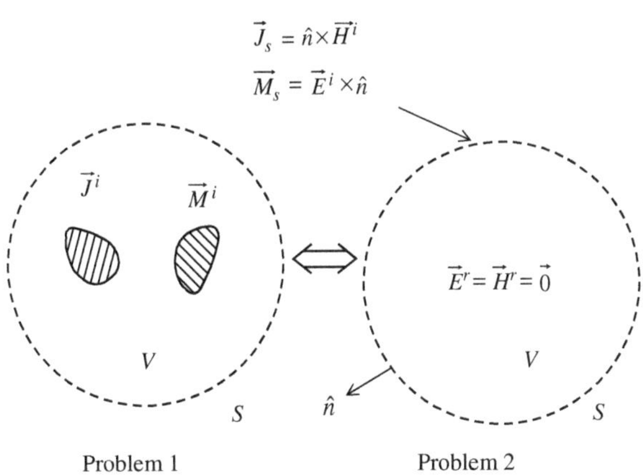

Figure 5.1 Outside the volume V, problems 1 and 2 are equivalent

In particular, if the incident field $(\vec{E}^i, \vec{H}^i)$ is a plane wave, which corresponds to sources rejected at infinity, we can replace the incident wave inside a closed domain V by distributions of fictitious electric and magnetic currents, located on the surface S delimiting the domain and having the following expression:

$$\vec{J}_s = \hat{n}' \times \vec{H}^i = -\hat{n} \times \vec{H}^i, \qquad \vec{M}_s = \vec{E}^i \times \hat{n}' = -\vec{E}^i \times \hat{n}, \tag{5.5}$$

where $\hat{n}$ is the normal unit vector oriented towards the region exterior to the domain V. Outside V, the field radiated by the currents $(\vec{J}_s, \vec{M}_s)$ is identically zero.

5.2.2 *Corollary 2*

If in addition to the sources, some obstacles are present in the domain considered previously, the incident field on S is modified by these obstacles. Then, we have

$$\vec{E}^i = \vec{E}^o + \vec{E}^D, \qquad \vec{H}^i = \vec{H}^o + \vec{H}^D, \tag{5.6}$$

where $(\vec{E}^o, \vec{H}^o)$ is the field due to the primary sources and $(\vec{E}^D, \vec{H}^D)$ is the field diffracted by the obstacles. If we designate by $(\vec{J}_s, \vec{M}_s)$ the fictitious currents defined by the new incident field $(\vec{E}^i, \vec{H}^i)$ given by (5.6), and by the normal unit vector $\hat{n}'$, the field radiated by these sources in the interior domain is still equal to $(\vec{E}^i, \vec{H}^i)$ and the field radiated outside the domain by $(\vec{J}_s, \vec{M}_s)$ is identically zero (see Figure 5.2).

5.2.3 *Other forms of the equivalence theorem*

We substitute to the volume V in Figure 5.2 a perfect conductor having the same surface S. The inclusion of this body transforms the field $(\vec{E}^i, \vec{H}^i)$ radiated by the sources in the presence of the obstacles outside V in the field:

$$\vec{E} = \vec{E}^i + \vec{E}^D, \qquad \vec{H} = \vec{H}^i + \vec{H}^D \tag{5.7}$$

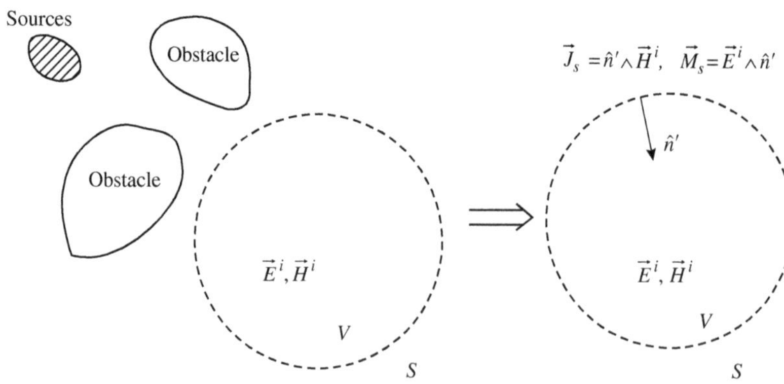

Figure 5.2 *The field radiated by the equivalent currents $(\vec{J}_s, \vec{M}_s)$ alone is equal to the field radiated by the primary sources in the presence of the obstacles*

where $(\vec{E}^D, \vec{H}^D)$ is the field diffracted by the perfect conductor occupying volume V. On the surface S, the tangential component of the electric field is zero, hence $\vec{M}_s = \vec{0}$. Let $\vec{H}_s$ be the value of the total magnetic field on S. Then $\vec{J}_s = \hat{n} \times \vec{H}_s$ is the electric current induced on the surface S of the conductor. The diffracted field $(\vec{E}^D, \vec{H}^D)$ outside of V also corresponds to the field radiated by the current $\vec{J}_s$ in the presence of the obstacles. Moreover, since the electromagnetic field inside a perfect conductor is identically zero, the field radiated by the currents $\vec{J}_s$ in the interior domain of V is equal to the opposite of the incident field $(\vec{E}^i, \vec{H}^i)$. As a consequence, we obtain the equivalence shown in Figure 5.3.

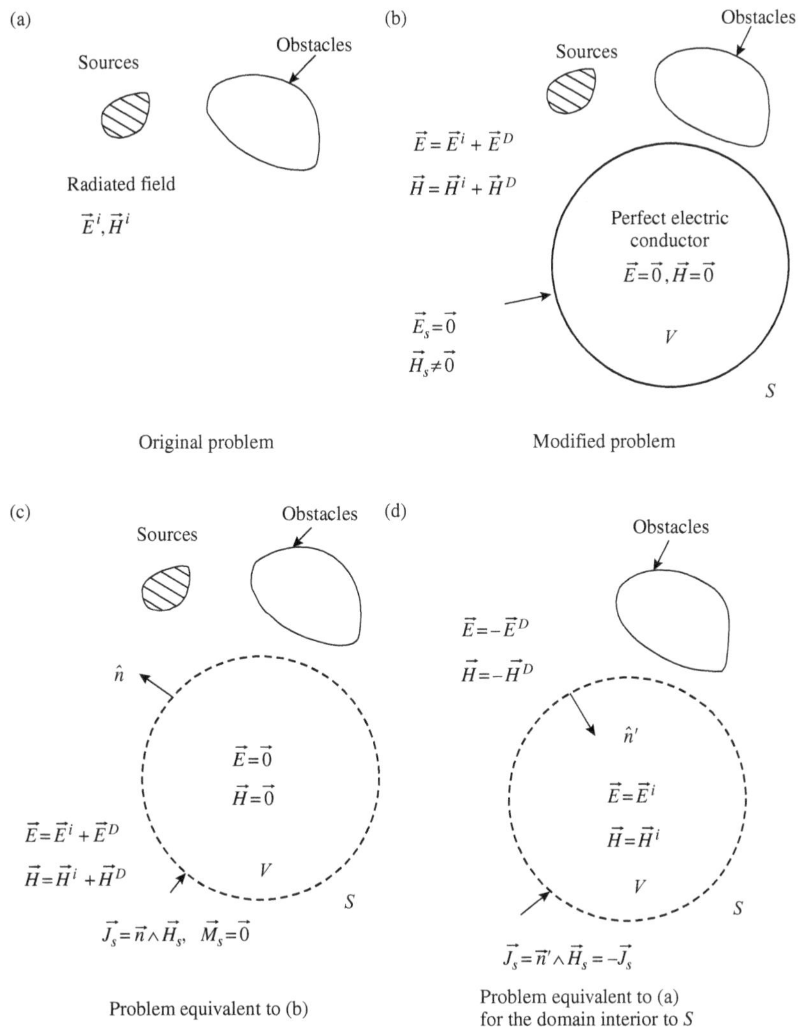

Figure 5.3 Equivalences leading to the third Schelkunoff principle

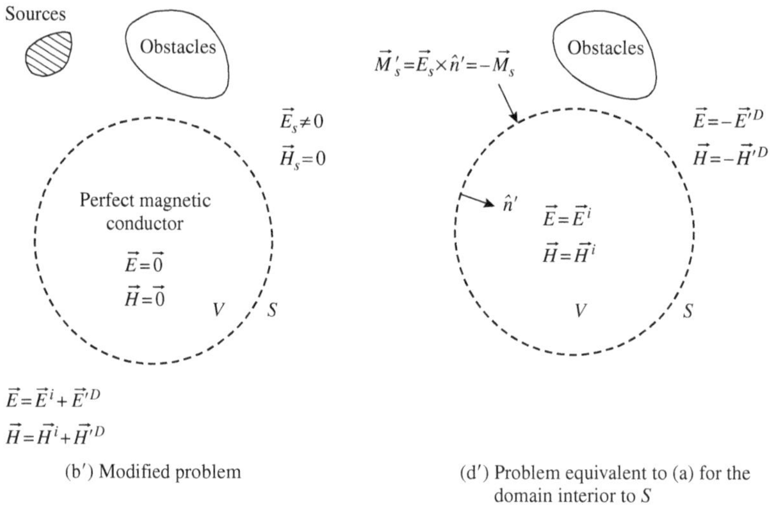

Figure 5.4 *Equivalence leading to the fourth Schelkunoff principle, dual of the third principle*

As the current $\vec{J}_s' = -\vec{J}_s$ in problem (d), it radiates in the presence of the obstacles, the field $(-\vec{E}^D, -\vec{H}^D)$ in the exterior domain of V and $(\vec{E}^i, \vec{H}^i)$ in the interior domain of V. It follows therefore that the sum of the solutions of problems (b) and (d) gives again the solution of problem (a).

This result corresponds to the third Schelkunoff principle [31]. It can, of course, be extended to a perfect magnetic conductor by replacing problems (b) and (d) by problems (b′) and (d′) defined in Figure 5.4.

The sum of the solutions of problems (b′) and (d′) gives again the solution of problem (a). These theorems can be directly generalised to the case when conducting or dielectric obstacles are inside the volume V. The inclusion of such obstacles has no effect on the solution of problems (b) or (b′). Problems (d) or (d′), however, are modified and in order to obtain problems which are equivalent to problem (a) we have to superpose problems (b″) and (d″) or (b‴) and (d‴) of Figure 5.5.

5.3 Application of the equivalence theorem to the hybridisation of the methods

The different forms of the equivalence principle presented in Figures 5.3–5.5 have in common the fact that the sheet of currents which radiates towards the interior domain of the volume V, gives again the incident field created by the sources in the presence of the obstacles located outside V. This property may be turned to perform the hybridisation of a numerical method with asymptotic techniques. We will present two examples.

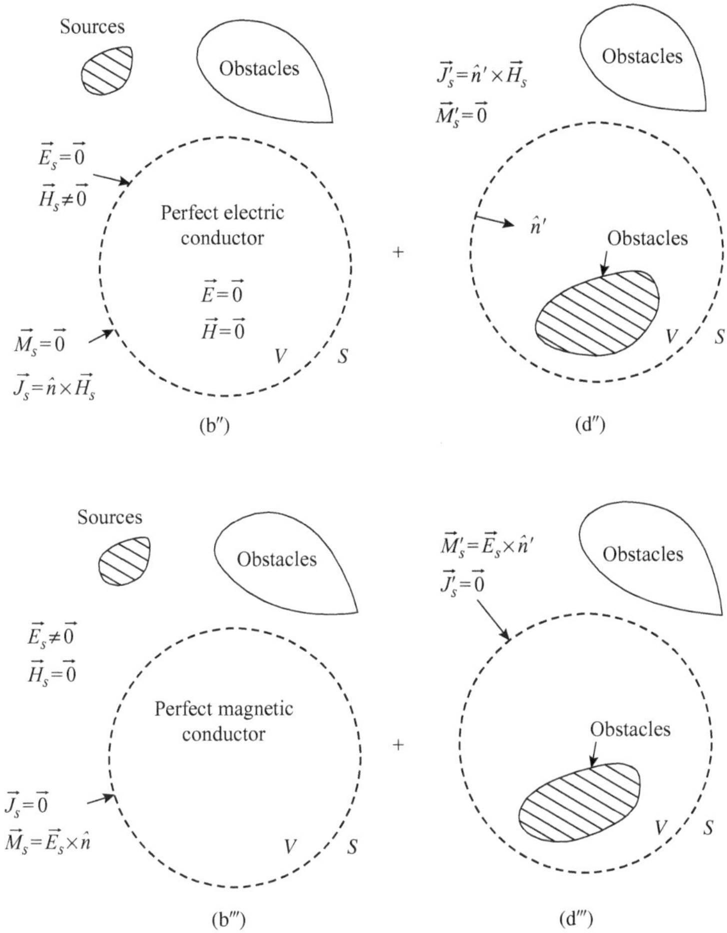

Figure 5.5 Extension of the equivalences shown in Figures 5.3 and 5.4

5.3.1 Cavity in a smooth perfectly conducting surface

Let us consider a cavity bored in a perfectly conducting body and suppose that the surface of the body away from the cavity is smooth and delimited by edge type singularities. More generally, the surface away from the cavity is a regular piece of surface belonging to a large object having a complex geometry.

The original problem is the diffraction of an incident plane wave by the complete structure which is schematised in Figure 5.6 by the cavity and the surface in its vicinity, which has a sharp edge and some separated obstacles.

The original problem is modified by introducing into the cavity a perfectly conducting body B, the contour of which follows exactly the walls of the cavity and which regularises the exterior surface of the structure (see Figure 5.7). The body B plays the part of the perfect conductor occupying the volume V in Figure 5.3(b).

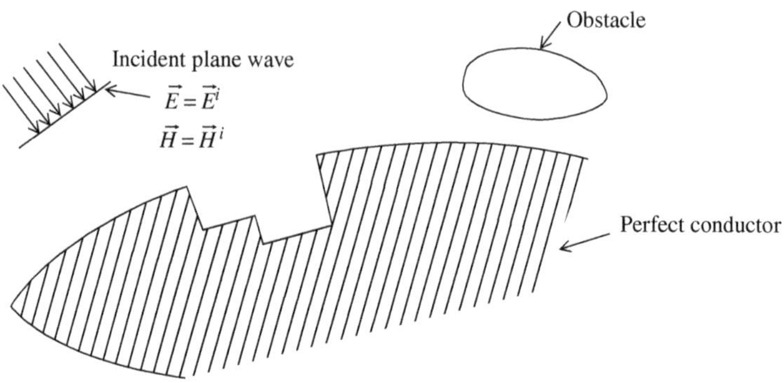

Figure 5.6 Original problem

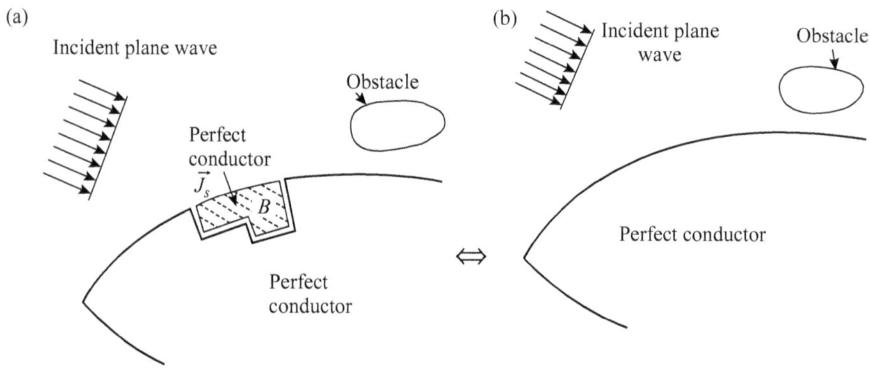

Figure 5.7 Modified problem

Let $\vec{J}_s$ be the current induced on the surface of B in the presence of the complete structure. On the walls of B facing those of the cavity, $\vec{J}_s$ is equal to zero since the space separating these walls from the cavity is supposed infinitely thin. Only the current on the exterior surface of B is different from zero. The modified problem reduces therefore to the diffraction of a plane wave by the structure, the cavity of which has been short-circuited by a perfectly conducting surface (Figure 5.7(b)). Owing to the fundamental equivalence stated in Section 5.2, the complementary problem is the radiation of a sheet of current $-\vec{J}_s$ in the presence of the complete structure with its cavity and obstacle (Figure 5.8).

The modified problem (Figure 5.7(b)) is solved by the asymptotic methods by taking into account all waves (or rays) reaching a point M in the region exterior to the structure and the obstacle: direct waves, reflected waves, edge-diffracted waves, creeping waves. In particular, if M is on the surface which closes smoothly the cavity as in Figure 5.7(a), the asymptotic methods give the currents $\vec{J}_s$ which play the role of the source in the complementary problem.

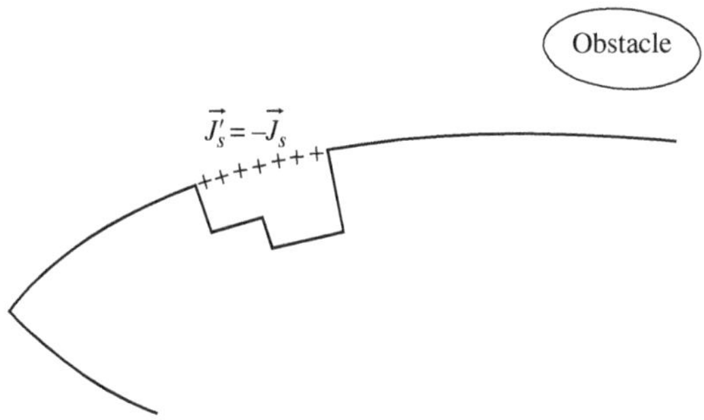

Figure 5.8 Complementary problem

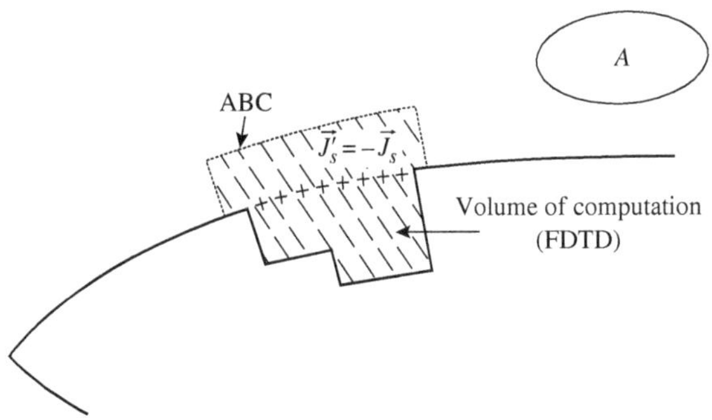

Figure 5.9 Solution of the complementary problem (ABC: Absorbing Boundary Condition)

The complementary problem can be solved by the finite difference method in the time domain (FDTD), applied to a sinusoidal wave. For this purpose we limit the volume in which this method is applied to the interior of the cavity and its near exterior environment (Figure 5.9).

The interactions of the field radiated by $\vec{J}'_s$ with the other parts of the structure are taken into account indirectly by an iteration procedure the principle of which is now given in detail.

Let Σ be a surface infinitely close to the current sheet $\vec{J}'_s$ on its exterior side and let $\vec{J}_\Sigma$ and $\vec{M}_\Sigma$ be the electric and magnetic currents on Σ calculated by the FDTD. According to the induction theorem [23], the currents $\vec{J}_\Sigma$ and $\vec{M}_\Sigma$ radiate a field which is equal to zero inside the volume limited by the cavity and the surface Σ (Figure 5.10(a)).

We can therefore replace the interior medium of the cavity by a perfect conductor and transform the radiation problem outside the volume in which the FDTD is applied, into the radiation of a sheet of current in front of a perfectly conducting surface (Figure 5.10(b)). Since $\vec{J}_\Sigma$ does not radiate in front of a perfect conductor, only $\vec{M}_\Sigma$

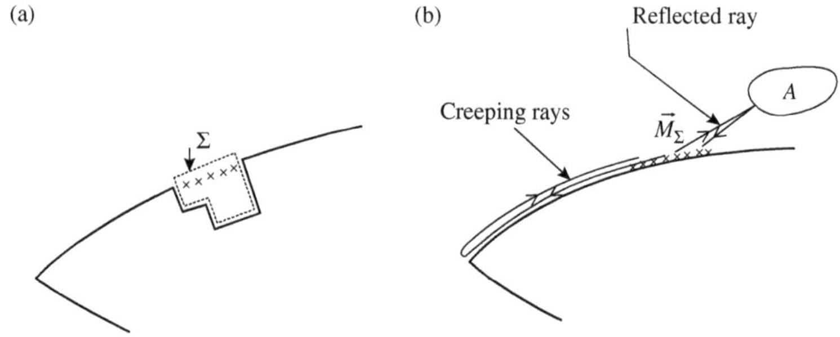

Figure 5.10 Application of the induction theorem

radiates. The problem illustrated in Figure 5.10(b) can again be solved by the asymptotic methods, by calculating the field radiated by $\vec{M}_\Sigma$ and reflected or diffracted by those parts of the structure which lie outside the FDTD computation volume.

It is important to mention that the radiated field at large distance, on a monostatic device can be obtained by applying the reciprocity theorem to the asymptotic solutions giving the current $\vec{J}_s$ on the regularised surface, induced by the incident plane wave. The rays and the diffraction coefficients therefore remain unchanged. In a bistatic device, the ray-searching algorithm of the direct problem can be applied by choosing an incident direction opposite to the observation direction.

The field back-scattered on the 'connection surface' short-circuiting the cavity, however, needs the search of new rays. The associated field gives a correction $\vec{J}_s^{(1)}$ of the excitation current $\vec{J}_s$. By applying again the FDTD algorithm to the sources $\vec{J}_s^{(1)'} = -\vec{J}_s^{(1)}$, we obtain the first reiterated term for the field in the cavity. This procedure can be repeated, but in principle it converges rapidly owing to the divergence in space of the radiated or diffracted waves and the reduction of their amplitude when they undergo a diffraction. Numerical experiences on typical shapes have shown that in the absence of reflected rays on the obstacles A, the contribution of the first reiterated term on the final result is negligibly small.

In summary, the application of the hybrid method which has been described involves the computation of the current $\vec{J}_s$ on the connection surface and the computation of the field $\vec{E}^R$ radiated by the magnetic currents $\vec{M}_\Sigma$ in the presence of exterior obstacles. When the surface containing the cavity is smooth and the singularities delimiting it as well as the exterior obstacles are all situated at a distance from the cavity larger than the wavelength, $\vec{J}_s$ and $\vec{E}^R$ can be calculated by the asymptotic solutions presented in Chapter 4.

The method described so far remains valid when the cavity is partially or totally filled by a dielectric material or if its walls are covered by absorbing material. Indeed, it is always possible to substitute to the interior domain of the cavity which contains the dielectric or absorbing material and extends to the metallic walls, a perfectly conducting body B the limits of which face the internal metallic walls and smooth the exterior surface as shown in Figure 5.11. As before, the currents induced on the surfaces of B which face the metallic walls of the cavity are zero and only the current

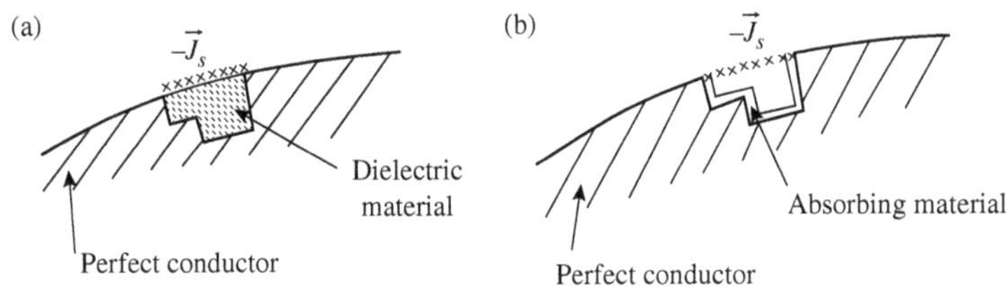

Figure 5.11 Equivalence theorem applied to a cavity

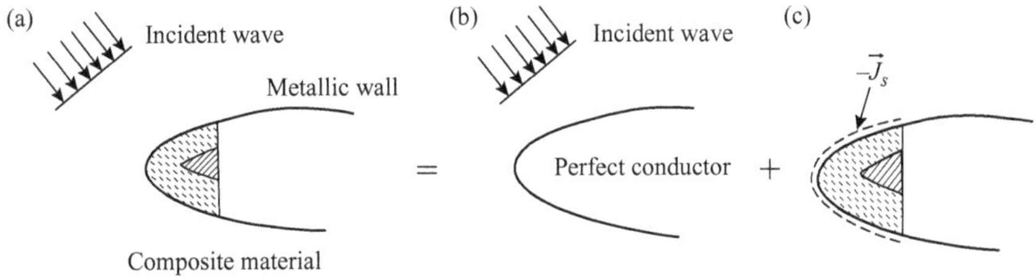

Figure 5.12 Equivalence theorem applied to the front edge of a wing

$\vec{J}_s$ induced on the surface of B which short-circuits the cavity, remains. By applying the equivalences shown in Figure 5.3, we see that the modified problem, resulting from the inclusion of B is reduced as before to the diffraction of a plane wave by the regularised structure, the exterior surface of which is perfectly conducting, whereas the complementary problem is the radiation of the sheet of currents $-\vec{J}_s$ in the presence of the complete structure comprising the cavity containing dielectric or absorbing materials and the exterior obstacles (Figures 5.11(a) and (b)).

In Figure 5.11(b), the excitation current sheet extends to the limits of the metallic walls of the cavity.

The principle of the hybrid method can be applied to geometries other than a cavity. An example is the diffraction by the front edge of a wing constituted by composite materials. This problem can be decomposed as shown in Figure 5.12.

In general, the method is well adapted to a class of problems characterised by a convex regularisation surface without singularities, which can be matched without discontinuity to the neighbouring surface and which surrounds that part of the structure which cannot be treated by the asymptotic methods. But the method can also be applied to a truncated wedge which can be transformed by regularisation into a full wedge for which an asymptotic solution is available (Figure 5.13).

This technique has been applied to truncated or rounded wedges by Hunter [25].

5.3.2 *Protrusion standing out of a smooth perfectly conducting surface*

The method described in the preceding paragraph can, in principle, also be applied to a protrusion standing out of a smooth surface. However, very frequently the dimensions of the surface that regularises the structure are either too large in terms of

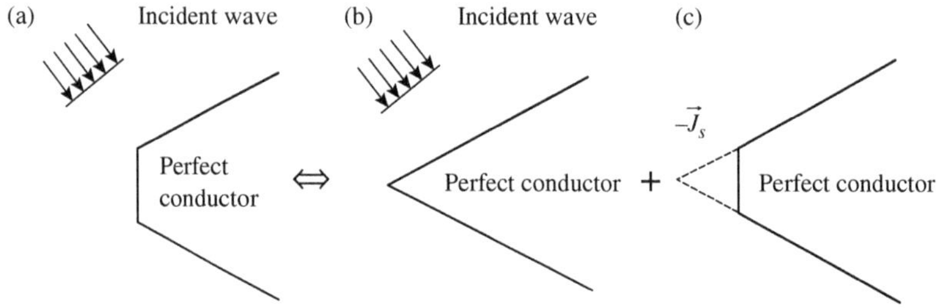

Figure 5.13 Equivalence theorem applied to a truncated wedge

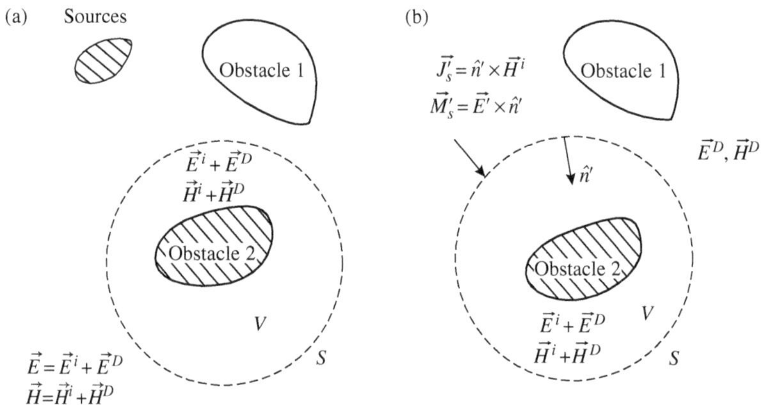

Figure 5.14 Extension of corollary 2

the wavelength so that the application of the numerical techniques takes too much CPU time or too complex for the application of asymptotic methods.

A typical example is a wire antenna standing widely out of its support. For this case, it is preferable to change the method. Instead of regularising the structure with a fictitious surface including the region which cannot be treated by asymptotic methods, we perform this regularisation with a fictitious surface crossing this region.

Suppose that in corollary 2 of Section 5.2, we introduce a second obstacle into the fictitious volume V with exterior surface S. Now, let $(\vec{E}^i, \vec{H}^i)$ be the field radiated by the primary sources in the presence of obstacle 1, but in the absence of obstacle 2 denote by $\vec{J}_s' = \hat{n}' \times \vec{H}^i, \vec{M}_s' = \vec{E}^i \times \hat{n}'$ the corresponding currents on S, where $\hat{n}'$ is the unit vector normal to S and oriented towards the inside of V. Then corollary 2 leads to the result shown in Figure 5.14.

From corollary 2, we know that the field radiated outside the volume V by the currents $(\vec{J}_s', \vec{M}_s')$ is identically zero. The incident field on obstacle 2 is $(\vec{E}^i, \vec{H}^i)$. It gives rise to a diffracted field which radiates outside V and interacts with obstacle 1, which again radiates on obstacle 2. The result of the multiple interactions of obstacles 1 and 2 is the diffracted field $(\vec{E}^D, \vec{H}^D)$ defined as the modification of the incident field $(\vec{E}^i, \vec{H}^i)$ due to the primary sources in the presence of obstacle 1, when obstacle 2 is introduced in the problem (Figure 5.14(a)). We can see that problems (a) and (b) of Figure 5.14 are only equivalent inside the volume V. Outside V, the equivalence is obtained if we add $(\vec{E}^i, \vec{H}^i)$ to problem (b). An equivalence that is valid in the space outside both obstacles is obtained by letting the fictitious surface S tend to the surface of obstacle 2. Under these conditions we obtain the equivalence shown in Figure 5.15.

A variant of the equivalence shown in Figure 5.15 can also be established for an obstacle separated into two fictitious parts constituting two different obstacles. In order to establish this variant, we choose a volume V having a part s of its surface in contact with the surface of obstacle 1 as shown in Figure 5.16.

Along s, the electric field due to the radiation of the primary sources and the diffraction by obstacle 1 is orthogonal to s when obstacle 1 is perfectly conducting. It follows that the magnetic current $\vec{M}_s'$ is zero along s and hence only the electric current $\vec{J}_s'$ remains. Moreover, according to corollary 2, since obstacle 1 is exterior to V, the field radiated at an arbitrary point of this obstacle by the currents $(\vec{J}_s', \vec{M}_s')$ on S is zero when obstacle 2 is absent. Consequently, obstacle 1 does not give any contribution by diffraction to the field radiated by the fictitious currents $(\vec{J}_s', \vec{M}_s')$ at a point inside V.

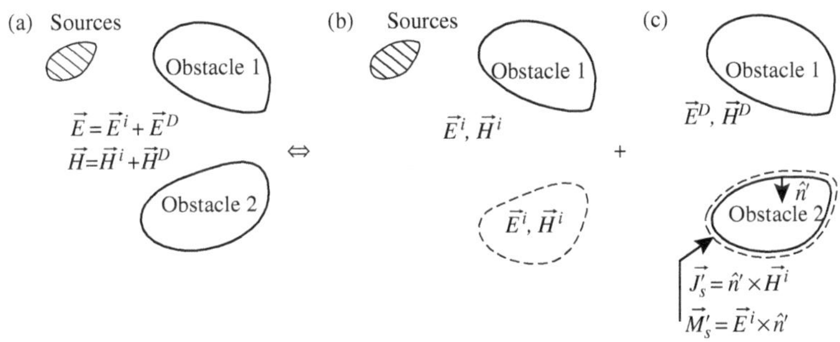

Figure 5.15 Equivalence valid for separated obstacles

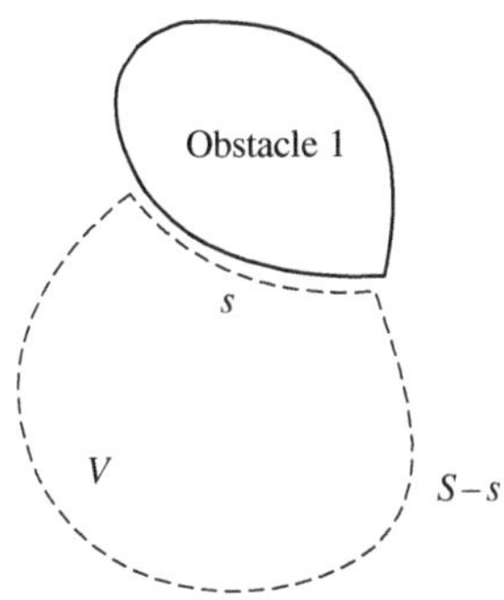

Figure 5.16 Fictitious volume in contact with an obstacle

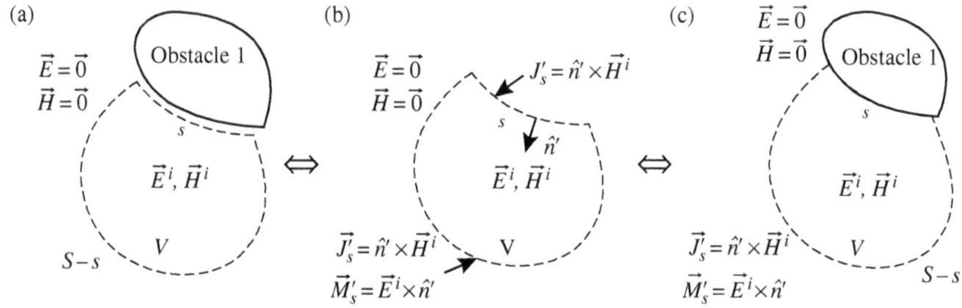

Figure 5.17 Equivalences for a fictitious volume in contact with an obstacle

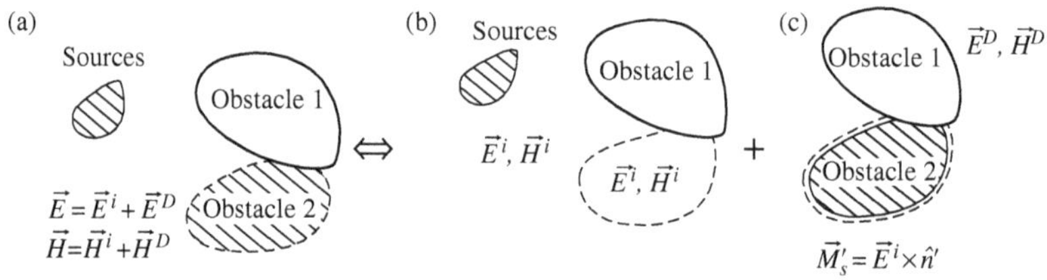

Figure 5.18 In the domain exterior to obstacles 1 and 2, the solution of problem (a) is equivalent to the sum of the solutions of problems (b) and (c)

We have therefore the equivalence of Figure 5.17 between problems (a) and (b). Owing to the linearity of Maxwell's equations, the total radiated field interior to V is the sum of the field due to the sources $(\vec{J}_s', \vec{M}_s')$ on $S - s$ and the field due to $\vec{J}_s'$ on (s). As the currents $\vec{J}_s'$ in the presence of a perfectly conducting obstacle do not radiate, we can replace the sheet of currents on S by the sheet of currents on $S - s$ in the presence of obstacle 1. It then follows the equivalence between problems (b) and (c) of Figure 5.17.

If a second obstacle is inserted inside V in the problem of Figure 5.17(c), the field inside V is modified in $(\vec{E}^i + \vec{E}^D, \vec{H}^i + \vec{H}^D)$ and we again obtain a situation similar to that in Figure 5.14 where the two problems (a) and (b) are equivalent inside V. If obstacle 2 is in contact with obstacle 1, and if S tends to the exterior surface of obstacle 2, we obtain the equivalence of Figure 5.18, which generalises that of Figure 5.15.

As obstacle 2 is supposed to be perfectly conducting, and according to the induction theorem [23], it is sufficient to impose in problem (c) the currents $\vec{M}_s' = \vec{E}^i \times \hat{n}'$. The surface separating obstacles 1 and 2 is arbitrary. This surface is chosen in order to simplify the solution of problem (b) by asymptotic methods. Usually, we impose the continuity of the tangent plane and of the curvature at the junction between this surface and the surface of obstacle 1.

A typical application of the equivalence illustrated in Figure 5.18 is a protrusion on a convex regular surface illuminated by a plane wave. In this case, the equivalence of Figure 5.18 takes the special form shown in Figure 5.19, where $(\vec{E}^i, \vec{H}^i)$ is

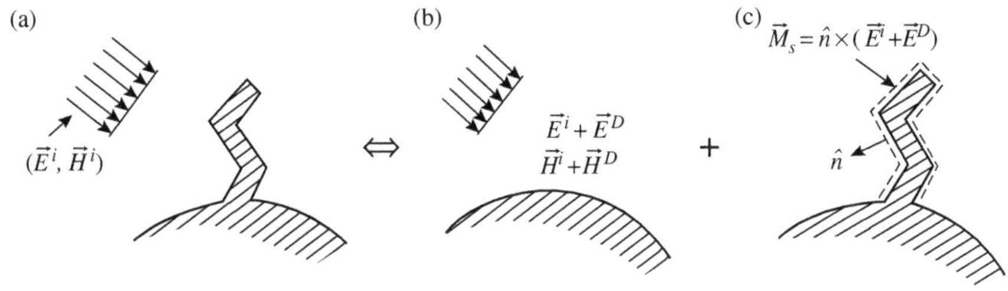

Figure 5.19 Equivalence adapted to a hybrid asymptotic/FDTD method

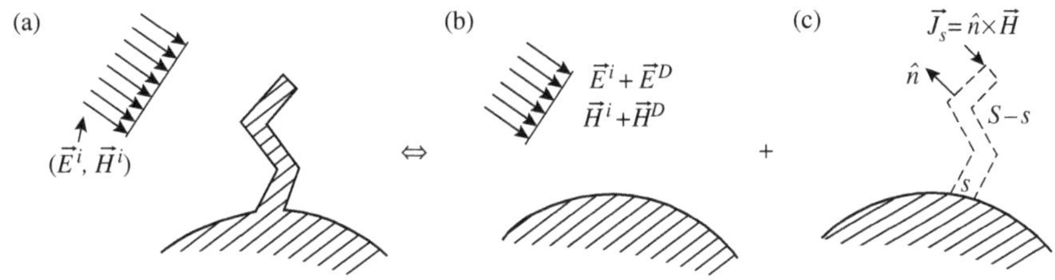

Figure 5.20 Equivalence adapted to a hybrid asymptotic/integral equation method

the field of the incident plane wave and $(\vec{E}^D, \vec{H}^D)$ the field diffracted by the smoothed surface.

The equivalence of Figure 5.19 is well adapted to a hybridisation between asymptotic methods and FDTD. Indeed the solution of problem (b) and the calculation of the excitation current $\vec{M}_s$ in problem (c) are worked out by asymptotic methods whereas the resolution of problem (c) is performed by the FDTD method.

The same equivalence can also be applied to a hybrid method where the asymptotic solutions are combined with an integral equation solution (MM solution, for instance). In this case, the excitation field $\vec{E}^i + \vec{E}^D$ is still given by the asymptotic near field solution of problem (b) (see Figure 5.20) and the unknowns are the currents $\vec{J}_s = \hat{n} \times \vec{H}$ on the surface $S - s$. These currents radiate in the presence of the regularised obstacle and therefore the Green's function of this obstacle has to be introduced in the integral equation instead of the free space Green's function. If the regularised surface is one of the classical shapes which fits into a separable coordinate system, its Green's function is well known and can be introduced in the hybrid technique. Hybrid solutions for wires in the presence of cylinders or spheres have been presented by many authors. But, in general, it is necessary to approximate the Green's function by an asymptotic Green's function.

In the hybrid asymptotic/FDTD method described before, the computation domain is limited by absorbing boundary conditions. However, if some other obstacles exist outside the regularised surface, the diffraction of the radiated field by these obstacles has to be taken into account through a reiteration procedure similar to that described in Section 5.3.1 for the cavity problem.

5.4 Generalisation to coated objects

The third Schelkunoff principle can be easily extended to an object covered with a dielectric or an absorbing coating. Indeed, we observe that the perfect electric conductor introduced in Figure 5.3(b) can be replaced by any other object, the boundary conditions of which are known, such as for instance a magnetic conductor or more generally a body with a known surface impedance, without modifying the superposition theorem.

In Figure 5.21, we consider the case where the original problem is modified by a perfect electric conductor partially coated by an absorbing material.

It is important to note that the obstacle B is not present in the modified problem. We have supposed that it is located inside the volume V occupied by the perfect electric conductor and therefore does not intervene in the diffraction problem 5.21(b).

Let $(\vec{E}_s, \vec{H}_s)$, $(\vec{E}_1, \vec{H}_1)$, $(\vec{E}_c = 0, \vec{H}_c)$, respectively, be the field on the exterior surface of the coating, on the surfaces of the truncations of the coating and on the bare region of the perfect conductor and let $(\vec{J}_s, \vec{M}_s)$, $(\vec{J}_1, \vec{M}_1)$, $\vec{J}_{sc}$ be the corresponding surface currents. In order to recover the solution of the original problem (Figure 5.21(a)), we must add to the solution of the modified problem (Figure 5.21(b)), the solution of the complementary problem of Figure 5.22.

We remark that the primary sources are absent in Figure 5.22 and that only the field in the domain exterior to the coating has to be taken into account in the solution of the modified problem (Figure 5.21(b)).

5.4.1 *Cavity in a coated smooth surface*

The application of the equivalence illustrated in Figures 5.21 and 5.22 to the problem of a cavity in a coated smooth surface is straightforward. Consider the original problem shown in Figure 5.23(a).

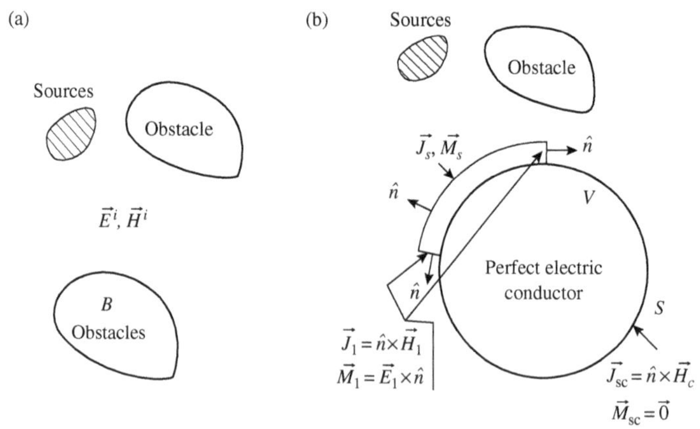

Figure 5.21 Modification of the original problem by a partially coated electric conductor

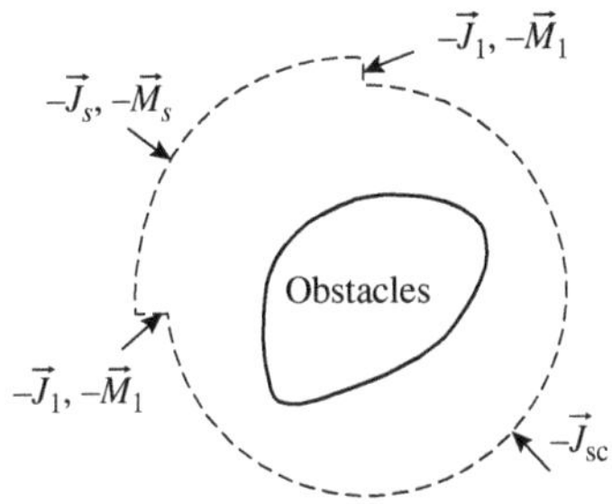

Figure 5.22 Complementary problem for a coated surface

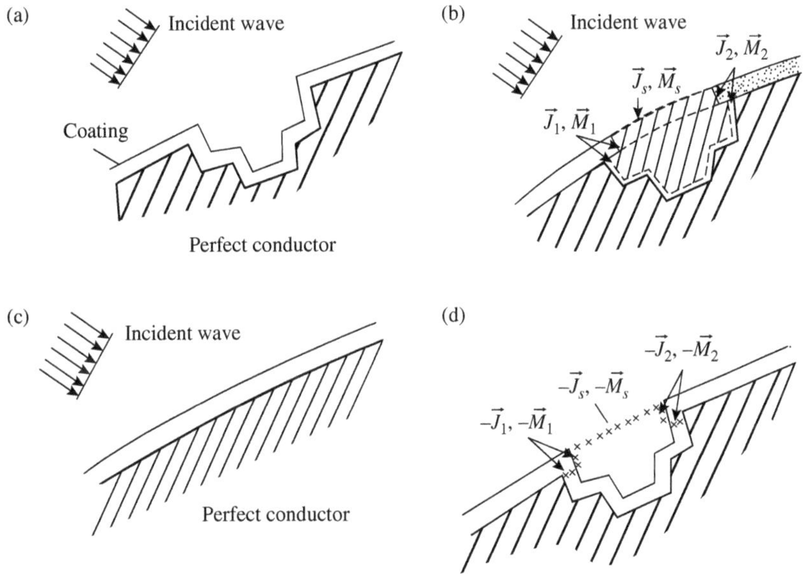

Figure 5.23 Generalisation of the third Schelkunoff principle to a cavity in a coated conducting surface

In a first step the surface is regularised by including a partially coated conductor occupying the hatched volume in Figure 5.23(b). Problem 5.23(c) is equivalent to problem 5.23(b).

The currents $(\vec{J}_s, \vec{M}_s)$ are obtained by calculating the field on the exterior surface of the coating in the problem of Figure 5.23(c), using asymptotic methods. The currents $(\vec{J}_1, \vec{M}_1)$ and $(\vec{J}_2, \vec{M}_2)$ along the truncations of the coating are also calculated in problem (c) since the asymptotic methods also give the field inside the coating. The fictitious surface which supports the equivalent currents in problem (c) is called the 'connection surface'. When the surface is a perfect conductor, the connection surface is simply the surface which regularises the structure. For a coated structure,

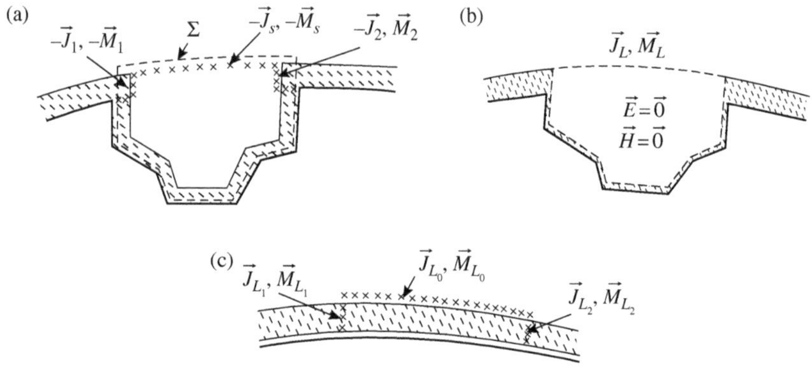

Figure 5.24 *For the exterior domain, problem (b) is equivalent to (a) and problem (c) is equivalent to (b)*

the connection surface comprises on the one hand the surface regularising the exterior surface of this structure and on the other hand the surfaces corresponding to the truncation of the coating on the hatched volume in problem (b) plus the bare surfaces of the inclusion which do not face a perfect conductor in problem (b).

The iterative method presented in Section 5.3.1, which allows us to solve the complementary problem, can also be extended to a cavity in a coated smooth surface. Indeed the induction theorem applies also to this case. If we consider that a closed fictitious surface Σ which follows the conducting walls of the cavity, crosses the coating and closes the cavity as shown in Figure 5.24(a), and if $(\vec{E}_L, \vec{H}_L)$ and $(\vec{J}_L, \vec{M}_L)$ denote, respectively, the field on Σ and its associated equivalent currents due to the radiation of the excitation current sheets $(-\vec{J}_s, -\vec{M}_s)$, $(-\vec{J}_1, -\vec{M}_1)$ and $(-\vec{J}_2, -\vec{M}_2)$, then the field radiated by $(\vec{J}_L, \vec{M}_L)$ inside the volume V defined by Σ is identically zero. Hence, we do not change the field in the exterior domain by introducing into the volume V an arbitrary object. If we choose a partially coated object which regularises the structure as shown in Figure 5.24(c), the complementary problem is reduced to the radiation of the current sheets $(\vec{J}_L, \vec{M}_L)$, $(\vec{J}_{L_1}, \vec{M}_{L_1})$, $(\vec{J}_{L_2}, \vec{M}_{L_2})$ in the presence of a smooth coated surface.

At order zero of the iteration procedure, the FDTD method is applied to the cavity and its vicinity excited by the sheets of currents $(-\vec{J}_s, -\vec{M}_s)$, $(-\vec{J}_1, -\vec{M}_1)$ and $(-\vec{J}_2, -\vec{M}_2)$ determined by solving problem (c) of Figure 5.23 by the asymptotic methods. We call these currents the excitation currents of order zero. With the FDTD method, the field $(\vec{E}_L, \vec{H}_L)$ and the associated currents $(\vec{J}_L, \vec{M}_L)$ are calculated on that part of Σ which is not faced by a perfect conductor namely the current sheets $(\vec{J}_{L_o}, \vec{M}_{L_o})$, $(\vec{J}_{L_1}, \vec{M}_{L_1})$ and $(\vec{J}_{L_2}, \vec{M}_{L_2})$. The radiation of these currents in the domain outside the cavity is calculated by solving problem (c) by asymptotic techniques.

At order 1, we take into account the modification of the field $(\vec{E}_L, \vec{H}_L)$ and the currents $(\vec{J}_L, \vec{M}_L)$ due to the interaction of the field radiated by the excitation currents of order zero with exterior obstacles. As in the case of a perfectly conducting structure, we determine first the excitation currents of order 1: $(-\vec{J}_L^{(1)}, -\vec{M}_L^{(1)})$, $(-\vec{J}_{L_1}^{(1)}, -\vec{M}_{L_1}^{(1)})$ and $(-\vec{J}_{L_2}^{(1)}, -\vec{M}_{L_2}^{(1)})$ by applying the uniform GTD to the diffraction of the field radiated by the excitation currents of order zero with the exterior obstacles. Then we repeat the operations by applying the FDTD method with the excitation currents of order 1 and follow this iteration procedure until its convergence is obtained.

5.4.2 Protrusion standing out of a coated regular surface

The method presented in Section 5.3.2 for a protrusion standing out of a perfectly conducting surface can be generalised to a protrusion on a coated surface. If obstacle 1 in Figure 5.17 is coated by a dielectric or an absorbing material, the fictitious surface S has to be extended inside the coating so that (s) be in contact with the underlying perfectly conducting surface (Figure 5.25(a)). By following the same type of reasoning as in Section 5.3.2 we obtain a new form of equivalences displayed in Figure 5.25, where $(\vec{E}^i, \vec{H}^i)$ is the field radiated by the primary sources in the presence of the coated obstacle 1 but in the absence of obstacle 2.

If obstacle 2 is a perfect conductor and is in contact with the perfectly conducting core of obstacle 1 (see Figure 5.26(a)), then the equivalence of Figure 5.18 can be generalised as shown in Figure 5.26.

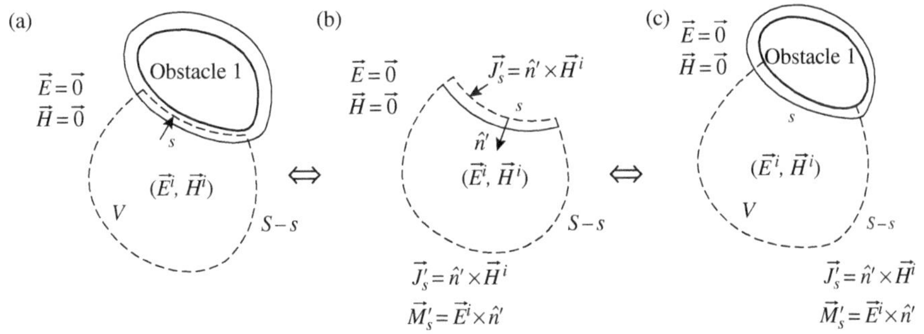

Figure 5.25 Equivalences generalising those of Figure 5.17

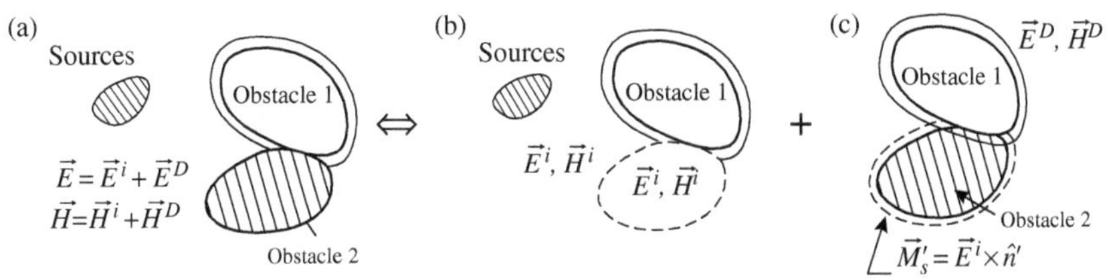

Figure 5.26 Outside obstacles 1 and 2, the solution to the original problem (a) is equivalent to the sum of solutions to problems (b) and (c)

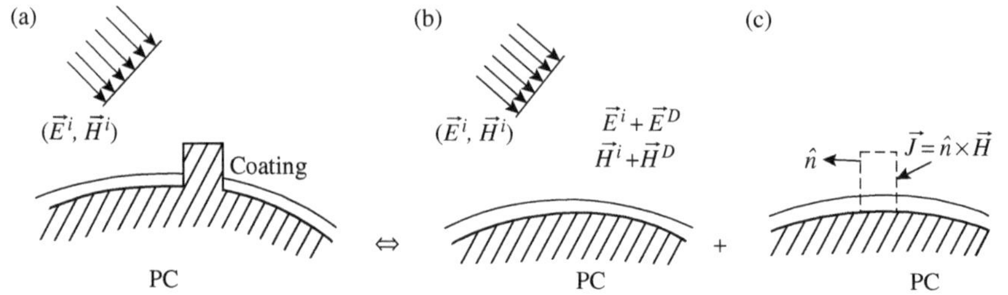

Figure 5.27 Equivalence adapted to a hybrid asymptotic/integral equation method

We see that in problem (b) the field $(\vec{E}^i, \vec{H}^i)$ has to be determined not only outside the coating but also inside the coating. Problem (c) is well adapted to the FDTD method whereas problem (b) may be solved by asymptotic methods.

The equivalence shown in Figure 5.26 can also be adapted to combine the asymptotic methods with an integral equation solution. In this case, we can decompose the original problem in the way shown in Figure 5.27, where the unknown currents $\vec{J}$ in problem (c) are obtained by solving the EFIE with an excitation term given by the opposite of the tangential component of $\vec{E}^i + \vec{E}^D$. The latter is calculated inside and outside the coating by solving problem (b) with the asymptotic methods. The EFIE is restricted to the exterior surface of the protrusion standing out of the metallic part of the surface supporting it. Since the current $\vec{J}$ radiates in the presence of the smoothed coated surface, we have to introduce in the EFIE the Green's function corresponding to the coated regularised surface for a source and an observation point arbitrarily located inside or outside the coating. This Green's function can be approximated by asymptotic methods.

5.5 Brief review of asymptotic solutions adapted to the development of hybrid methods

The techniques that have been presented in the preceding chapters for combining numerical methods with asymptotic solutions can be applied to a real problem by using the asymptotic solutions presented in this book. As we have mentioned before, PO alone or simple GTD formulae which are only valid at large distance from the scattering object are not sufficient in most applications. More sophisticated solutions like those presented in this book are needed. Especially, the field on the regularised structure which gives the excitation currents used as inputs in the numerical method for a cavity, involves the asymptotic currents presented in Chapter 4.

In the case of a protrusion, the field also has to be calculated outside the regularised surface and especially, a uniform solution is needed when the observation point is moved away from the surface. Asymptotic solutions for this problem have been developed in References 26 and 27 for a smooth convex surface and an observation point located in the lit region or in the shadow region. A technique somewhat more time consuming but more robust and which gives directly a uniform solution, consists in calculating the near field by integrating the radiation of the asymptotic currents with the aid of the radiation integral.

In the hybrid asymptotic/integral equation method, an asymptotic approximation of the Green's function is also needed to apply this method. A critical situation arises when both the source and the observation point are close to the surface.

Asymptotic solutions to this problem may be established by extending the method presented in Reference 28 to a source slightly above the surface. It is also possible to integrate the radiation of the asymptotic currents induced by a point source situated on the surface or close to it. A uniform asymptotic solution for a 2D convex object has been obtained in Reference 29 by extending Ivanov's procedure [30] to the case where both the source and the observation point are close to the surface.

5.6 Numerical results

We present two examples of application of the hybrid method described in Section 5.3.1. The first one concerns a 2D structure with an ogival cross-section, having two slots parallel to its generatrices. The second one concerns a 3D structure having the form of a rectangular cavity, representing an inlet.

5.6.1 *Slotted ogival cylinder*

The transverse dimensions of the structure are given in wavelengths in Figure 5.28.

The structure is perfectly conducting. We have used the formulae for the asymptotic currents on the smoothed surface and for the field radiated at large distance by the equivalent magnetic currents given in Chapter 4. The complementary problem has been solved by applying the FDTD method. No iteration procedure has been used.

The curves of Figure 5.29 give the bistatic radar cross-section as a function of the observation angle θ^d defined in Figure 5.28, for an incident plane wave propagating in the direction normal to the generatrices with an angle of 20° with the symmetry

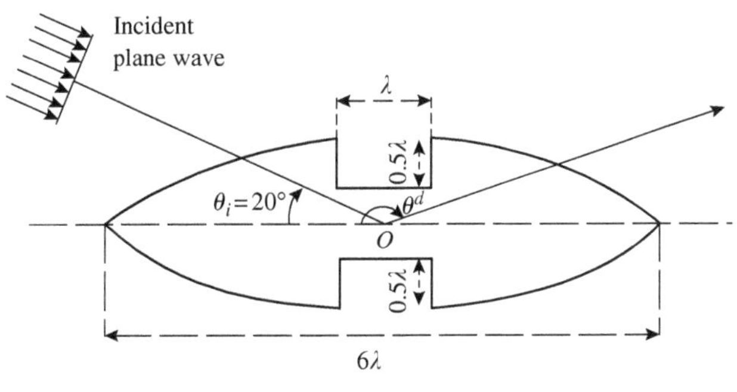

Figure 5.28 Cross-section of the structure and angles of illumination and observation

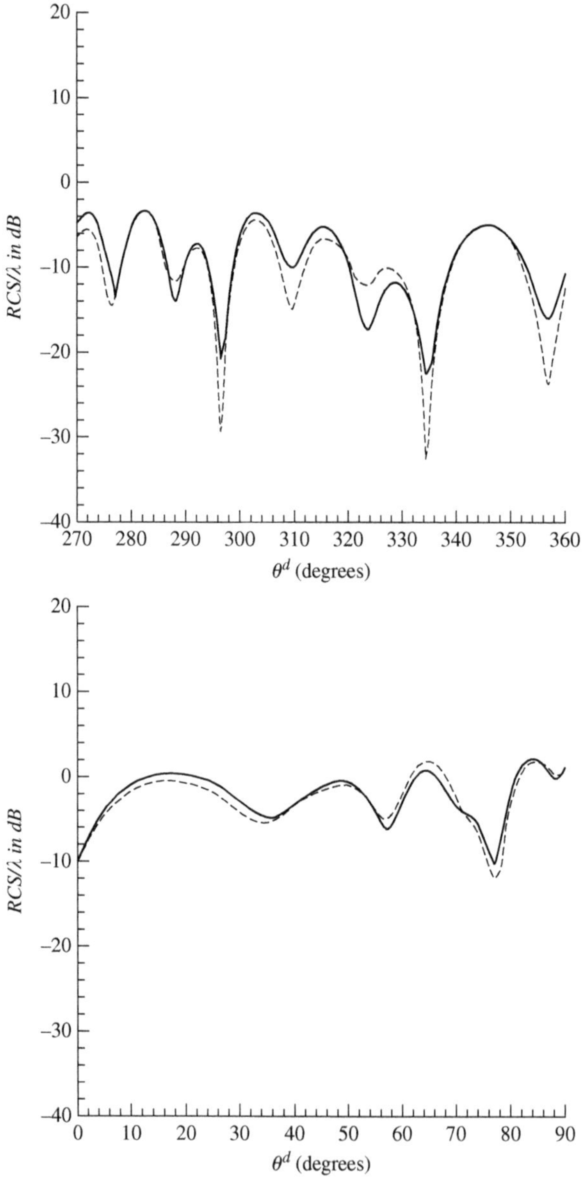

Figure 5.29 Bistatic radar cross-section of a slotted ogival cylinder: ———, hybrid asymptotic/FDTD method; – – – – – –, reference curve (FDTD)

plane of the structure. The incident electric field is polarised perpendicular to the generatrices.

The solid curve corresponds to the reference results obtained by applying the FDTD method to the complete slotted structure. The agreement between both results

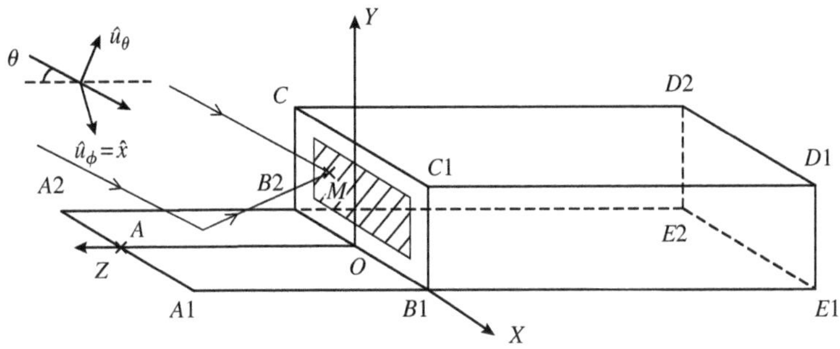

Figure 5.30 Rectangular cavity illuminated by an incident plane wave

is very satisfactory despite the fact that without performing any iteration on the excitation current, the field radiated by the equivalent magnetic currents and scattered back to the slots by the edges of the ogive, is neglected.

5.6.2 Rectangular cavity

The structure is shown in Figure 5.30. Its lower wall is extended in order to simulate the fuselage on which the inlet is attached. The hatched rectangle represents the opening of the rectangular cavity.

The dimensions of the cavity and of the structure are specified in Figure 5.31. They have been chosen such that the scattered field by the complete structure can also be calculated by the FDTD method alone at a frequency of 5 GHz.

The incident field is a plane wave propagating in the direction $\vec{k}^i$. Only the case where the opening is situated in the region illuminated by the incident field is considered. Apart from the direct and reflected rays represented in Figure 5.30, other rays reach the opening like singly edge-diffracted rays, reflected rays which are diffracted by an edge, direct and reflected rays which are doubly diffracted by edges and higher-order edge-diffracted rays.

In order to determine the rays, the structure is represented in a Cartesian coordinate system the axes of which are parallel to the sides of the box. The origin O is located at the centre of the edge B_1B_2 as shown in Figure 5.30.

We have restricted our analysis to the case where the direction of propagation of the incident field lies in the symmetry plane ZOY of the structure. So, it is defined by a single angle θ^i. The direction of observation lies in the same plane and is defined by the angle θ^d. For a fixed value of θ^i, θ^d varies from 0 to 90°.

In order to calculate the field diffracted by this structure, the hybrid method described in Section 5.3.1 combining the asymptotic techniques with the FDTD method has been applied.

In the first step, the opening of the cavity is short-circuited. This modified problem is then solved by the asymptotic methods for the induced currents $\vec{J}_s$ on the perfectly conducting surface closing the cavity and for the scattered far field. In the second step, the field radiated by the sheet of currents $-\vec{J}_s$ in the presence of the structure

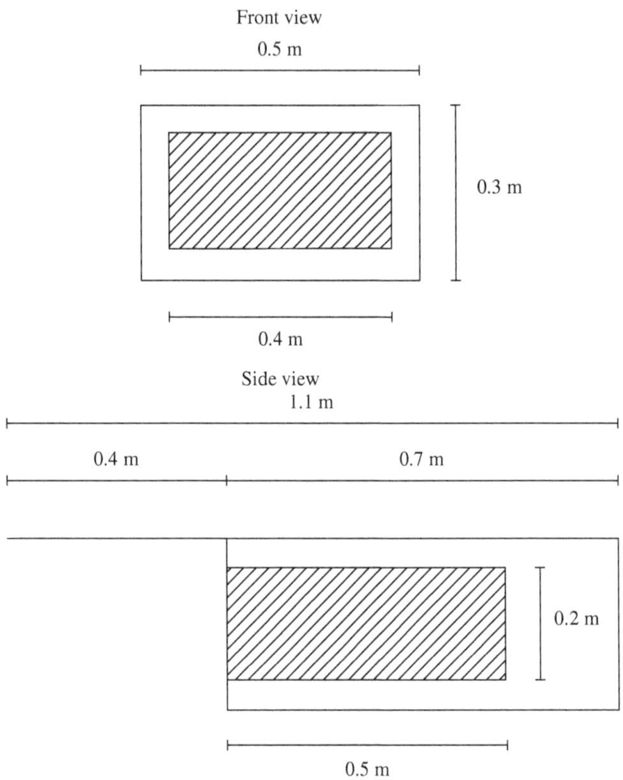

Figure 5.31 *Dimensions of the structure and of the cavity*

with its open cavity is calculated by the FDTD. This complementary problem has been solved in a volume comprising the cavity and its surroundings limited to a part of the exterior surface of the box containing the cavity, corresponding to one tenth of its length. A check of the validity of this approximation has been made by applying the FDTD to the whole structure and comparing the results to those obtained with a truncated cavity. The differences between both results were negligibly small. The solution of the complementary problem by the FDTD method gives the electric field $\vec{E}_\Sigma$ in the opening of the cavity and the corresponding magnetic current sheet $\vec{M}_\Sigma = \vec{E}_\Sigma \times \hat{n}$. In the third step, the far field radiated by this current sheet in the presence of the complete structure has been calculated by the asymptotic methods. A preliminary numerical analysis of the convergence of the method with respect to the sampling of the 'connecting' surface which closes the opening of the cavity, has been performed. It has been observed that the convergence is obtained by a sampling of $\lambda/12$ in both directions, where λ is the wavelength.

The far field $\vec{E}^R$ radiated by the magnetic currents $\vec{M}_\Sigma$ determined by the FDTD method, in the presence of the structure with its closed cavity, has been calculated by the asymptotic method which in this particular application is the classical GTD.

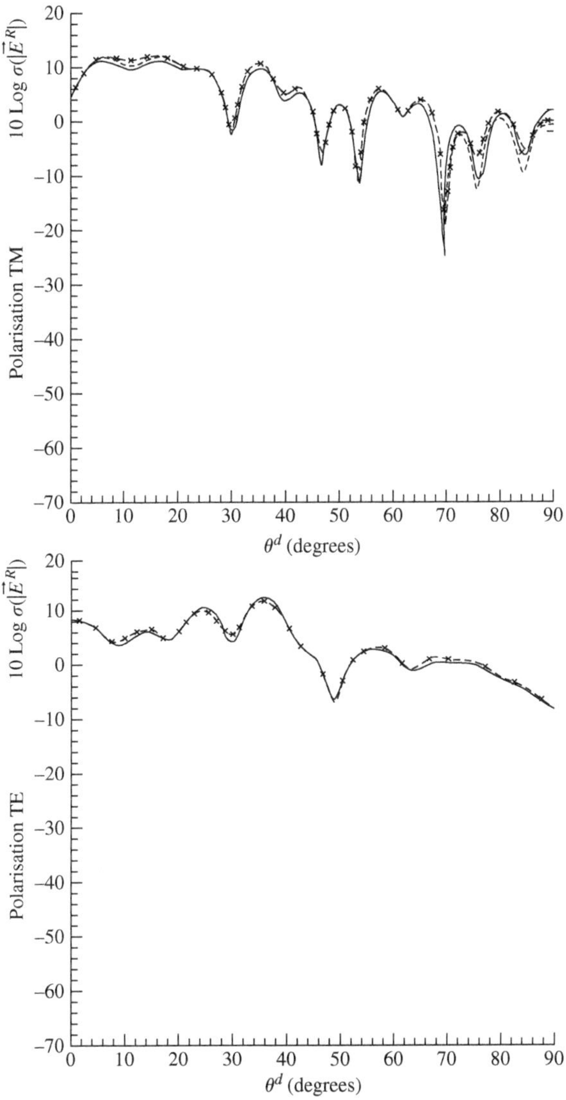

Figure 5.32 *Far field radiation (in dB) of the magnetic current $\vec{M}_{\Sigma}$ as a function of the observation angle θ^d for $\theta^i = 20°$, $F = 5$ GHz. ——— GTD results, - - - - - - FDTD results with $\vec{J}_s$ calculated by GTD × × × × × FDTD results with $\vec{J}_s$ calculated by FDTD*

The results obtained have been checked by comparing them to a direct calculation, by the FDTD method, of the far field radiated by the sheet of current $-\vec{J}_s$ in the presence of the open cavity. In this procedure the calculation has been performed with two different inputs for $\vec{J}_s$, one obtained by the asymptotic method and the other by a direct application of the FDTD method.

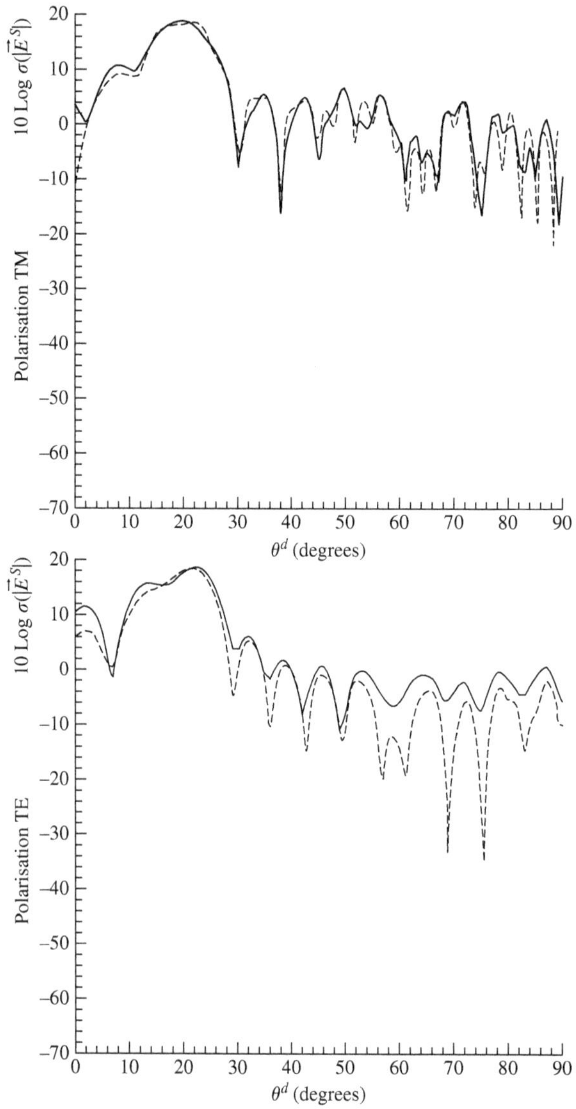

Figure 5.33 *Far field (in dB) scattered by the structure with its closed cavity (modified problem) as a function of the observation angle θ^d for $\theta^i = 20°$, $F = 5\,GHz$.* ——— *GTD results,* ------ *reference solution (FDTD)*

The curves in Figure 5.32 give the results in decibels (dB) of this comparison for the quantity $\sigma(|\vec{E}^R|) = 4\pi r^2 |\vec{E}^R|^2/|\vec{E}^i|^2$, for r large, as a function of the observation angle for $\theta^i = 20°$ fixed. It can be observed that the differences between the GTD results and the two other methods are small (less than 1 dB) and are of the order of the differences between the two FDTD results.

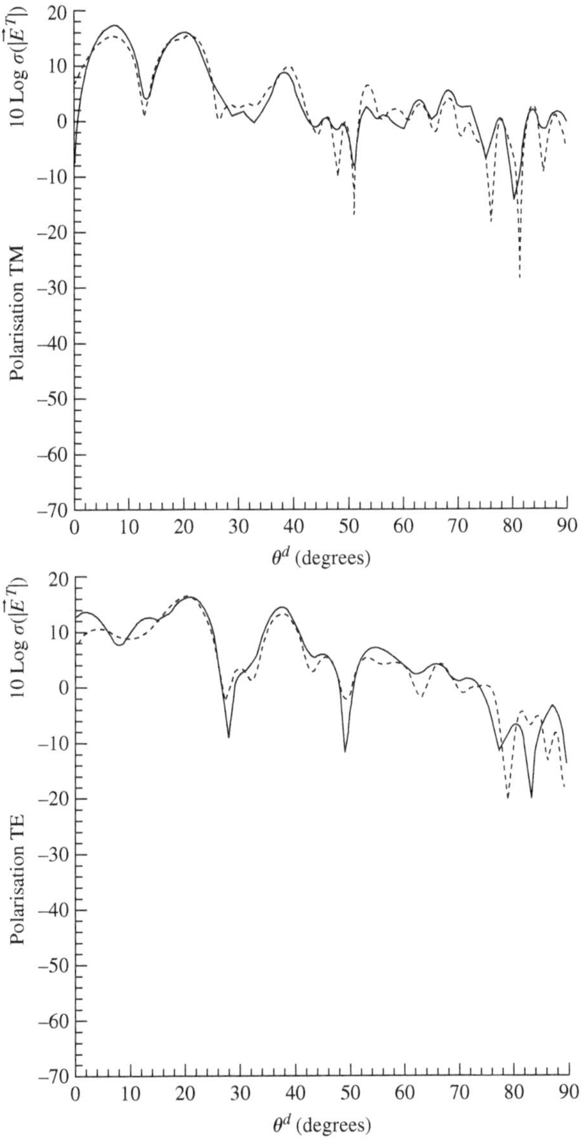

Figure 5.34 *Far field (in dB) scattered by the structure with its open cavity (original problem) as a function of the observation angle θ^d for $\theta^i = 20°$, $F = 5\,GHz$. ——— Hybrid asymptotic/FDTD method, – – – – – – reference solution (FDTD)*

The field $\vec{E}^s$ scattered by the structure, the cavity of which being closed, is represented in Figure 5.33. The curves give the quantity $\sigma(|\vec{E}^S|) = 4\pi r^2 |\vec{E}^S|^2/|\vec{E}^i|^2$ in dB for r large, as a function of the observation angle for $\theta^i = 20°$. The results obtained by GTD are compared to those given by the reference method, which in the present case is just the FDTD applied to the same structure. Some differences appear between the results given by both methods which are mainly due to the fact that in the GTD model, double diffractions by edges have not been taken into account. The total field $\vec{E}^T$ scattered by the complete structure with its open cavity (original problem) is equal to the sum $\vec{E}^R + \vec{E}^S$. Figure 5.34 shows the quantity $\sigma(|\vec{E}^T|) = 4\pi r^2 |\vec{E}^T|^2/|\vec{E}^i|^2$ in dB for r large, as a function of the observation angle, for $\theta^i = 20°$. The solid line corresponds to the results obtained by the hybrid method. The reference solution, represented by the dashed line, corresponds to the FDTD method applied directly to the original problem. The agreement between both methods is rather good despite the simple GTD model used in all steps in the hybrid procedure which was limited to double reflections and to single edge-diffraction phenomena associated with zero or one reflection. It can be seen by comparing the results of Figure 5.33 to those of Figure 5.34, that the largest deviations from the reference curve in Figure 5.34 are located at the same angular positions as those observed in Figure 5.33. Hence, by improving the GTD model used for solving the modified problem (closed cavity), the results of the hybrid method could also be improved. However, it is not necessary to augment the actual GTD model for the determination of the currents $\vec{J}_s$, by adding double edge-diffraction contributions, in the case where the opening of the cavity is directly illuminated by the incident wave, since then the contributions of the direct rays and the rays reflected by the plate $A_1A_2B_2B_1$ (see Figure 5.30) are dominant. The same GTD model is also sufficient for the calculation of the far field radiated by the sheet of magnetic currents $\vec{M}_\Sigma$ determined by the FDTD method, when the observation point is seen from any point of the opening.

The remark concerning the calculation of the current $\vec{J}_s$ by GTD, remains valid for a more realistic problem, like an inlet on a real air-plane which is supposed to be directly illuminated by an incident wave. Again direct and reflected waves are dominant in the determination of the sheet of currents $\vec{J}_s$ which act as secondary sources for the FDTD method. Hence, a simple GTD model limited to directed, reflected and single diffracted rays is sufficient for obtaining an accurate solution of the complementary problem in the angular domain where both the source and the observation point are seen from any point of the opening of the inlet. For the modified problem, however, in order to prevent some inaccuracies especially in the axial direction, a more refined GTD model including higher-order diffraction phenomena like double diffraction by edges, creeping waves and edge-diffracted creeping waves is necessary.

5.7 References

1 BOUCHE, D., MOLINET, F., and MITTRA, R.: ‘Asymptotic and hybrid techniques for electromagnetic scattering’, *Proceedings of the IEEE*, 1993, **81** (12), pp. 1658–84

2 NEWMAN, E.: ‘An overview of the hybrid MM/Green’s function method in electromagnetics’, *Proceedings of the IEEE*, 1988, **76** (3), pp. 270–82

3 MEDGYESI-MITSCHANG, L. N., and WANG, D. S.: 'Review of hybrid methods on antenna theory', *Annales des Télécommunications*, 1989, **44** (9–10) pp. 445–55

4 MEDGYESI-MITSCHANG, L. N., HURST, M. P., PUTMAN, J. M., and WANG, D. S.: 'New hybrid methods for antenna analysis'. Invited Conference, JINA, Nice, November 8–10, 1994

5 THIELE, G. A.: 'Overview of selected hybrid methods in radiating system analysis', *Proceedings of the IEEE*, 1992, **80** (1), pp. 66–78

6 MEDGYESI-MITSCHANG, L. N., and PUTNAM, J. M.: 'Hybrid formulation for arbitrary 3-D bodies'. 1994 ACES Symposium Record, vol. II, March 1994, pp. 267–74

7 JAKOBUS, U., and LANDSTORFER, F.: 'Improved PO-MM hybrid formulation for scattering from three-dimensional perfectly conducting bodies of arbitrary shape', *IEEE Transactions on Antennas and Propagation*, 1995, **AP-43** (2), pp. 162–9

8 HODGES, R. E., and RAHMAT-SAMII, Y.: 'An iterative current-based hybrid method for complex structures', *IEEE Transactions on Antennas and Propagation*, 1997, **AP-45** (11), pp. 265–76

9 KAYE, M., MURTHY, P. K., and THIELE, G. A.: 'An iterative method for solving scattering problems', *IEEE Transactions on Antennas and Propagation*, 1985, **AP-33**, pp. 1272–9

10 MURTHY, P. K., HILL, K. C., and THIELE, G. A.: 'A hybrid-iterative method for scattering problems', *IEEE Transactions on Antennas and Propagation*, 1986, **AP-34**, pp. 1173–80

11 DESPRÉS, B.: 'Sur une formulation variationnelle de type ultra-faible', *CRAS série I*, 1994, **318**, p. 939

12 CESSENAT, O., and DESPRÉS, B.: 'Application of an ultra-weak variational formulation of elliptic PDE to the 2D Helmholtz problem', *SIAM Journal of Numerical Analysis*, 1998, **35** (1), pp. 255–99

13 HUTTUNEN, T., MONK, P., and KAIPIO, J.: 'Computational aspects of the ultra-weak variational formulation', *Journal of Computational Physics*, 2002, **182**, pp. 27–46

14 PERREY-DEBAIN, E., LAGHROUCHE, O., BETTESS, P., and TREVELYAN, J.: 'Plane wave basis finite elements and boundary elements for 3D wave scattering'. Philosophical Transactions of the Royal Society, vol. 362, no. 1816, 2004, pp. 561–77

15 LAGHROUCHE, O., BETTESS, P., PERREY-DEBAIN, E., and TREVELYAN, J.: 'Plane wave basis finite elements and boundary elements for 3D wave scattering', *Communications in Numerical Methods in Engineering*, 2003, **19**, pp. 715–23

16 LAGHROUCHE, O., BETTESS, P., and ASTLEY, R. J.: 'Modelling of shortwave diffraction problems using approximating systems of plane waves', *International Journal of Numerical Methods in Engineering*, 2002, **54**, pp. 1501–33

17 GILADI, E., and KELLER, J.: 'A hybrid numerical asymptotic method for scattering problems', *Journal of Computational Physics*, 2001, **174**, pp. 226–47

18 KWON, D. H., BURKHOLDER, R., and PATHAK, P. H.: 'Efficient MoM formulation for large PEC scattering problems using asymptotic phase front extraction (APE)', *IEEE Transactions on Antennas and Propagation*, 2001, **49** (4), pp. 583–91

19 DARRIGRAND, E.: 'Coupling of fast multipole method and microlocal discretization for the 3D Helmholtz equation', *Journal of Computational Physics*, 2002, **181**, pp. 126–54

20 TIBERI, G., MONORCHIO, A., MANARA, G., and MITTRA, R.: 'Hybridizing asymptotic and numerically rigorous techniques for solving electromagnetic scattering problems using the characteristics basis functions (CBFs)'. 2003 IEEE Antennas and Propagation Society International Symposium, URSI Dig., Columbus, Ohio, June 22–27, 2003, p. 519

21 TIBERI, G., ROSACE, S., MONORCHIO, A., MANARA, G., and MITTRA, R.: 'Electromagnetic scattering from large faceted conducting bodies by using analytically-derived characteristic basis functions', *IEEE Antennas and Wireless Propagation Letters*, 2003, vol 2, issue 20, pp. 290–3

22 NETO, A., MACI, S., VECCHI, G., and SABBADINI, M.: 'A truncated Floquet wave diffraction method for the full wave analysis of large phased arrays. (I) Basis principles and 2-D cases, (II) Generalisation to 3-D cases', *IEEE Transactions on Antennas and Propagation*, 2000, **AP-48** (4), pp. 594–611

23 HARRINGTON, R. F.: 'Time-harmonic electromagnetic fields' (McGraw-Hill, New York, 1961), Chap. 3

24 CHEN, K. M.: 'A mathematical foundation of the equivalence relation', *Physical Reviews*, 1958, **109**, pp. 1863–7

25 HUNTER, J. D.: 'Aspects of electromagnetic scattering'. PhD Thesis, University of Canterbury, Christchurch, New Zealand, 1980

26 CHOU, H. T., PATHAK, P. H., and HSU, M.: 'Extended uniform geometrical theory of diffraction solution for the radiation of antennas located close to an arbitrary, smooth, perfectly conducting, convex surface', *Radio Science*, 1997, **32** (4), pp. 1297–317

27 BOUCHE, D., MOLINET, F., and MITTRA, R.: 'Asymptotic methods in electromagnetics' (Springer-Verlag, Berlin, 1997), Chap. 5.5.3

28 MOLINET, F.: 'Asymptotic current method applied to the near field radiation of antennas mounted on a perfectly conducting convex surface'. 26th ESA Antenna Technology Workshop on *Satellite Antenna Modelling and Design Tools*, ESTEC, Noordwijk, The Netherlands, November 12–14, 2003, pp. 83–90

29 MOLINET, F., and TORT, S.: 'Improved asymptotic solutions for the near field radiated by antennas mounted on a perfectly conducting convex surface'. Paper presented at the ICEEA Conference, Torino, 1999

30 IVANOV, V. I.: 'Uniform asymptotic behaviour of the field produced by a plane wave reflection at a convex cylinder', *USSR Journal of Computational Mathematics and Mathematical Physics*, 1971, **2**, pp. 216–32

31 SCHELKUNOFF, S.A.: 'Field Equivalence Theorem' Communications in Pure and Applied Maths, vol. 4, June 1951, pp. 43–59

Index

www.ingramcontent.com/pod-product-compliance
Ingram Content Group UK Ltd.
Pitfield, Milton Keynes, MK11 3LW, UK
UKHW020803290926
15470UKWH00003B/939

9 780863 414473